PROTEIN KINASE CK2 – FROM STRUCTURE TO REGULATION

Protein Kinase CK2 – From Structure to Regulation

Edited by

KHALIL AHMED

Cellular and Molecular Biochemistry Research Lab
University of Minnesota
Veterans Affairs Medical Center
One Veterans Drive
Minneapolis, MN 55417
USA

JORGE E ALLENDE

Program de Biologia Celular y Molecular
Instituto de Ciencias Biomedicas
Facultad de Medicina
Universidad de Chile
Casilla 70086
Santiago 7, Chile

OLAF-GEORG ISSINGER

Department of Biochemistry and Molecular Biology
University of Southern Denmark
Campusvej 55
5230 Odense
Denmark

Reprinted from *Molecular and Cellular Biochemistry*, Volume 227 (2001)

Kluwer Academic Publishers

Dordrecht / Boston / London

Distributors for North, Central and South America:
Kluwer Academic Publishers
101 Philip Drive
Assinippi Park
Norwell, Massachusetts 02061 USA
Telephone (781) 871-6600
Fax (781) 681-9045
E-mail <kluwer@wkap.com>

Distributors for all other countries:
Kluwer Academic Publishers Group
Distribution Centre
Post Office Box 322
3300 AH Dordrecht, The Netherlands
Telephone 31 78 6392 392
Fax 31 78 6546 474
E-mail <services@wkap.nl>

Electronic Services <http://www.wkap.nl>

A C.I.P. Catalogue record for this book is available from the Library of Congress

ISBN 0-7923-7666-8

Printed on acid-free paper

Printed in the United States of America

Molecular and Cellular Biochemistry:

An International Journal for Chemical Biology in Health and Disease

CONTENTS VOLUME 227, Nos. 1 & 2, November 2001

PROTEIN KINASE CK2 – FROM STRUCTURE TO REGULATION
Drs. Khalil Ahmed, Jorge E. Allende and Olaf-Georg Issinger

Molecular and Cellular Biochemistry **227:** 1–2, 2001.

Preface

Signal transduction through phosphorylation and dephosphorylation of proteins in the cell is now a well recognized mechanism involved in countless physiological and pathological processes. Consequently, the enzymes, known as protein kinases, which catalyze the phosphorylation of proteins are critical regulators of cellular events. One of these protein kinases is the protein kinase CK2 (also known as casein kinase 2) that has been implicated in multiple functions including control of cell growth and proliferation. CK2 is a protein serine/threonine kinase which is a highly conserved and ubiquitous protein kinase. It is localized in the cytoplasmic and nuclear compartments which accords with its multiple functional activities in the cell. Pertinent to this is also the recognition that a large number of putative substrates for this kinase have been identified in various compartments of the cell. New evidence from several laboratories has further reinforced the involvement of CK2 in signal transduction related to many cellular functions thus underscoring the significance of its functional role in normal and abnormal cell growth and proliferation.

The state of the knowledge concerning this intriguing protein kinase has been discussed at specially organized international meetings held every three years. The Third International Meeting on this subject entitled 'Protein Kinase CK2: From Structure to Regulation' was held in San Esteban, Chile, on January 8–10, 2001, bringing together a number of major investigators engaged in research in this area. The previous two meetings on CK2 were held in Heidelberg (1994) and Villard de Lans near Grenoble (1997); the meeting in Chile was timely as it highlighted the impressive progress that has since been made in this area. Key developments during the past three years pertain to the elaboration of the crystal structure and definition of novel functions of the kinase such as its role as an inhibitor of apoptosis. Additionally, the shuttling of the kinase to various compartments in response to physiological and stress stimuli appears to be a key feature of the functional regulation of its activity in the cell.

The overall organization of the meeting involved seven sessions for invited talks, two sessions dealing with general discussion, and a Poster Discussion for young investigators. A brief description of the talks in the various sessions highlights the recent developments in the studies of CK2 in relation to its function. The meeting was opened with a welcome address by the organizer of the meeting Dr. Jorge E. Allende (Santiago, Chile) who outlined the status of the current knowledge and unresolved issues regarding CK2. The first scientific session, chaired by C. Cochet (Grenoble, France), was entitled 'Structural and Functional Analysis'. In the first talk, O.-G. Issinger (Odense, Denmark) presented the latest information on the crystallization and characterization of crystals of human protein kinase CK2 holoenzyme. The second talk was by L.A. Pinna (Padova, Italy) who gave an account of the structural features underlying constitutive activity and specific inhibition of protein kinase CK2.

The second session, chaired by K. Ahmed (Minneapolis, USA), focused on the topic of the 'Interacting Partners of CK2'. D. Litchfield (London, Canada) presented his findings on functional specialization of CK2 isoforms and isoform-binding partners. J. Jiménez (Madrid, Spain) described new data on macromolecular interactions of the protein kinase CK2 component subunits studied by surface plasmon resonance. The talk by J.E. Allende (Santiago, Chile) was concerned with the functional analysis of CK2 expression in studies employing transfection of CK2 wild type and mutant subunits into COS-7 cells.

The third session chaired by O. Filhol-Cochet (Grenoble, France) dealt with the topic of 'Genes and Gene Expression'. In this session, the presentation by B. Boldyreff (Odense, Denmark) was on the physiological function of the protein kinase CK2 beta subunit involving studies of interacting partners and knockout mice. W. Pyerin (Heidelberg, Germany) introduced data on the transcriptional regulation of the human protein kinase CK2 genes alpha and beta. This was followed by a talk by K. Ackermann (Heidelberg, Germany) on genes targeted by protein kinase CK2 as examined by a genome-wide expression array analysis in yeast.

The theme of the fourth session chaired by O.-G. Issinger (Odense, Denmark) was 'Cell Localization'. In this session, the first talk presented by K. Ahmed (Minneapolis, USA) presented an overview of the functional consequences of CK2 signaling to the nuclear matrix. M. Montenarh (Homburg, Germany) discussed his work on subcellular localization of individual subunits of CK2 and its interaction with other cellular proteins. The final talk in this session was by C. Cochet (Grenoble, France) who presented data on dynamics of protein kinase CK2 subunits in living cells.

The fifth session, chaired by M.C. Sogayar (Sao Paulo, Brazil), considered the 'Biological Functions' of CK2. The first talk was by M. Schultz (Alberta, Canada) who gave an account of his studies showing that TATA binding protein-associated CK2 transduces DNA damage signals to the RNA polymerase III transcriptional machinery. D. Meek (Dundee, UK) presented studies on phosphorylation of the p53 regulatory partner, MDM2, at serine 267 by protein kinase CK2 *in vitro* and in cultured cells. Likewise, A. Bidwai (Morgantown, USA) described that *Drosophila* CK2 interacts with and phosphorylates bH1H proteins derived from the E(spl)C and hairy locus, and discussed implications of these data for neurogenesis and segmentation. The last talk in this session was by M.I. Vera (Santiago, Chile) who focused on studies dealing with environmental reprogramming of the expression of CK2 beta subunit in the carp fish.

The sixth session chaired by M.I. Vera (Santiago, Chile) was on the topic of 'Plant and Yeast CK2'. This session recognized the acquisition new information on the potential involvement of CK2 in plant cell biology. S. Passeron (Buenos Aires, Argentina) described data on cloning and expression of β and β' subunits of *Candida albicans* CK2. M. Riera (Barcelona, Spain) gave an account of the maize protein kinase CK2 with focus on its regulation and functionality. The final talk in the session was by L. Holuigue (Santiago, Chile) describing the involvement of CK2 in the transcriptional activation of genes by salicylic acid in tobacco.

The seventh session chaired by A. Quest (Santiago, Chile) was devoted to 'Involvement in Disease'. This session brought forth the significant new observations on the role of CK2 in certain disease processes in addition to neoplasia. The first talk by L. Banks (Trieste, Italy) was on the emerging role of CK2 in viral biology, and summarized the studies on CK2 regulation of the HPV-16 E7. The talk by F. Meggio (Padova, Italy) detailed the possible implications of protein kinase CK2 in infectious and neurodegenerative diseases describing phosphorylation of HIV-Rev transactivator and up-regulation by prion protein. D. Seldin (Boston, USA) reviewed studies of the protein kinase CK2 in Wnt signaling and mammary gland tumorigenesis. The final talk in the session was by K. Ahmed (Minneapolis, USA) who presented data describing the response of cancer cells to interruption of the CK2 signal by employing antisense oligodeoxynucleotides.

An important new feature of this meeting was the two 'General Discussion' sessions. The first session led by L.A. Pinna brought forth the intriguing questions relating to the stoichiometry of the various subunits of CK2. The issues relating to the nature of the equilibrium between the holoenzyme and its subunits remains to be fully understood. It would appear that the structural features of the subunits may play a role in their interactions. Further, evidence has emerged suggesting that various subunits can indeed exist individually. The second discussion session led by D.W. Litchfield related to issues such as the role of CK2 in cell proliferation, its regulation during proliferation, and targets of CK2 during proliferation. Consideration of the aggregate information supports the conclusion that CK2 is essential for cell viability, growth and proliferation. The various features that may impact on the regulation of CK2 (such as covalent modification, regulatory interactions with other small molecules and proteins, etc.) need further evaluation. Future prospects of CK2 studies may deal with the study of novel functional roles (such as in apoptosis and stress response), development of strategies such as mouse knockouts, and application of genomics and proteomics. Important questions remain concerning a precise classification of substrate protein targets for CK2. Thus, these two general discussion sessions not only highlighted the current status of our knowledge on CK2 but also brought forth several important issues that need to be resolved in the future.

The 21 papers in this focused issue of *Molecular and Cellular Biochemistry* are based on the materials presented by the authors at the meeting. We believe they provide a view of the latest progress on protein kinase CK2. To that end, we express our sincere gratitude to all the participants in the meeting for their efforts.

The major credit for the organization of the meeting goes to Jorge E. Allende and Catherine C. Allende who, with the most able and tireless assistance of Ms. Monique Fortin, took care of all the organizational details making the meeting a resounding success. The meeting was generously sponsored by ICGEB (International Centre for Genetic Engineering and Biotechnology, Trieste, Italy), CONICYT (National Commission for Scientific and Technological Research, Santiago, Chile), IUBMB (International Union of Biochemistry and Molecular Biology, Symposium and Interest Groups Committee IG200), ICBM (Institute of Biomedical Sciences, Faculty of Medicine, University of Chile, Santiago, Chile) and CORPORATION RELAB (Red Latinoamericana de Ciencias Biologicas, Santiago, Chile). We are deeply appreciative of their support as without it the meeting would have not been possible.

Guest Editor: Khalil Ahmed, Minneapolis, USA
Guest Co-editors: Jorge E. Allende, Santiago, Chile
Olaf-Georg Issinger, Odense, Denmark

Molecular and Cellular Biochemistry **227**: 3–11, 2001.

Characterization of CK2 holoenzyme variants with regard to crystallization

Barbara Guerra,[1] Karsten Niefind,[2] Inessa Ermakowa[2] and Olaf-Georg Issinger[1]

[1]*Institut für Biochemie, Universität zu Köln, Köln, Germany;* [2]*Institut for Biokemi og Molekylær Biologi, Syddansk Universitet, Odense, Denmark*

Abstract

A search for strategies was conducted in order to obtain a human protein kinase CK2 preparation which would be suitable for crystallization, despite the fact that the recombinant enzyme is abundant and can be readily purified to homogeneity. This seemingly contradiction is based on the fact that the catalytic subunit moiety of the human CK2 holoenzyme is not stable neither as a free subunit nor in the tetrameric complex. All attempts to prevent degradation failed. Hence, alternative approaches were designed in order to avoid this degradation, which was expected to hamper any crystallization efforts severely. One of the approaches chosen was the production of a chimeric holoenzyme made up from a human regulatory subunit and a catalytic subunit from *Z. mays*. The plant catalytic subunit, in contrast to the human counterpart is very stable and does not undergo this kind of degradation. The second strategy to tackle the problem of instability was to produce the homologous recombinant human CK2 holoenzyme and then, instead of trying to avoid degradation, attempt to accelerate degradation until all catalytic subunit material was converted to the degraded form, i.e. a 40 kDa polypeptide. (Mol Cell Biochem **227**: 3–11, 2001)

Key words: CK2, crystallization, truncated protein kinases, protein purification

Introduction

Protein kinase CK2 is a tetrameric enzyme which consists of two catalytic and two regulatory subunits.

Both subunits have been cloned and expressed from various organisms such as yeast [1], plants [2], and mammals [3]. There is a great evolutionary conservation from yeast to mammals (for review, [4]), however, one of the major differences found so far affects polypeptide length which differs distinctly between plants and mammals. It is the carboxy terminal end that is involved in this shortening. Earlier observations showed that CK2 holoenzyme isolated from different mammalian organ material had the tendency to break down in lower molecular mass products irrespective of the presence or absence of protease inhibitors. As a matter of fact it was observed that the purified native enzyme's catalytic subunit slowly converted from the 44 kDa form to a ca. 38 kDa form as judged by SDS-PAGE [5]. A comparison of the kinase activity before and after breakdown did not show any significant changes suggesting that the 6 kDa polypeptide did not affect the phosphoryl transferase activity used to phosphorylate casein or the synthetic peptide *in vitro*. Hence, it did not come to a surprise when the first recombinant catalytic subunit of CK2 was cloned and expressed to observe a similar behavior, i.e. short after induction of the expression by IPTG appeared already a cleavage product of ca. 40 kDa. Again the addition of protease inhibitors did not have any effect on this cleavage process. Once the full length polypeptide was converted to the 40 kDa form no further breakdown was observed.

By contrast, using the catalytic subunit from *Z. mays* no such cleavage was observed. However, as already pointed out above, the CK2 gene product from maize is ca. 6 kDa shorter than the human counterpart. This observation is intriguing and suggests that the removal of the carboxy terminal end might affect some feature other than kinase activity which

Address for offprints: Olaf-Georg Issinger, Syddansk Universitet, Institut for Biokemi og Molekylær Biologi, Campusvej 55, 5230 Odense, Denmark (E-mail: ogi@bmb.sdu.dk)

may be specific for mammals but not needed in plants. In any case the lack of the C-terminus does not lead to a measurable change of activity towards the test substrates casein and the synthetic peptide.

This observation prompted us to use the stable catalytic subunit from maize for crystallization experiments. Indeed the stable maize enzyme was purified and crystallized [6] and the structure was solved [7]. Until recently (M. Pages, pers. comm.) all attempts to find a regulatory subunit in *Z. mays* were unsuccessful, even leading to the belief that there may not be a regulatory subunit at all in maize. A similar assumption was made for CK2alpha from *Dictyostelium discoides* [8]. Hence, in order to tackle the structure of the tetrameric holoenzyme we opted for the analysis of a chimeric tetramer consisting of maize α and human β, not considering recombinant human α for the reasons mentioned already in detail above.

Material and methods

Protein kinase assay

CK2 activity was tested 5 min at 37°C in a mixture containing 25 mM Tris/HCl pH 8.5, 200 mM NaCl, 5 mM $MgCl_2$, 1 mM DTT, 50 μM [γ-^{32}P]ATP (specific activity: 1000–2000 cpm $pmol^{-1}$), 190 μM synthetic peptide substrate (RRRDDDSDDD) in a total vol. of 50 μl. One unit of enzyme is defined as the amount of activity necessary to transfer 1 μmol phosphate min^{-1} into the synthetic peptide substrate at 37°C. In the experiments where the behavior of rhCK2 and the partial degraded form were compared with respect to salt, heat, urea, trypsin, 0.02 mU of enzyme were routinely used. The reaction was terminated by setting the samples on ice and by spotting 30 μl on P81 phosphocellulose paper as described previously [9]. The ^{32}P incorporated was measured with a liquid scintillation counter (Packard).

Phosphorylation of calmodulin and casein by rhCK2 and rhCK2$^{\Delta}$

Phosphorylation of calmodulin and casein by rhCK2 before and after partial degradation was performed in a reaction mixture containing 25 mM Tris/HCl pH 8.5, 150 mM NaCl, 5 mM $MgCl_2$, 1 mM DTT, 50 μM [γ-^{32}P]ATP (specific activity 1000 cpm $pmol^{-1}$) in a 30 μl final vol. Calmodulin (100 ng) or casein (50 ng) were phosphorylated by 0.03 mU of rhCK2 before and after partial degradation for 30 min at 37°C. When present 50 ng polylysine or 1.2 ng heparin were used in the phosphorylation assay mix. The samples were analyzed by SDS-PAGE and the radioactivity incorporated was measured in an Instant Imager Apparatus (Packard).

Gel-filtration chromatographic analysis

Analytical gel filtration chromatographic analysis was performed after overnight incubation at 4°C of recombinant chimeric CK2 with different amounts of SDS, as indicated in the figures, in buffer A containing 25 mM Tris/HCl pH 8.5 and 300 mM NaCl. Samples were loaded onto a Superdex 200 column mounted on a SMART chromatographic system (Amersham-Pharmacia) equilibrated in buffer A.

Results and discussion

Comparison of CK2 holoenzyme from bovine testes (CK2), recombinant human CK2α (rhCK2α) and recombinant CK2α from maize (rmCK2α)

We have analyzed native CK2 from bovine testes, together with recombinant human and plant CK2α on immunoblots in order to detect also trace amounts of degraded material, which might escape detection by the lesser sensitive Coomassie staining. Figure 1 shows that in the case of the native CK2 holoenzyme (only α-subunit detection is shown) from bovine testes and in the case of the recombinant human CK2α degradation product is readily visible whereas in the case of the CK2α from *Z. mays* no such product is detectable confirming earlier observations of an unstable mammalian CK2α subunit [5, 10].

Purification of a chimeric tetrameric CK2 holoenzyme consisting of CK2α subunit from maize and CK2β subunit from man

The proteins were expressed separately as described earlier [11]. The bacterial pellets expressing either the one or the other subunit were mixed in a 1:1 ratio, lysed and the purification carried out according to the scheme already described for the homologous recombinant human tetrameric CK2 holoenzyme [11]. In brief, the first steps of the purification in-

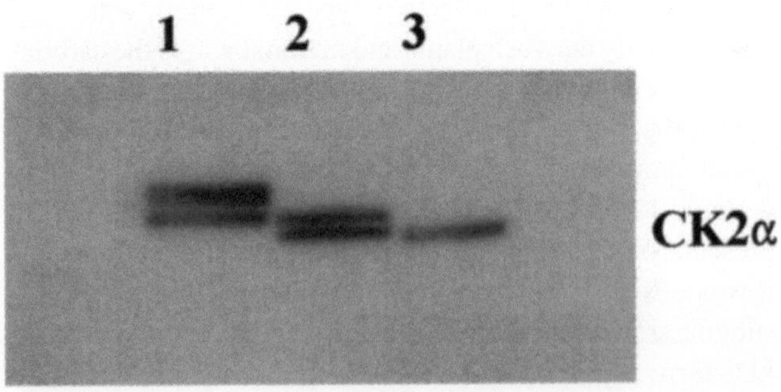

Fig. 1. Immunoblot showing the native CK2α from bovine testes (lane 1), recombinant human CK2α and recombinant CK2α from *Z. mays*.

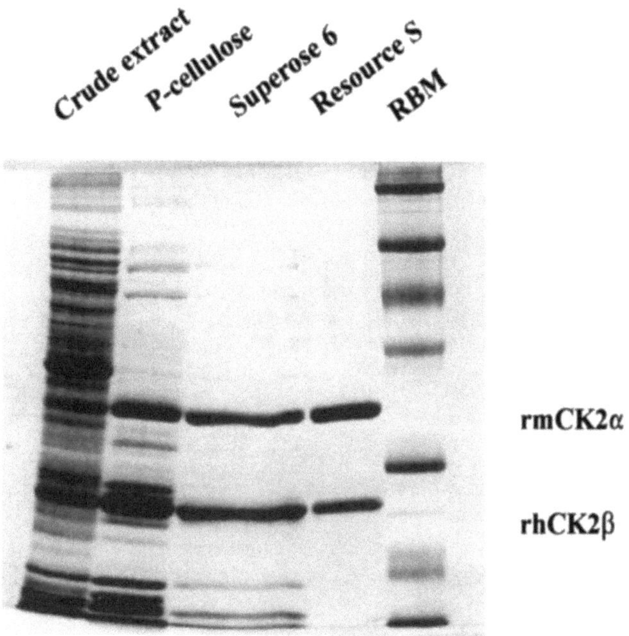

Fig. 2. 12.5% SDS-PAGE analysis from combined bacterial extracts containing the chimeric recombinant CK2 holoenzyme assembled from human CK2β and maize CK2α. Samples of CK2 collected after each purification step were loaded. Protein applied: lane1, 30 μg; lanes 2 and 3, 20 μg; lane 4, 6 μg; lane 5, rainbow marker.

volved phosphocellulose chromatography, gel filtration on Superose 6 (Amersham-Pharmacia) with the final purification step using Resource S (Amersham-Pharmacia). The final product was virtually purified to homogeneity (Fig. 2). Table 1 shows the degree of purification achieved during the various chromatographic steps involved throughout the purification. Starting from a total of 2.4 g bacterial extract containing the expressed recombinant chimeric CK2 from maize and man the final yield was roughly 24 mg of homogenous chimeric holoenzyme. A 20-fold purification was achieved. The specific activity obtained (2.5 U/mg) is well comparable to that of the homologous recombinant human CK2 holoenzyme but also to the catalytic subunit of maize CK2α alone (Table 2).

Table 1. Purification scheme of recombinant chimeric CK2

	Volume (ml)	Protein (mg)	Total activity (U)	Specific activity (U/mg)	Purification (fold)
Crude extract	680	2448	293.76	0.12	1
P-cellulose	714	350	105	0.30	2.5
Mini P-cell	35	234.5	150	0.64	5.33
Superose	170	69.2	134.9	1.95	16.25
Resource S	4.5	24.4	62.9	2.58	21.5

Table 2. Kinase activity comparison of different forms of recombinant protein kinase CK2

	Specific activity (μmol/min/mg)
Recombinant human CK2 Mw 140,170	3.34
Recombinant human CK2 (partially degraded) Mw 128,000	2.9
Recombinant chimeric CK2 Mw 128,344	1.98
Recombinant human CK2α Mw 45,143	0.59
Recombinant maize CK2α Mw 39,230	1.12
Recombinant human CK2β Mw 24,942	0

The influence of various biochemical and biophysical challenges on the recombinant human CK2 holoenzyme and the chimeric molecule

Influence of monovalent salt ions

Over a range from 0–500 mM NaCl we did not observe significant differences in the peak activity distribution. In the case of the recombinant human holoenzyme CK2 activity was highest at 150 mM whereas in the chimeric holoenzyme the activity peak shifted slightly to 100 mM NaCl (Fig. 3A).

Quenching effect of GTP

In order to test for the typical duo-cosubstrate specificity of protein kinase CK2, i.e. to use ATP as well as GTP as a cosubstrate, enzyme activity was measured in the presence of (i) ATP as the sole phosphoryl donor and (ii) after adding a second non-radioactive phosphoryl donor to the assay mix, i.e. GTP. The addition of GTP led to a reduction of CK2 activity suggesting a competing effect of this nucleotide. The losses in activity were in both cases, very comparable for the human holoenzyme and the chimera (Fig. 3B).

The effect of polylysine

Polylysine is known to stimulate CK2 activity for reasons so far not completely understood. By contrast to the denaturating challenge exerted to the two CK2 forms, the stimulation of the protein kinase activity for both the chimeric enzyme and the recombinant human holoenzyme were indistinguishable (Fig. 3C).

Formation of chimeric enzyme in vitro

In an experiment, designed to check the influence of increasing amounts of CK2β on protein kinase activity while the amount of CK2α was kept constant maximum activity was obtained close to equimolar amounts of subunits in the case of human CK2α whereas in the case of the CK2α from maize the number of CK2β molecules needed to obtain maximum activity was in the range of 1.5–2.0 (Fig. 3D).

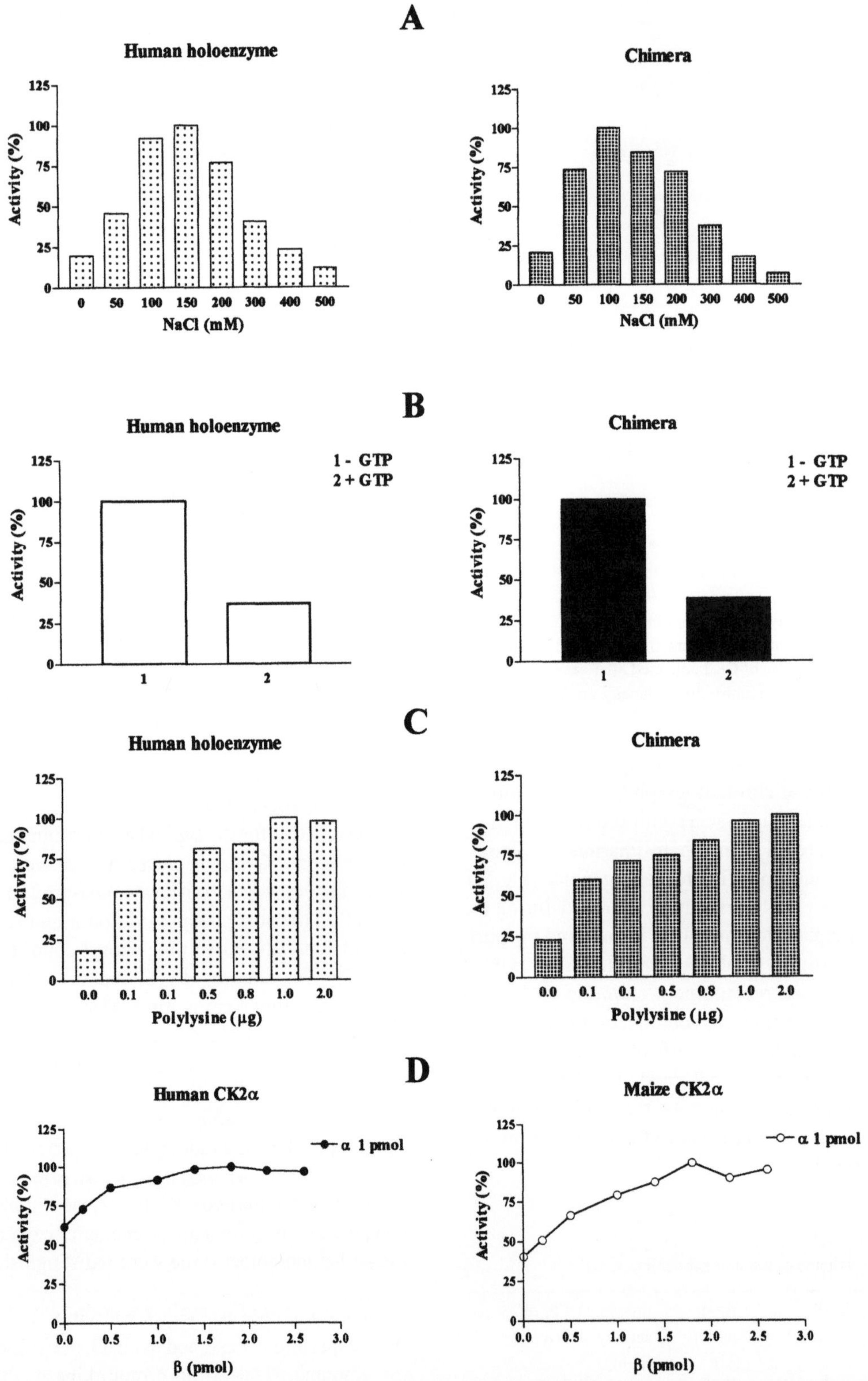

Fig. 3. Various challenges on enzyme activity of recombinant human holoenzyme and the chimeric CK2 from recombinant human CK2β and recombinant CK2α from maize. (A) Influence of increasing amounts of NaCl. (B) Influence of GTP as a competitor to ATP. (C) Effect of polylysine on enzyme activities. (D) Reconstitution experiments were carried out by keeping the amount of CK2α constant and adding increasing amounts of CK2β.

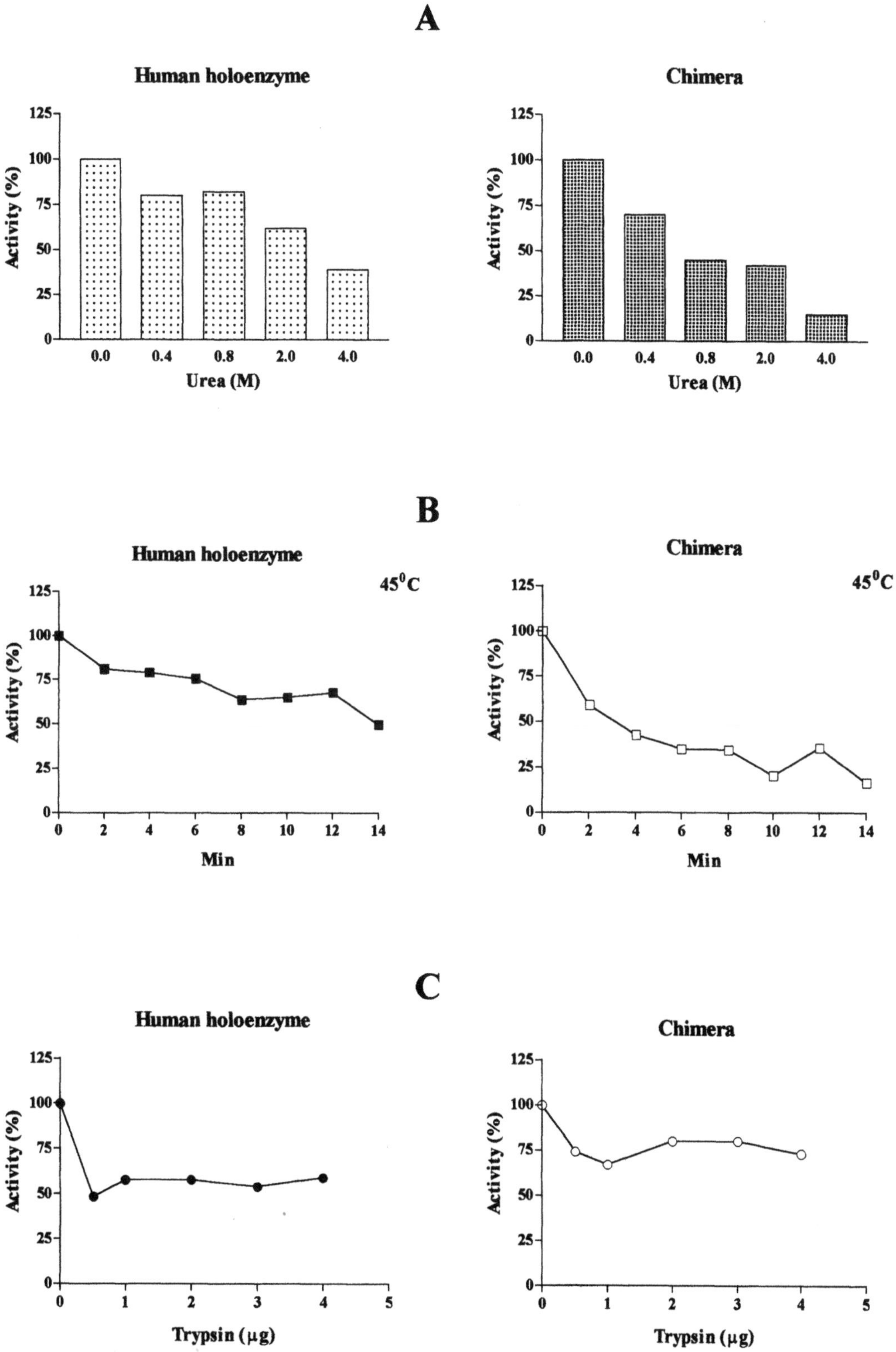

Fig. 4. Comparative analysis of rhCK2 and recombinant chimeric CK2 with respect to different treatments. (A) Incubation of the enzymes for 10 min at room temperature in the presence of different urea concentrations (0–4 M). (B) The influence of 45°C over a time period up to 14 min on enzyme activity for both the recombinant human holoenzyme and the chimera was analyzed. (C) Partial trypsin digestion was done by incubating the enzymes with increasing amounts of trypsin for 2 min at 37°C. The samples were immediately cooled on ice and the digestion was stopped by adding 4 mol excess soybean trypsin inhibitor. In all experiments CK2 activity is expressed in percent. The untreated sample was set 100% in all cases.

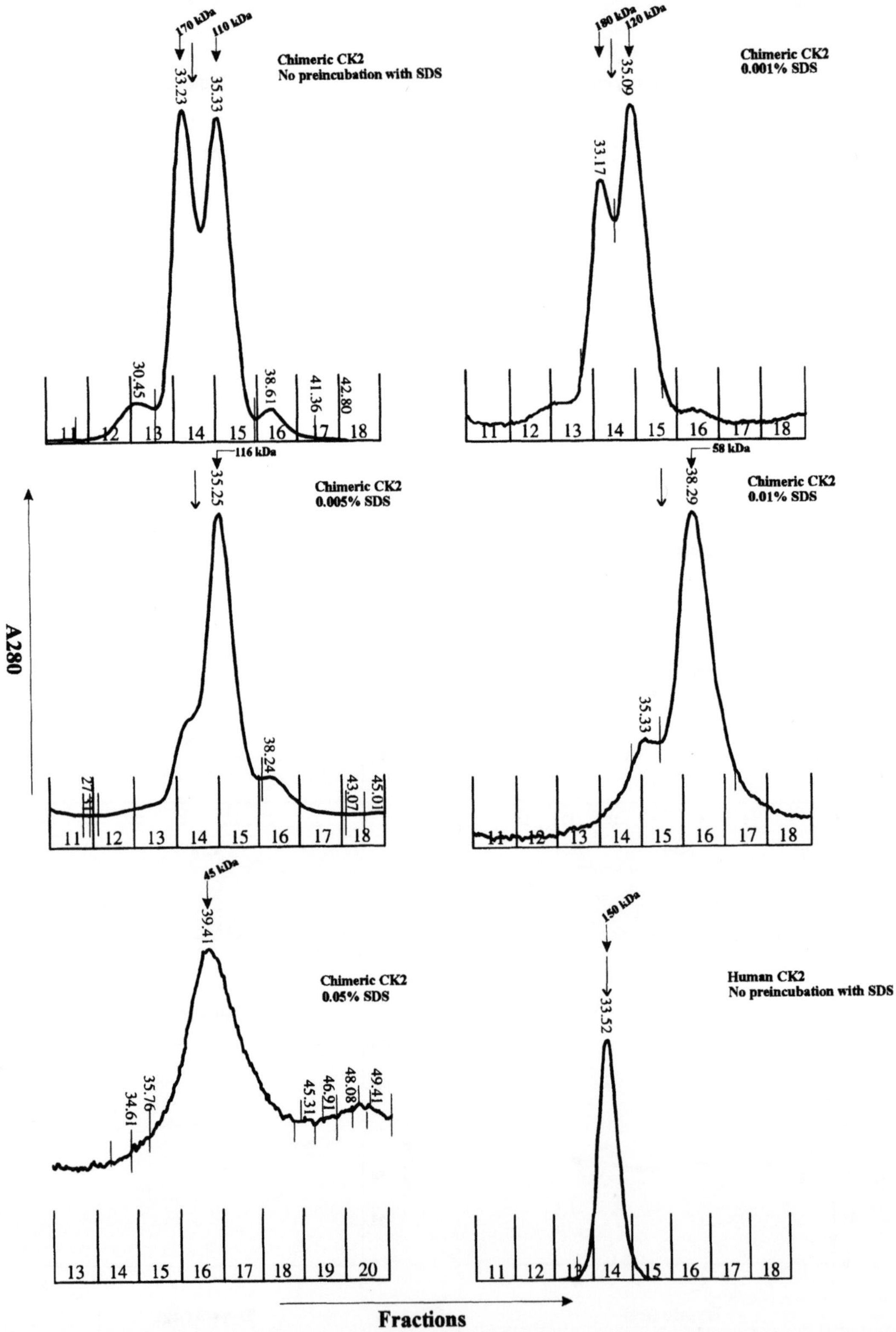

Fig. 5. Analytical gel-filtration chromatographic analysis of recombinant chimeric CK2 holoenzyme and rhCK2. 12 μg of either rhCK2 or recombinant chimeric CK2 after dialysis overnight with different amounts of SDS in buffer A, as indicated in the figure, were applied on a Superdex 200 PC 3.2/30 column (SMART system, Amersham-Pharmacia) equilibrated in buffer A.

Treatment with urea
The influence of urea treatment on CK2 activity is shown in Fig. 4A. Recombinant human holoenzyme and chimeric CK2 were challenged with increasing amounts of urea starting from 0.4–4 M urea. Figure 4A shows the effect of the various urea concentrations on the activity of the two enzymes. In the case of the recombinant human holoenzyme the effect of urea lead to a ca. 50% reduction in activity between 2–4 M urea. A more than 50% reduction was already obtained at a concentration of 0.8 M in the case of the chimeric enzyme. These data clearly show that the chimeric enzyme, in agreement with the heat denaturation results (see below), is not that stable than the recombinant human holoenzyme.

Heat inactivation effect
When the homologous human holoenzyme was subjected to 45°C over a time period of 14 min the CK2 holoenzyme activity dropped to 50%. This degree of inactivation was already observed after 2 min in the case of the chimeric enzyme suggesting that this complex was not so tight in order to protect the enzyme from denaturation. 14 min incubation of the chimeric enzyme led to a >80% loss of activity (Fig. 4B).

Partial trypsin digestion effect
Treatment of the human holoenzyme and the chimeric enzyme with 0.5–4 μg trypsin for 2 min at 37°C showed comparable inactivation patterns, yet it looked as if the chimeric enzyme was even more resistant to partial tryptic digestion as the recombinant human holoenzyme (Fig. 4C).

Gel filtration analysis of the chimeric CK2 holoenzyme
In order to characterize the chimeric holoenzyme also by gel filtration analysis the chimeric enzyme was analyzed in parallel to the recombinant human holoenzyme (Fig. 5). As it turned out the obtained sedimentation profile revealed a double peak in the case of the chimeric enzyme as compared to the homologous holoenzyme from human. This was a surprising result because we had expected, from what we saw on the SDS PAGE analysis result, also one homogenous peak during gel filtration. Since we assumed an unspecific association leading to higher molecular mass complexes we added increasing amounts of SDS (0.001–0.05%) (Fig. 5). As one can readily see from the profiles in Fig. 5 this treatment did not solve the problem, by contrast the chimeric CK2 started to deteriorate at SDS concentrations > 0.01%. In fact the peak of activity indicated by the open arrow did not follow the peak of absorbance indicated by the closed arrow (Fig. 5). At 0.05% SDS the tetrameric complex was almost completely dissociated into its subunits and the activity was not any more detectable.

Dynamic light scattering
DLS, using a DynaPro-801 instrument, revealed that the chimeric CK2 was strongly heterogenous by aggregation in solution in spite of the perfect SDS PAGE analysis. The DLS experiments further led to a polydispersity of 62% and to an average molecular mass of 310 kDa. For this experiment a protein concentration of 0.2 mg/ml was used. The polydispersity and the molecular mass values indicate a strong heterogeneity in solution which did not change by varying the NaCl concentration in the buffer from 0.15–1 M. The lowest relative polydispersity found was 47% which is still far from homogeneity, i.e. < 20%.

Production of a stable recombinant human CK2 holoenzyme (rhCK2)

As already mentioned above, the recombinant human CK2 holoenzyme was not homogenous with respect to a single polypeptide as judged from the SDS PAGE analysis (Fig. 6A). The amount of the breakdown component varied from roughly 1 to >50%. We did not find a correlation between CK2 activity and the percentage of breakdown product present in our enzyme preparation.

When looking for an alternative to the chimeric maize/human holoenzyme we decided to speed up the degradation of the homologous human recombinant CK2α subunit artificially and try to convert all native full length protein into the final breakdown compound. This was achieved by storage of the preparation on the shelf at room temperature for about 10 days. By this time the CK2α was estimated to be >99% converted into the breakdown product (Fig. 6B). The

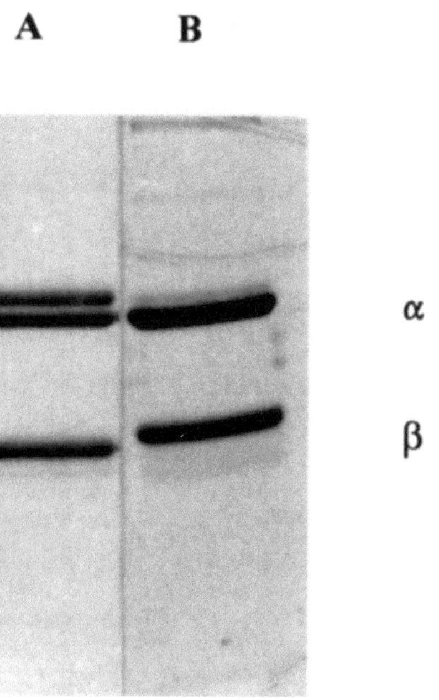

Fig. 6. 12.5% SDS-PAGE analysis of rhCK2 before (lane a) and after (lane

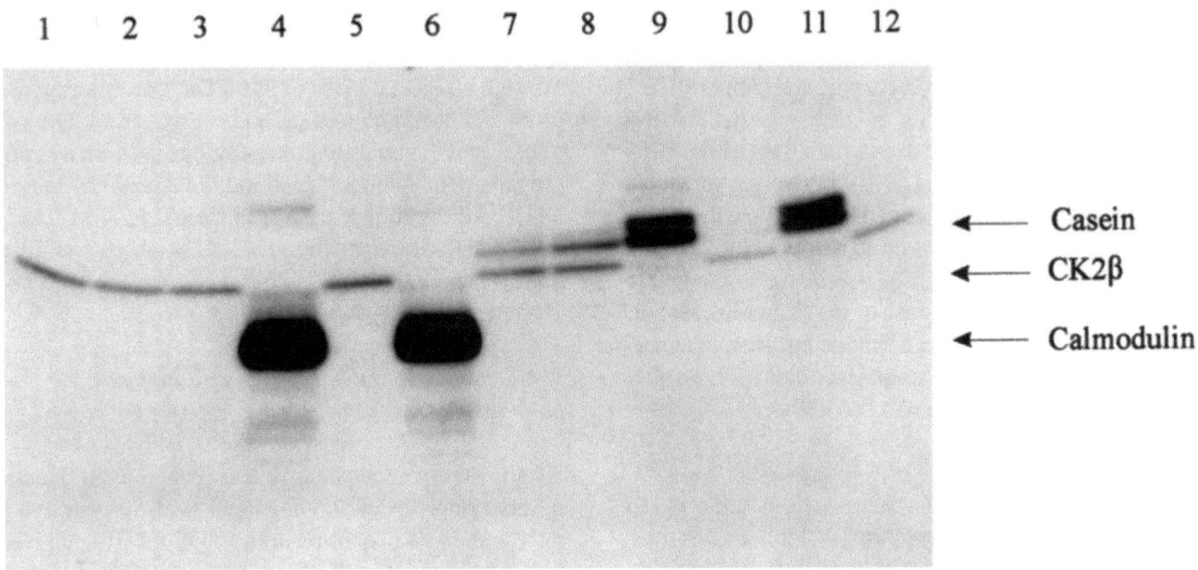

Fig. 7. Autoradiograph showing the phosphorylation of calmodulin and casein using rhCK2 before and after degradation. Lanes 1 and 2 show the autophosphorylation of the β-subunit of rhCK2 before and after degradation, respectively. Lanes 3 and 4 show phosphorylation of calmodulin in the absence and presence of polylysine catalysed by rhCK2 before degradation. Lanes 5 and 6 compare to lanes 3 and 4 in the presence of the partially degraded rhCK2. Lanes 7 and 8 show phosphorylation of casein catalysed by rhCK2 before and after degradation, respectively. Lane 9 and 11 show phosphorylation of casein in the presence of polylysine by rhCK2 before and after degradation, respectively. Lanes 10 and 12 shows phosphorylation of casein in the presence of heparin by rhCK2 before and after degradation, respectively. This figure is from Fig. 3 of [12].

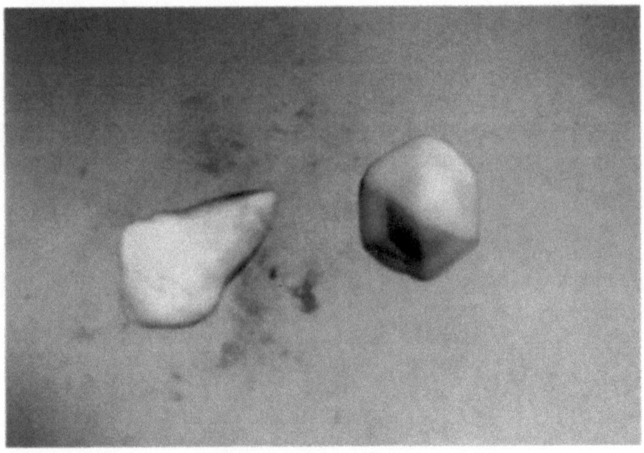

Fig. 8. Crystal of the partial degraded form of rhCK2 ($rhCK2^{\Delta}$)

so prepared homogenous CK2 holoenzyme from human source was perfectly stable. The activity was comparable to the activity found for the 'unmatured' holoenzyme (Table 2).

Hence, one important criteria, i.e. kinase activity, was shown to be unaltered by the 10 days maturation process at room temperature.

Crystallization experiments with rhCK2

As already described by [12] two crystal forms were obtained, e.g. (i) crystals with a cuboid like morphology which did not diffract beyond the 10 Å resolution level in a synchrotron beam and (ii) crystals of the second form shown in Fig. 6, where a diffraction data set to 3.5 Å resolution was obtained [12].

Acknowledgments

K. Niefind obtained an EMBO short term fellowship (ASTF 9305). I. Ermakowa is funded by the Professor Dr. Werner-Petersen-Stiftung, Kiel, Germany. The generous support by Prof. Dietmar Schomburg, Cologne, is gratefully acknowledged. O.G.I. is supported by the Danish Cancer Society (grant No. 96 100 40) and the Danish Research Council (grant No. 9601695).

References

1. Padmanabha R, Glover CVC: Casein kinase of yeast contains two distinct alpha polypeptides and an unusually large beta subunit. J Biol Chem 262: 1829–1835, 1987
2. Dobrowolska G, Boldyreff B, Issinger OG: Cloning and sequencing of CK2α subunit from *Z. mays*. Biochem Biophys Acta 1129: 139–140, 1991
3. Meisner H, Heller-Harrison R, Buxton J, Czech MP: Molecular cloning of the human casein kinase II α subunit. Biochemistry 28: 4072–4076, 1989

4. Guerra B, Issinger OG: Protein kinase CK2 and its role in cellular proliferation, development and pathology. Electrophoresis 20: 391–408, 1999
5. Münstermann U, Fritz G, Seitz G, Yiping L, Schneider HR, Issinger OG: CKII is elevated in solid tumors and rapidly proliferating non-neoplastic tissue. Eur J Biochem 189: 251–257, 1990
6. Guerra B, Niefind K, Pinna LA, Schomburg D, Issinger OG: Expression, purification and crystallization of the catalytic subunit of protein kinase CK2 from *Zea mays*. Acta Cryst D54: 143–145, 1998
7. Niefind K, Guerra B, Ermakowa I, Issinger OG: Crystal structure of the catalytic subunit of protein kinase CK2 from *Zea Mays* at 2.1 Å resolution. Embo J 1: 2451–2462, 1998
8. Kikkawa U, Mann SK, Firtel RA, Hunter T: Molecular cloning of casein kinase II alpha subunit from *Dictyostelium discoideum* and its expression in the life cycle. Mol Cell Biol 12: 5711–5723, 1992
9. Kuenzel EA, Krebs EG: A synthetic peptide substrate specific for casein kinase II. Proc Natl Acad Sci 82: 737–741, 1985
10. Boldyreff B, Meggio F, Dobrowolska G, Pinna LA, Issinger OG: Expression and characterization of a recombinant maize CK2 alpha. Biochem Biophys Acta 1173: 32–38, 1992
11. Grankowski N, Boldyreff B, Issinger OG: Isolation and characterization of recombinant human casein kinase II subunits and from bacteria. Eur J Biochem 198: 25–30, 1991
12. Niefind K, Guerra B, Ermakowa I, Issinger OG: Crystallization and preliminary characterization of crystals of human protein kinase CK2. Acta Cryst D56: 1680–1684, 2000

Molecular and Cellular Biochemistry **227**: 13–19, 2001.

Generation of mutants of CK2α which are dependent on the β-subunit for catalytic activity

Stefania Sarno,[1] Paola Ghisellini,[2] Luca Cesaro,[1] Roberto Battistutta[3] and Lorenzo A. Pinna[1]

[1]*Dipartimento di Chimica Biologica and Centro per lo Studio delle Biomembrane del CNR, Universita' di Padova, Padova;* [2]*Istituto di Biofisica, Universita' di Genova, Genova;* [3]*Dipartimento di Chimica Organica e Centro di Ricerca Biopolimeri del CNR, Universita' di Padova, Padova, Italy*

Abstract

To shed light on the structural features underlying high constitutive activity of protein kinase CK2 a number of mutants of the human CK2α–subunit altered in the interactions between the N-terminal segment and the activation loop have been generated and shown to be defective in catalytic activity. In particular the truncated mutant Δ2–12 displays under standard conditions an almost complete loss of catalytic activity accounted for by a dramatic rise in its Km for ATP (from 10 to 206 μM) and a reduced Kcat. Such a drop in efficiency is paralleled by conformational disorganization, as judged from Superdex 75 gel filtration profile. Both catalytic properties and gel filtration behaviour similar to those of wild type CK2α were restored upon association with the regulatory β-subunit, suggesting that constitutive activity is conferred to CK2α and to CK2 holoenzyme through different molecular mechanisms. In the holoenzyme an assumable release of tension at the backbone of Ala-193 (as seems to be indicated by a comparison of the crystal structures of maize CK2α alone vs. a CK2α–β peptide complex) may result in the ability of the activation loop to adopt its proper conformation independently of interactions with the N-terminal segment. (Mol Cell Biochem **227**: 13–19, 2001)

Key words: protein kinase CK2, α catalytic subunit, N-terminal domain, casein kinase 2, CK2α mutants

Introduction

One of the most remarkable features of protein kinase CK2 is high constitutive activity of both its holoenzyme and the isolated catalytic subunits. This is not the case of most protein kinases, whose activity has to be triggered by specific effectors and/or molecular events, notably phosphorylation. In particular, the few protein kinases displaying like CK2 a quaternary structure, e.g. PKA (cyclic AMP-dependent protein kinase) and CDKs (cyclin dependent protein kinases), these latter belonging to the same CMGC group as CK2 [1], are strictly controlled by the association-dissociation of their regulatory subunits (see Fig. 1). In contrast, the phosphotransferase activity of the catalytic subunits of CK2 (α and α′) is altered, but neither suppressed by nor entirely relying on association with the regulatory β-subunits. Although in fact there are sub-sets of CK2 protein substrates whose phosphorylation is either critically promoted or, conversely, inhibited by the β-subunit (see Fig. 1) the α/α′-subunits definitely are catalytically competent both in the absence and presence of the β-subunit as judged from the phosphorylation of peptide substrates.

At variance with the catalytic subunits of CDKs whose activation is stringently dependent on association with the regulatory cyclin, and whose N-terminus is just upstream from the glycine rich loop, the constitutively active catalytic subunits of CK2, either α or α′, are endowed with an extra N-terminal segment of about 30 residues (see Fig. 2A). This region makes extensive contacts with elements of the catalytic core, notably the activation loop and the first part of helix-C [2]. These interactions, which are reminiscent of those occurring between the activatory cyclin A and the catalytic subunit of CDK2 [3], have been postulated to be responsible for the constitutive activity of CK2α [2].

Address for offprints: L.A. Pinna, Dipartimento di Chimica Biologica, Viale G. Colombo, 3 35121 Padova, Italy

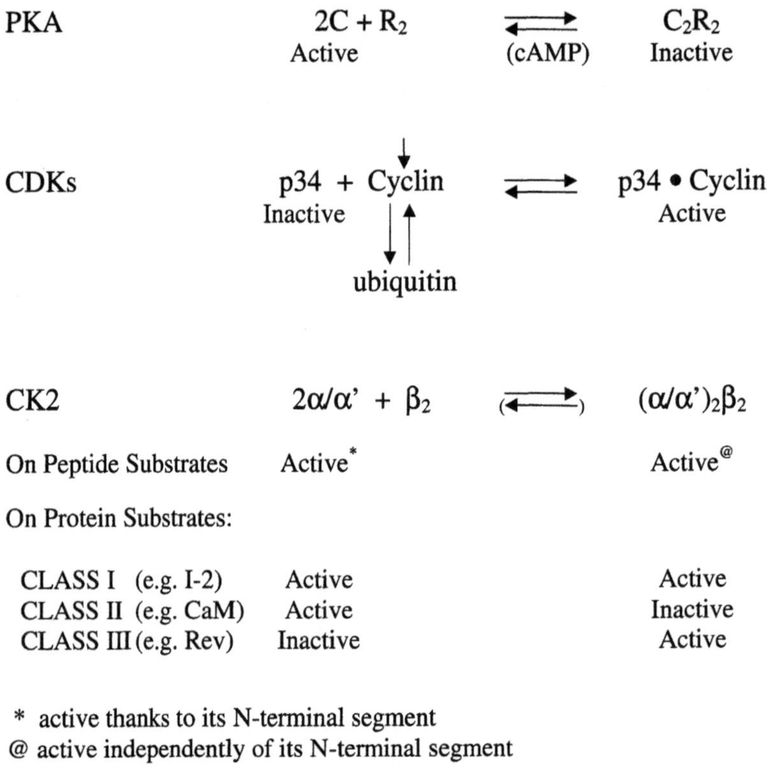

Fig. 1. Influence of quaternary structure on the activity of three oligomeric protein kinases. (I-2 = inhibitor-2 of protein phosphatase-1; CaM = calmodulin). For details see text.

To probe this hypothesis and to check whether the β-subunit might surrogate the activatory role of the N-terminal segment, we have designed a number of human CK2α mutants in which the contacts between the N-terminal segment and the activation loop are expected to be disrupted or weakened. Here we describe the generation and expression of these mutants and we report on the biochemical characterization of one of them (Δ2–12) whose catalytic activity under standard assay conditions is lost, but can be entirely restored upon association with the β-subunit.

Materials and methods

Mutagenesis and expression of CK2 α-subunits

The αY26F and αΔ2–6 mutants were generated as previously described [4]. The αE180A, αY182F and the N-terminal deleted mutants αΔ2–12, αΔ2–18 and αΔ2–24 were obtained using the 'QuickChange site-directed mutagenesis' kit (Stratagene). Each mutant was generated using two synthetic oligonucleotide primers, each complementary to opposite strands of human α cDNA inserted in pT7-7 vector, 5′-CTGGGGTTTGGCTGCGTTTTATCATCCTG-3′,5′-GGGTTTAGCAGAGTTT-TTCCATCCTGGCCAAGAG-3′ for E180A and Y182F respectively. 5'-GGAGATATACATATGACAGATGTTAATACACACAGACCTCG-3′, 5′-GGAGATATACATATGAGACCTCGAGAATACTGGGATTACG-3′ and 5′-GGAGATATACATATGGATTACGAGTCACATGTGGTGG-3′ were the primers used to generate sequential deletions αΔ2–12, αΔ2–18 and αΔ2–24 respectively, using as template the α cDNA of the previous deletion. All the mutants were completely sequenced and used to transform BL21(DE3) competent bacteria cells. Protein expression and purification was performed as previously described [5]. α catalytic subunits were dialyzed against 25 mM Tris pH 7.5 and 50% of glycerol and stored at –20°C.

Assay of CK2 activity

Phosphorylation of the specific synthetic peptide substrate RRRADDSDDDDD [100 μM] was performed by incubating CK2α-subunits alone (0.44 pmol), or in the presence of the β-subunit, for 10 min at 30°C in a final volume of 30 μl containing 50 mM Tris-HCl, pH 7.5, 12 mM $MgCl_2$, 100 mM NaCl and 50 mM [γ-^{32}]ATP (specific activity 300 cpm/pmol). Kinetic analyses were performed by incubating CK2α wild type and αΔ2–12 (0.22 and 2.2 pmol, respectively) with variable amounts of ATP (5–500 μM) at a fixed peptide substrate

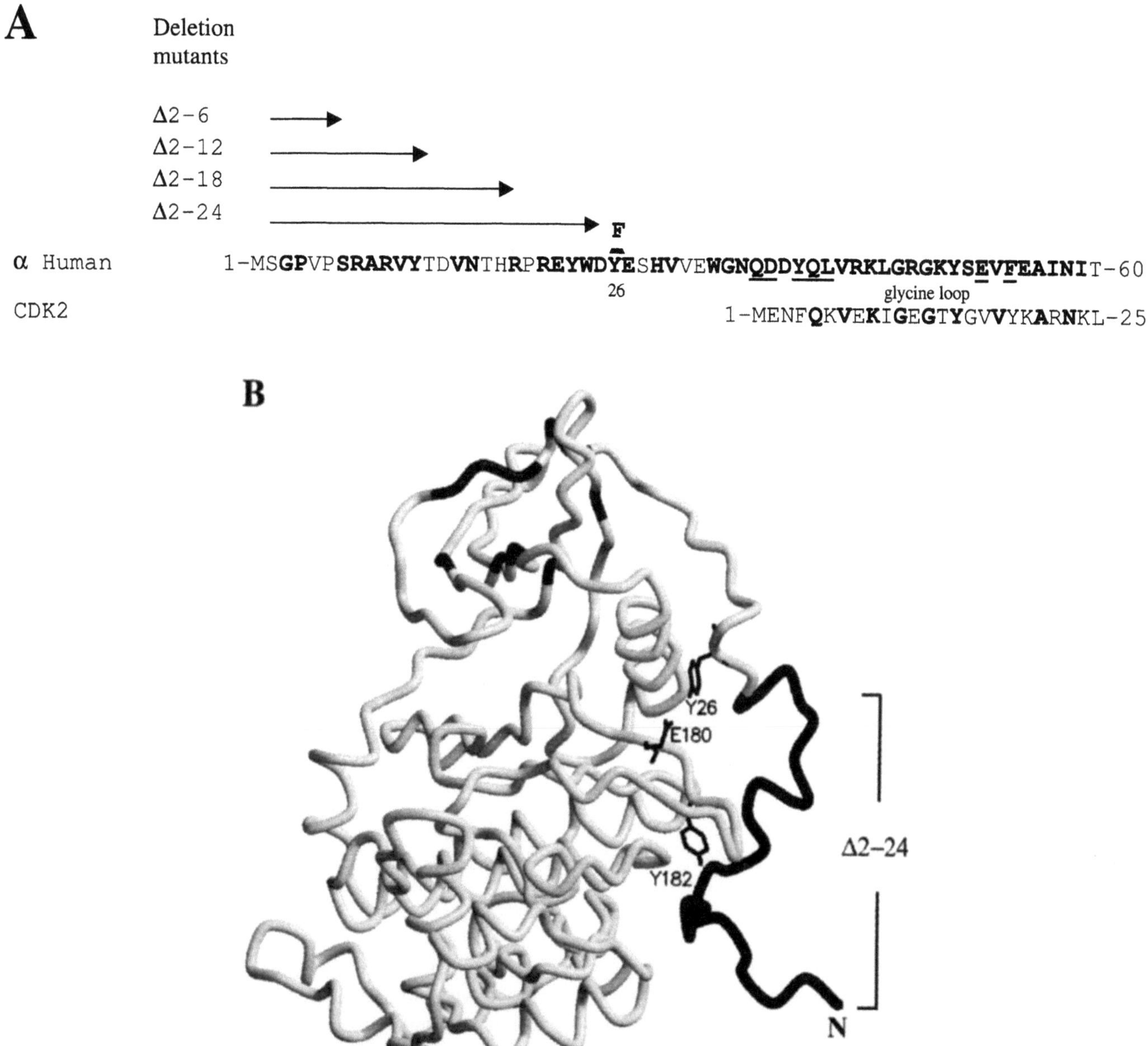

Fig. 2. Mapping of CK2α mutations affecting the interaction between the N-terminal segment and the activation loop. (A) The N-terminal region of human CK2α is aligned with that of CDK2 catalytic subunit. Deletions from the N-terminal side are indicated by arrows; point mutation of Tyr-26 to phenylalanine is denoted by a triangle. Residues making contacts with β in the maize CK2α/β 181–203 peptide complex [8] are underlined. (B) A modelization of the backbone of human CK2α based on the crystal structure of maize CK2α [2] (>70% identical) is presented, showing the N-terminal segment largest deletion (Δ2–24), the residues mutated in the activation loop (E-180, Y-182) and in the N-terminal segment (Y-26) in black. Contacts with the β peptide 181–203 [8] are denoted by dark gray color. The figure is drawn using Raster3D and MOLSCRIPT programs [10, 11]

concentration (200 μM), or varying the peptide substrate concentration between 10–250 μM at a fixed concentration of ATP (50 μM), without NaCl. ^{32}P incorporation was evaluated by the phosphocellulose paper procedure [6]. Kinetic constants were determined by double-reciprocal plots constructed from initial measurements fitted to the Michaelis–Menten equation.

Gel electrophoresis

Different aliquots of purified α mutants were subjected to 12% SDS-PAGE according to Laemmli [7] and stained with Coomassie Brilliant Blue.

Gel filtration analysis

Purified α-subunits (200 μl) either alone or assembled with equimolar amounts of the β-subunit, were subjected to Superdex 75 HR 10/30 (Pharmacia) FPLC (fast protein liquid chromatography). Before applying the sample, the column was equilibrated with 50 ml of 25 mM Tris-HCl pH 7.5, 7 mM 2-mercaptoethanol and 0.5 M NaCl. The elution profile was analyzed by OD monitor at 280 nm. Carbonic anhydrase (29 kDa), bovine serum albumin (66 kDa) and β-amylase (200 kDa) were run separately at the same conditions as molecular mass markers.

Results and discussion

Generation of mutants defective in the interactions between the N-terminal segment and the activation loop

The main interactions occurring between the N-terminal segment and the activation loop in the crystal structure of maize CK2α [2] are summarized in Table 1. In order to assess the actual relevance of these interactions in conferring constitutive activity to human CK2α, seven mutations have been performed, as schematically summarized in Figs 2A and 2B. Four mutants bear deletions of increasing extension on their N-terminal segment, lacking 5 (Δ2–6), 11 (Δ2–12), 17 (Δ2–18) and 23 (Δ2–24) residues, respectively. The other mutants (Y26F, E180A, Y182F) underwent point mutations of residues whose side chains in the crystal structure of maize CK2α [2] participate with hydrogen bonds between the activation loop and the N-terminal segment (see Table 1). These residues, like most of the residues in the N-terminal segment and all those in the activation loop, are conserved between maize and human CK2.

All the mutants were expressed in *E. coli* and purified to near homogeneity (Fig. 3) as described in the experimental section. Under standard assay conditions all mutants proved more or less severely defective in catalytic activity, corroborating the concept that the interactions between the N-terminal segment and the activation loop play a crucial role in conferring constitutive activity to CK2α. A detailed account of the properties of mutant Δ2–12 is provided below.

Table 1. Main interactions between the N-terminal segment and the activation loop of CK2α

N-terminal segment residue	Activation loop residue
Ala9-N	Tyr182-OH
Asn16-NΔ2	Tyr182-O
Tyr23-CE2	Phe181-CE1
Trp24-CB	Phe181-CD1
Tyr26-OH	Glu180-OE2

Drawn from [2].

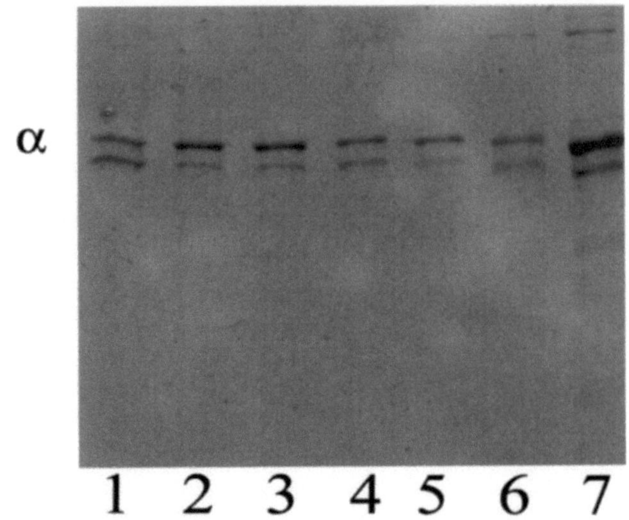

Fig. 3. SDS-PAGE analysis of six CK2α mutants. The Coomassie-stained gel is shown. CK2α mutants were purified as described under Materials and methods. Variable amounts (0.5–1 μg) of each mutant were loaded on the gel. Lane 1, α wild type; lane 2, αY26F; lane 3, αY182F; lane 4, αE180A; lane 5, αΔ2–6; lane 6, αΔ2–12; lane 7, αΔ2–18. All mutants, as well as CK2α wild type give rise to a doublet due to proteolytic degradation of α. Both bands are recognized by anti-CK2α antibodies (not shown).

Biochemical characterization of mutant Δ2–12

A kinetic analysis of mutant Δ2–12 revealed that its negligible activity under standard assay conditions is accounted for by both a 20-fold higher Km value for ATP (206 μM instead of 10 μM) and a reduced Kcat value (see Table 2). Also the Km value for the phosphoacceptor peptide substrate (calculated in the presence of 50 μM ATP) is substantially increased. These data are consistent with a scenario whereby, by disrupting the interactions between the N-terminal segment and the activation loop, the structure of the active site is deeply altered, resulting in a reduced aptitude to bind the substrates. Gel filtration experiments (Fig. 4) suggest that the 2–12 deletion destabilises the whole structure of CK2α, giving rise to partially unfolded conformers and/or to aggregated forms which are eluted before the peak of CK2α wild type.

Interestingly, upon addition of equimolar amounts of the β-subunit, these differences in gel filtration profile between

Table 2. Kinetic constants of CK2 α w.t. and αΔ2–12 for ATP and the peptide substrate RRRADDSDDDDD

	Km (ATP; μM)	Km (peptide; μM)	Kcat (min^{-1})
α wt.	10	26	64
αΔ2–12	206	140	44

Experimental conditions are detailed in Materials and methods. The data are the mean of at least three independent experiments with standard errors not exceeding 10%.

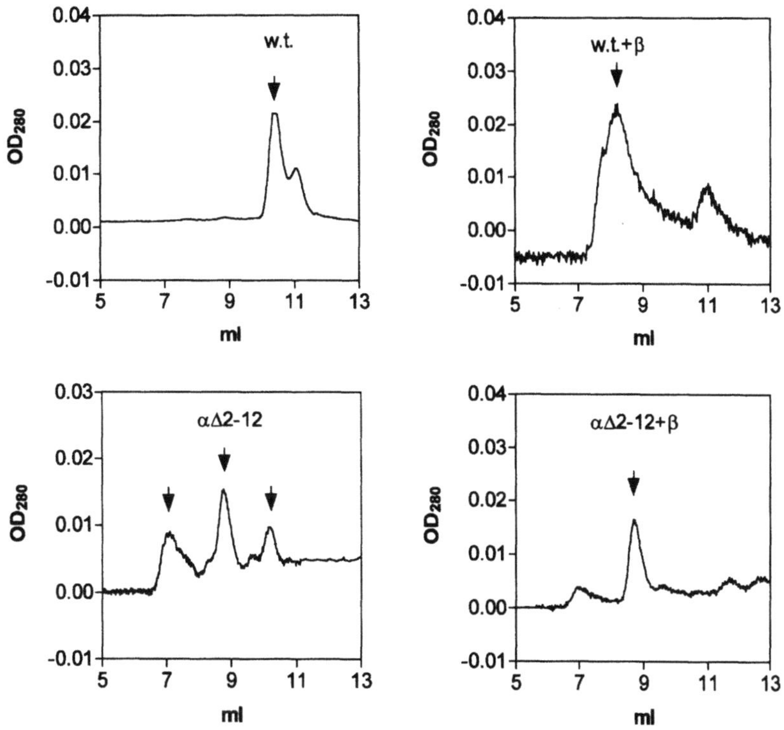

Fig. 4. Gel filtration profiles of CK2α wild type and Δ2–12, alone (left panels) and in combination with equimolar amount of β-subunit (right panels). Purified α wild type (1 pmol) and αΔ2–12 (1.2 pmol) were loaded and run through a Superdex 75 column as described in Materials and methods. CK2 holoenzymes were obtained by incubating together at 4°C for 1 h equimolar amount of the α- and β-subunits (1 and 0.6 pmol of α wild type and Δ2–12, respectively). The identification of the α-subunits and CK2 holoenzyme was confirmed by immunoreaction with anti-α- and anti-β-subunit antiserum in the peaks denoted by arrows (not shown).

Δ2–12 and wild type disappear (Fig. 4), indicating that in both cases the regular heterotetrameric holoenzyme has been formed. Consequently also the catalytic activity of the deleted mutant Δ2–12, which is almost undetectable under standard assay conditions, is restored to wild type level upon addition of equimolar amounts of the β-subunit (Fig. 5).

Interaction with a β-peptide induces conformational changes into the activation loop of maize CK2α

The data above would indicate that not only does the binding of β not require the integrity of the N-terminal segment of CK2α, but also that upon binding of β the catalytic subunit undergoes activation by a mechanism different from the one conferring constitutive activity to the isolated α-subunit.

As the structure of CK2 holoenzyme is not yet available, the precise mechanism by which CK2α associates with and is activated by the β-subunit remains a matter of conjecture. Some information, however, can be drawn from the structure of a chimeric complex between maize CK2α and a synthetic fragment of human β which includes the main residues involved in physical interaction with α [8]. The structural data are consistent with the concept that the occurrence of α–β interactions does not require the integrity of the N-terminal segment of α, since the physical contacts between the two subunits, though concentrated in the N-terminal lobe of α, are not involving any of the residues of the N-terminal segment (see Figs 2A and 2B).

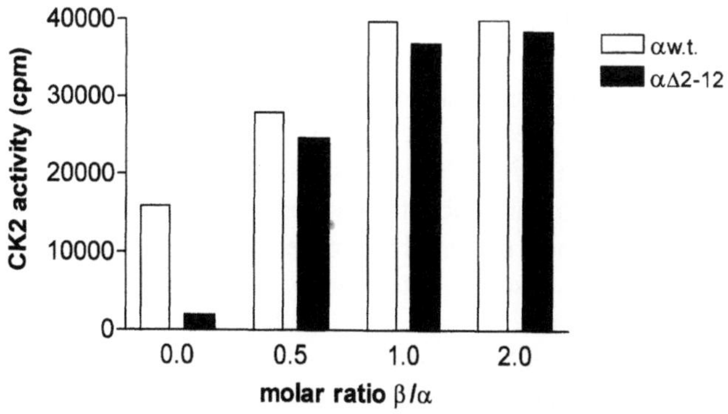

Fig. 5. Stimulation of CK2α and αΔ2–12 catalytic activity by the β-subunit. The activity of CK2α and αΔ2–12 (0.44 pmol) was tested toward the synthetic peptide substrate RRRADDSDDDDD (100 μM) either in the absence or in the presence of increasing amounts of the β-subunit. Phosphorylation conditions and evaluation of ^{32}P incorporated are described in the experimental section.

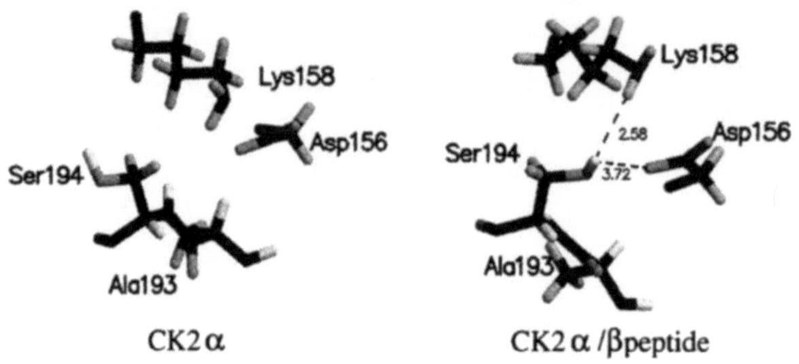

Fig. 6. Conformational changes occurring at the activation loop of maize CK2α upon binding of the β 181–203 peptide. Based on the coordinates of maize CK2α [9] (left side) and of the maize CK2α/β peptide complex [8] (right side). For details see text.

On the other hand, the overall structure of the α-subunit seems to be only marginally altered by its association with the β-peptide, as judged from the almost perfect superimposition of the two chains with rmsd on corresponding Cα atoms of about 0.55 Å [8]. In the crystal structure of maize CK2α, Ala-193, at the end of the activation loop, is the only residue falling in a disallowed region of the Ramachandran plot [9]. The structural tension of the backbone around Ala-193 is suspected to reflect the tendency of the activation loop to change its orientation unless it is locked in its 'active' conformation [2]. The tension at Ala-193 in the isolated CK2α could be released in the tetrameric holo-enzyme where the interactions with the N-terminal segment could be no longer required in order to confer to the activation loop its proper conformation. This hypothesis is in accordance with the crystal structure of maize CK2α in complex with a fragment of the human β-subunit [8], in which residue Ala-193 moves to a permitted region with a consequent release of the conformational strain. Anyway, the relative low resolution of this complex (3.16 Å) prevents a definite proof of this hypothesis.

The rotation around the backbone of Ala-193 is accompanied by the displacement of the side chain of the adjacent residue, Ser-194, whose hydroxyl group makes interactions with two important residues of the catalytic loop, Asp-156 and Lys-158. These interactions in conjunction with the released tension at Ala-193 could stabilize the activation loop in its active conformation, making the intervention of the N-terminal segment dispensable in order to keep the active site accessible to substrates. Interestingly both Ala-193 and Ser-194 are conserved in the catalytic subunits (either α or α′) of all known CK2 enzymes, suggesting that they do play a crucial functional role in this kinase.

Conclusions

The data presented provide the clear-cut demonstration that the N-terminal segment present in the catalytic subunits of CK2 plays a crucial role in conferring high constitutive activity to the isolated subunits, by holding the activation loop in its proper conformation. In particular a deletion of 11 residues from the N-terminal end (Δ2–12) is sufficient to suppress catalytic activity under standard assay conditions. Such a drop in phosphorylation efficiency is accounted for by a dramatic raise in Km for ATP, indicating that accessibility to the catalytic site has been seriously impaired. This functional impairment is paralleled by a striking alteration in the gel filtration elution pattern, consistent with the occurrence of important conformational changes due to the N-terminal deletion. Both the activity and the regular gel filtration behaviour of αΔ2–12 are restored to wild type level by interaction with the β-subunits. In this respect the truncated mutant behaves like the catalytic subunits of CDKs (where the N terminal tail of the CK2 α/α′-subunits is absent) which are inactive unless they associate with the cognate cyclins (see Fig. 1). This discloses functional similarity between cyclins and the β-subunit of CK2. The mechanism by which the β-subunit preserves the catalytic activity of CK2α, rendering dispensable the activatory mechanism relying on the N-terminal segment, is still a matter of conjecture. A clue however is possibly provided by the crystal structure of maize CK2α in complex with a fragment of the human β-subunit [8], revealing subtle yet significant conformational alterations affecting two residues at the end of the activation loop. These are Ala-193 which falls in a disallowed region of the Ramachandran plot in the crystal structure of the isolated CK2α–subunit [2, 9], while in the complex it moves to a permitted region [8], and Ser-194 whose side chain in the complex undergoes a rotation and interacts with two important residues of the catalytic loop (see Fig. 6). The functional relevance of this observation is corroborated by the conservation of both Ala-193 and Ser-194 throughout the catalytic subunits of CK2 from different species.

References

1. Hanks SK, Hunter T: Protein kinases 6. The eukaryotic protein kinase superfamily: Kinase (catalytic) domain structure and classification. FASEB J 9: 576–596, 1995
2. Niefind K, Guerra B, Pinna LA, Issinger O-G, Schomburg D: Crystal structure of the catalytic subunit of protein kinase CK2 from Zea mays at 2.1 A resolution. EMBO J 17: 2451–2462, 1998
3. Jeffrey PD, Russo AA, Polyak K, Gibbs E, Hurwitz J, Massague J, Pavletich NP: Mechanism of CDK activation revealed by the structure of a cyclin A-CDK2 complex. Nature 376: 313–320,1995
4. Sarno S, Marin O, Ghisellini P, Meggio F, Pinna LA: Biochemical evidence that the N-terminal segments of the α subunit and the β subunit play interchangeable roles in the activation of protein kinase CK2. FEBS Letters 441: 29–33, 1998
5. Sarno S, Vaglio P, Meggio F, Issinger O-G, Pinna LA: Protein kinase CK2 mutants defective in substrate recognition. Purification and kinetic analysis. J Biol Chem 271: 10595–10601, 1996
6. Ruzzene M, Pinna LA: Assay of protein kinases and phosphatases using specific peptide substrates. In: D.G. Hardie (ed). Protein Phosphorylation. A Practical Approach. Oxford University Press, Oxford, UK, 1999, pp 221–253
7. Laemmli UK: Cleavage of structural proteins during the assembly of the head of bacteriophage T4. Nature (London) 227: 680–685, 1970
8. Battistutta R, Sarno S, De Moliner E, Marin O, Issinger, O-G, Zanotti G, Pinna LA: The crystal structure of the complex of Zea mays alpha subunit with a fragment of human beta subunit provides the clue to the architecture of protein kinase CK2 holoenzyme. Eur J Biochem 267: 5184–5190, 2000
9. Niefind K, Putter M, Guerra B, Issinger O-G, Schomburg D: GTP plus water mimic ATP in the active site of protein kinase CK2. Nat Struct Biol 6: 1100–1103, 1999
10. Merritt EA, Bacon DJ: Raster3D – photorealistic molecular graphics. Meth Enzymol 277: 505–524, 1997
11. Kraulis J: MOLSCRIPT: A program to produce both detailed and schematic plots of protein structures. J Appl Crystallogr 24: 946–950, 1991

Molecular and Cellular Biochemistry **227**: 21–29, 2001.

Functional specialization of CK2 isoforms and characterization of isoform-specific binding partners

David W. Litchfield, Denis G. Bosc, David A. Canton, Ronald B. Saulnier, Greg Vilk and Cunjie Zhang

Department of Biochemistry, University of Western Ontario, London, Ontario, Canada

Abstract

In mammals, protein kinase CK2 has two isozymic forms of its catalytic subunit, designated CK2α and CK2α′. CK2α and CK2α′ exhibit extensive similarity within their catalytic domains but have completely unrelated C-terminal sequences. To systematically examine the cellular functions of each CK2 isoform in mammalian cells, we have generated human osteosarcoma U2-OS cell lines with the expression of active or inactive versions of each CK2 isoform under the control of an inducible promoter [22]. Examination of these cell lines provides evidence for functional specialization of CK2 isoforms at the cellular level in mammals with indications that CK2α′ is involved in the control of proliferation and/or cell survival. To understand the molecular basis for functional differences between CK2α and CK2α′, we have undertaken studies to identify proteins that interact specifically with each isoform of CK2 and could contribute to the regulation of their independent functions. A novel pleckstrin-homology domain containing protein, designated CK2-interacting protein 1 (i.e. CKIP-1) was isolated using the yeast two hybrid system as a protein that interacts with CK2α but not CK2α′ [23]. When expressed in cells as a fusion with green fluorescent protein, CKIP-1 localizes to the cell membrane and to the nucleus. In this study, we present evidence from deletion analysis of CKIP-1 suggesting that a C-terminal region containing a putative leucine zipper has a role in regulating its nuclear localization. Collectively, our data supports a model whereby CKIP-1 is a non-enzymatic regulator of CK2α that regulates the cellular functions of CK2α by targeting or anchoring CK2α to specific cellular localization or by functioning as an adapter to integrate CK2α-mediated signaling events with components of other signal transduction pathways. (Mol Cell Biochem **227**: 21–29, 2001)

Key words: protein kinase CK2, CK2-interacting proteins, CKIP-1, functional specialization, inducible expression

Introduction

Protein kinase CK2 is a widely distributed protein serine/threonine kinase that is remarkably conserved between species [1–6]. In most organisms, CK2 has traditionally been viewed as a tetrameric enzyme composed of two catalytic and two regulatory subunits. Direct microsequence analysis of the individual subunits of purified bovine CK2 yielded the first definitive evidence for the existence of isozymic forms of the catalytic subunits of CK2 (α and α′) in mammals and led to the isolation of human cDNAs encoding each of the catalytic CK2 isoforms [7, 8]. The two isozymic forms of CK2 are the products of distinct genes located on different chromosomes [9, 10]. Both isoforms of CK2 are broadly expressed and although there are wide variations in absolute levels of each form of CK2 in different tissues, the two isozymes appear to be present in all tissues examined. CK2α and CK2α′ exhibit extensive similarity within their catalytic domains. Therefore, it is not surprising that the enzymatic characteristics (i.e. affinity for substrates and turnover rates) displayed *in vitro* by CK2α and CK2α′ are similar and that they appear to have some overlapping cellular properties [3, 11, 12].

Although the existence of isozymic forms of CK2 has been

Address for offprints: D.W. Litchfield, Department of Biochemistry, University of Western Ontario, London, Ontario N6A 5C1, Canada (E-mail: litchfi@uwo.ca)

known for some time, the majority of studies that have been directed towards the elucidation of the regulation or functions of CK2 in mammalian cells have failed to distinguish between these forms. This may result in part from the similar enzymatic characteristics of CK2α and CK2α′ which makes it difficult, if not impossible, to distinguish between CK2α and CK2α′ solely on the basis of enzymatic activity *in vitro*. In contrast to their highly similar catalytic domains, CK2α and CK2α′ have completely unrelated C-terminal domains [8, 13]. Notably, CK2α and CK2α′ exhibit extensive identity between species and the unique features of the C-terminal domains are highly conserved. For example, the deduced amino acid sequence of CK2α exhibits 98% identity between human and chicken sequences. Similarly, CK2α′ exhibits 97% amino acid identity between these species. The high conservation of the unique features of CK2α and CK2α′ suggests that these two forms of CK2 may have unique cellular functions.

Two isoforms of CK2 also exist in *S. cerevisiae* and genetic studies indicate that, while they can compensate for each other to maintain cell viability, they do have independent functions [3, 14, 15]. One isoform (designated CKA2) is involved in cell cycle progression whereas the other isoform (CKA1) is required for maintenance of cell polarity. There is also mounting evidence to suggest that the two isoforms of CK2 exhibit functional specialization in mammals. For example, differences in the subcellular localization of CK2α and CK2α′ have been observed at different stages in the cell cycle [16]. Also, CK2α and CK2α′ are differentially phosphorylated during cell division suggesting that the two proteins have independently regulated properties that are associated with cell cycle progression [17–20]. Additional support for the prediction that CK2α and CK2α′ exhibit functional specialization in mammals is derived from the observation that PP2A interacts with CK2α, but not CK2α′ [21]. Recent studies with mice with targeted disruption of CK2α′ suggest that CK2α′ has a unique role in spermatogenesis [12]. Although these results provide ample evidence for functional differences between CK2α and CK2α′ in mammals, the precise independent cellular functions of CK2α and CK2α′ remain poorly defined. Additionally, there is as yet virtually no information regarding the *in vivo* substrates that are phosphorylated by either CK2α or CK2α′ (i.e. what are their bona fide targets and do they have distinct cellular targets?). Thus, the molecular mechanisms responsible for specific cellular effects of CK2α and/or CK2α′ remain poorly defined. To address these issues, we have undertaken two complementary approaches that will be described here. First, we have established mammalian cell lines exhibiting regulated expression of either active or kinase-inactive forms of CK2α and CK2α′ [22]. With induced expression of active CK2α and CK2α′, it might be predicted that the phosphorylation of respective target proteins would be enhanced. By comparison, we expect that the induced expression of catalytically inactive CK2 isoforms would interfere with the phosphorylation of the respective targets of CK2α and CK2α′. As discussed further in this report, initial studies with these inducible cell lines provides evidence for functional specialization of CK2α and CK2α′ at the cellular level in mammalian cells. Secondly, to begin to understand the molecular basis for the functional specialization of CK2α and CK2α′, we have performed systematic studies to identify isoform-specific binding partners. These studies have yielded a novel CK2-interacting protein (designated CKIP-1) that interacts with CK2α, but not CK2α′ [23]. Studies performed to date that will be discussed here are consistent with a model in which CKIP-1 functions as a non-enzymatic regulator of CK2α in cells.

Materials and methods

Characterization of cell lines with regulated expression of protein kinase CK2

Human osteosarcoma U2-OS cell lines with regulated expression of CK2α, CK2αK^{68}/M, CK2α′, or CK2α′K^{69}/M together with CK2β have been previously described [22]. Assays of CK2 activity in cell extracts was performed using the synthetic peptide RRRDDDSDDD as described [22, 24]. Cell proliferation was monitored as described by counting cells at intervals following the removal of tetracycline [22]. Data summarized in Table 1 reflects the results of cell counts on

Table 1. Effects of induced expression of active or inactive CK2α or CK2α′

	CK2α	CK2αK^{68}/M	CK2α′	CK2α′K^{69}/M
CK2 activity	Increase (2.1-fold)	Increase (1.7-fold)	Increase (2.8-fold)	Increase (1.7-fold)
Proliferation rate	No change	No change	Decrease (~2-fold)	Decrease (~5-fold)
Specific cell cycle arrest	No	No	No	No
Sub-G0/G1 population on FACS profile	No change	No change	No change	Increase

A variety of parameters were examined following the induced expression of kinase-active or kinase-inactive (i.e. CK2αK^{68}/M or CK2α′K^{69}/M) forms of CK2α and CK2α′. Induced expression of each form of CK2α or CK2α′ coordinately with CK2β was achieved by tetracycline removal. Measurements were performed as described in [22] and in Materials and methods.

day 8 following tetracycline removal. Cell cycle analysis was performed by FACS after staining cellular DNA with propidium iodide. The percentage of cells within individual stages of the cell cycle was determined using ModFit LT Version 2.0 modelling software (Verity Software House, Topsham, ME, USA).

Constructs encoding EGFP-CKIP-1 fusion proteins

The construct encoding EGFP-CKIP-$1^{308-409}$ used in this study was previously described [23]. Constructs encoding full-length CKIP-1 (i.e. residues 1–409) and a deletion mutant of CKIP-1 encoding residues 1–346 as fusions with EGFP were generated using the polymerase chain reaction with Pfu polymerase (Stratagene). The following primers were used to amplify full length CKIP-1: 5′- TGA TCA GAA TTC CCA TGA TGA AGA AGA ACA ATT CCG CCA AG-3′ (forward) and 5′-GAA TTC GTC GAC CCC ACC CTG CCC TCA CAT CAG G-3′ (reverse). For, CKIP-1^{1-346}, the same forward primer was used in combination with the following reverse primer: 5′-GAA TTC GTC GAC TCA AGA ATC CGG CGG AGA CCG AGG G-3′. The amplification products were subcloned into pCR-Blunt and verified by sequencing. The coding regions of CKIP-1 were liberated from pCR-Blunt using EcoRI and SalI and subcloned into the respective sites of pEGFP-C3 (Clontech).

Subcellular localization of EGFP-CKIP-1 fusion proteins

Plasmids encoding EGFP or the respective EGFP-CKIP-1 fusion proteins were transfected into human osteosarcoma Saos-2 cells using Fugene 6 (Roche) according to the manufacturer's recommendations. On the day following transfection, cells were fixed with 3% paraformaldehyde in PBS without permeabilization. EGFP fluorescence was visualized with a Zeiss LSM 410 inverted confocal microscope [23]. The images were pseudocolored using built-in LSM software.

Results

Mammalian cells with tetracycline regulated expression of CK2α and CK2α′

To directly investigate the individual functions of CK2α and CK2α′ in mammalian cells, we developed cell lines with regulated expression of CK2α or CK2α′ [22]. To specifically interfere with the functions of each isoform, we also developed cell lines exhibiting inducible expression of kinase-inactive (i.e. dominant interfering) mutants of CK2α and CK2α′, designated CK2αK^{68}/M and CK2α′K^{69}/M, respectively. Prior studies using transient transfection indicated that, with the exception of their activity defects, the kinase-inactive mutants are indistinguishable from wild-type CK2α and CK2α′ in terms of prominent nuclear localization and complex formation with CK2β [25]. To develop the cell lines with regulated CK2 expression, human osteosarcoma U2-OS cells expressing the tetracycline-controlled transcriptional activator (tTA) were transfected with a bidirectional plasmid containing HA-tagged CK2α or CK2α′ cDNA together with a Myc-tagged CK2β cDNA to achieve coordinate overexpression of catalytic and regulatory CK2 subunits. The bidirectional plasmid was utilized because we, and others, had previously demonstrated that optimal CK2 expression required co-expression of CK2 subunits [25, 26]. Moreover, by inducing the co-ordinate expression of catalytic and regulatory subunits of CK2 in each of our cell lines, effects due to formation of hetero-tetrameric complexes (i.e. $\alpha\alpha'\beta_2$) rather than homo-tetrameric (i.e. $\alpha_2\beta_2$ or $\alpha'_2\beta_2$) complexes are minimized. Examination of the proliferative characteristics of these cell lines provided the most direct evidence to date for functional specialization of CK2α and CK2α′ at the cellular level. Notably, proliferation is dramatically attenuated by induced expression of CK2α′K^{69}/M but not by induced expression of CK2αK^{68}/M (Table 1). This result implies that CK2α′ has targets that are required for maintenance of optimal proliferation that are not phosphorylated by CK2α. Cells with induced expression of CK2α′ also exhibit attenuated proliferation, but unlike the effect that is observed with CK2α′K^{69}/M, the effects of CK2α′ are reversed by restoring tetracycline to the growth medium (data not shown). We expect that increased expression of CK2α′ will increase target phosphorylation whereas induced expression of CK2α′K^{69}/M will diminish target phosphorylation. Consequently, we believe that it is likely that the attenuated proliferation observed with CK2α′K^{69}/M arises through a different mechanism than that observed with CK2α′.

To characterize the mechanism responsible for the attenuated proliferation observed with induced expression of CK2α′ or CK2α′K^{69}/M, we performed FACS analysis to examine the cell cycle profile (summarized in Table 1). With this analysis, no specific cell cycle arrest was observed in any of the cell lines following altered expression of CK2 indicating that a single event associated with cell cycle progression was not responsible for the altered proliferation observed. FACS analysis did reveal an increase in the sub-Go/G1 fraction of cells following induction of CK2α′K^{69}/M. Since the sub-Go/G1 population may represent an apoptotic fraction, this result suggests that cell death may be one factor that contributes to the attenuated proliferation in cells expressing CK2α′K^{69}/M.

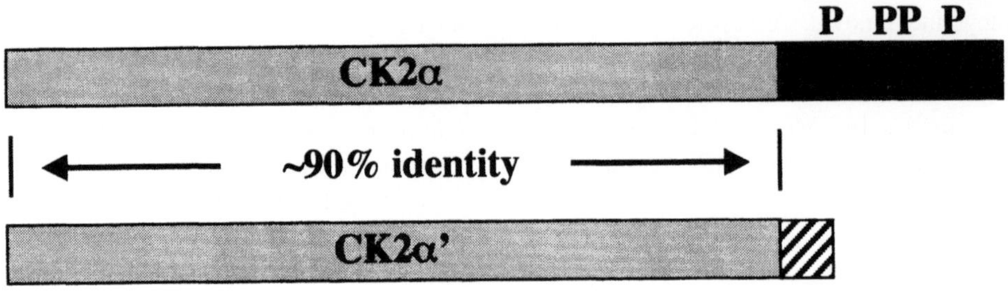

Fig. 1. Schematic representation of CK2α and CK2α'. Within their N-terminal 330 amino acids (based on the numbering of CK2α), the two catalytic isoforms of CK2, CK2α and CK2α', exhibit approximately 90% identity. All sub-domains found within all members of the protein kinase family are localized within this region of the two proteins. By comparison, the C-terminal domains of CK2α and CK2α', encoding 61 and 20 amino acids respectively, are completely unrelated. CK2α is phosphorylated in mitotic cells at 4 sites (Threonine-344, Threonine-360, Serine-362 and Serine-370) within its unique C-terminal domain.

Identification of CKIP-1: A novel CK2 interacting protein

There is ample evidence demonstrating the fundamental importance of protein:protein interactions in the transmission of many regulatory signals. Furthermore, we predict that the identification of isoform-specific interaction partners may yield insights into the functional specialization of CK2α and CK2α'. Consequently, we performed studies to systematically identify CK2 interacting proteins that have the capacity to regulate the unique aspects of CK2α and CK2α' [23]. Our studies have yielded two CK2-interacting proteins that exhibit isoform specificity. One of these proteins is Pin-1, a recently discovered peptide prolyl isomerase that is essential for mitosis [27, 28]. Interestingly, Pin-1 recognizes CK2α, but not CK2α', in a phosphorylation-dependent manner suggesting that Pin-1 could have a role in regulating isoform-specific mitotic functions of CK2 [29]. A second CK2-interacting protein, isolated from a human B cell cDNA library using the yeast two hybrid system is a novel protein that we have named CKIP-1 [23]. As illustrated in Fig. 2, CKIP-1 contains an N-terminal pleckstrin homology (PH) domain, P-X-X-P motifs reminiscent of SH3 binding sites, a potential leucine zipper and consensus phosphorylation sites for MAPK and CDK family members. Our initial studies indicate that CKIP-1 does not dramatically affect the catalytic activity of CK2. Furthermore, CKIP-1 is not effectively phosphorylated by CK2. Based on these observations and the fact that CKIP-1 contains a number of potential protein interaction motifs (including the P-X-X-P motifs, a potential leucine zipper and perhaps the PH domain as well), we hypothesize that CKIP-1 is a non-enzymatic regulator of CK2α. The PH domain appears to be necessary for membrane localization. This result was not unexpected since PH domains from other proteins have been shown to mediate protein:protein interactions or interactions with specific membrane phosphoinositides [30, 31]. Full length EGFP-CKIP-1 fusions are also detected within the nucleus of some, but not all, of the transfected cells suggesting that the subcellular distribution of CKIP-1 is dynamic, and perhaps regulated [23]. To further investigate the factors that regulate the subcellular distribution of CKIP-1, we have performed deletion studies with the EGFP-CKIP-1 fusion proteins. As shown in Fig. 3, we examined the subcellular distribution of a construct encoding an EGFP-CKIP-1 fusion that lacks the C-terminal domain of CKIP-1 (designated EGFP-CKIP-1^{1-346}) and a construct that expresses only the C-terminal domain of CKIP-1 (designated EGFP-CKIP-$1^{308-409}$). As is the case with full-length EGFP-CKIP-1, EGFP-CKIP-1^{1-346} localizes to the membrane. Furthermore, as shown in Fig. 3D, the majority of cells that express EGFP-CKIP-1^{1-346} exhibit nuclear localization of the EGFP fusion protein. In fact, deletion of the C-terminal domain of CKIP-1 appears to significantly enhance the nuclear complement of EGFP-CKIP-1 fusion proteins (compare Fig. 3D with Fig. 3C). By comparison, a construct encoding only the C-terminal domain of CKIP-1 (i.e. EGFP-CKIP-$1^{308-409}$) appears to be excluded from the nucleus (Fig. 3E). Collectively, these results suggest that sequences within the C-terminal 101 amino acids of CKIP-1 play a role in regulating its nuclear import or export or perhaps the cytoplasmic retention of CKIP-1.

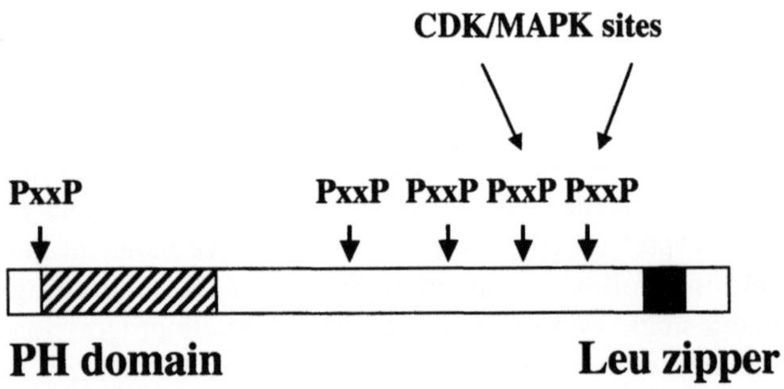

Fig. 2. Schematic representation of CKIP-1. The deduced amino acid sequence of CKIP-1 encodes a protein of 409 amino acids. A putative pleckstrin homology (PH) domain encompassing residues 21–132 is marked with hatched bars and a potential leucine zipper (Leu zipper) encompassing residues 347–373 is marked with the solid black bar. P-X-X-P motifs reminiscent of SH3-binding sites and consensus phosphorylation sites for MAP kinases or cyclin-dependent kinases are marked.

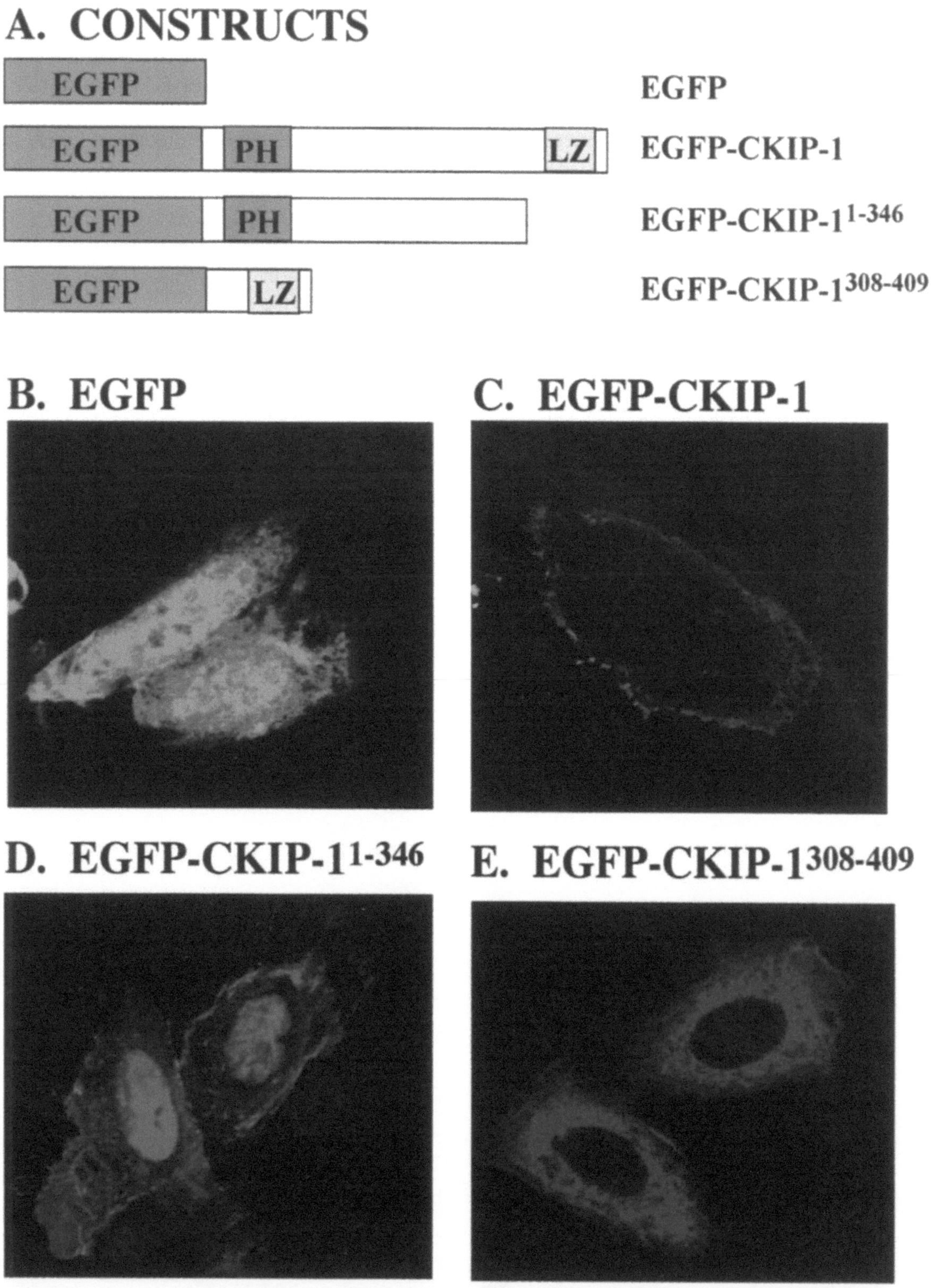

Fig. 3. Subcellular localization of EGFP-CKIP-1 fusion proteins. (A) Constructs encoding enhanced green fluorescent protein (EGFP) and EGFP fusions encoding full length CKIP-1 or deletions of CKIP-1. EGFP-CKIP-1$^{1-346}$ encodes a deletion product of CKIP-1 lacking its putative leucine zipper. EGFP-CKIP-1$^{308-409}$ encodes the C-terminal 101 amino acids of CKIP-1 including its putative leucine zipper. The pleckstrin homology (PH) domain and leucine zipper (LZ) of CKIP-1 are indicated. (B–E). Visualization of EGFP-CKIP-1 fusion proteins by confocal microscopy of Saos-2 cells transfected with EGFP or EGFP-fusion proteins as indicated. EGFP or EGFP-CKIP-1 fusion proteins are as indicated.

Discussion

Functional specialization of CK2 isoforms

There is a great deal of biochemical and genetic evidence to suggest that protein kinase CK2 is an important regulatory participant in a broad spectrum of cellular events [1–6]. However, to date, the majority of the studies that have been directed towards elucidation of the biological functions and regulation of CK2 have failed to distinguish between the different CK2 isoforms. Given the close enzymatic similarities between the different forms of CK2, there is a high likelihood that different forms of CK2 are capable of performing some degree of overlapping functions within cells. In fact, in the context of viability in yeast and in mice, disruption of the genes encoding only one form of CK2 is not lethal perhaps because of the overlapping functions or the potential for compensatory cellular actions from the remaining form of CK2 [3, 12, 14, 15]. These studies in yeast and in mice with targeted disruption of CK2α′ also provide evidence for functional specialization of the CK2 isoforms. The development of cell lines with regulated expression of CK2 has provided a unique opportunity for the systematic investigation of the cellular functions of CK2α and CK2α′ in mammalian cells. Initial studies with these cell lines demonstrated at the cellular level in mammalian cells that CK2α and CK2α′ have specialized functions since induced expression of catalytically inactive CK2α′ dramatically attenuates cell proliferation whereas induced expression of similar levels of inactive CK2α do not exert the same effect [22]. The attenuated proliferation that is observed following induction of inactive CK2α′ does not arise from a single specific cell cycle arrest. However, since CK2 appears to have functions associated with multiple stages during the cell cycle in mammalian cells and in yeast, this result may not be surprising [3, 32, 33]. More detailed examination of cells that have been synchronized at specific stages in the cell cycle will be necessary to determine whether CK2α′ is indeed required at specific points in the cell cycle.

In addition to the evidence for functional specialization of CK2 isoforms, these cell lines provide an opportunity for investigation of the role of each isoform of CK2 in specific cellular responses. For example, CK2 has been implicated in heat shock responses and in signaling pathways associated with UV responses [34–36]. By augmenting cellular levels of each CK2 isoform or by interfering with signaling pathways involving each CK2 isoform through the use of catalytically-inactive CK2 mutants, we expect to determine whether cellular responses to specific stimuli (e.g. UV, heat shock) involve either of the CK2 isoforms. Coupled with the emergence of genomic (i.e. microarrays) and proteomic (i.e. mass spectrometry) strategies for the identification of targets for specific signal transduction pathways, cell lines with regulated CK2 expression offer the unprecedented promise of identification of the bona fide cellular targets of CK2.

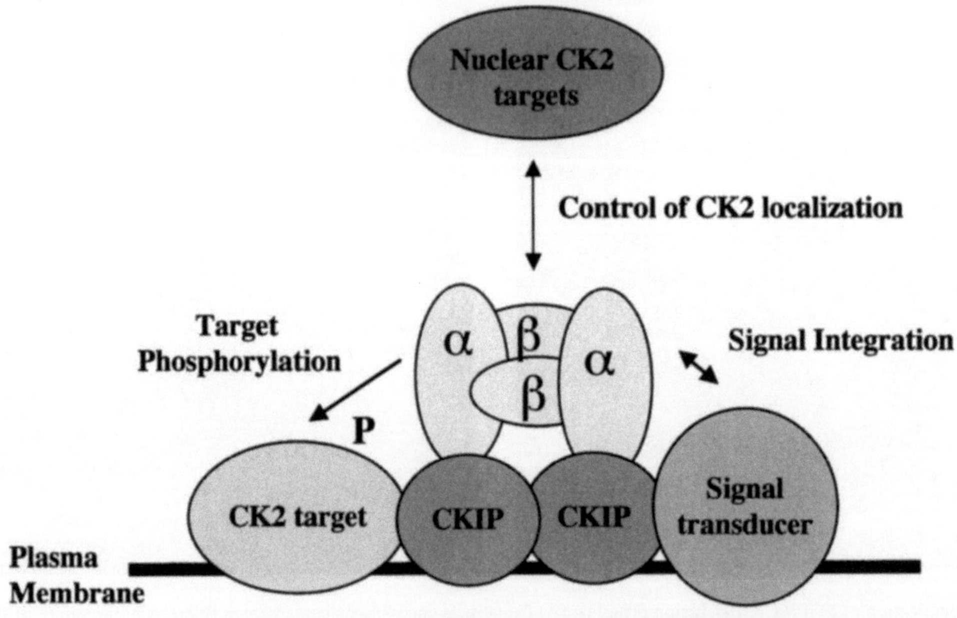

Fig. 4. Hypothetical model for regulation of CK2α by CKIP-1. Based on data described in the text, CKIP-1 appears to have the characteristics of a non-enzymatic regulator of CK2α. As a non-enzymatic regulator of CK2, CKIP-1 could control the ability of CK2 to phosphorylate its cellular targets by controlling the subcellular localization of CK2α. Alternatively, CKIP-1 could function as an adapter to integrate CK2-mediated signals with distinct signalling pathways. CKIP-1 is hypothetically illustrated as a dimer since many leucine zipper-containing proteins exist as dimers.

Isoform-specific binding partners

Protein:protein interactions play an essential role in regulating many signal transduction events [37, 38]. In particular, adapter or scaffold proteins that mediate indirect interactions between distinct signaling molecules and anchoring or targeting proteins that localize signaling molecules to specific cellular locations are important components of many signaling complexes. Some of these proteins may have the capacity to interact (perhaps even simultaneously) with a variety of distinct signaling molecules to facilitate coordinated and efficient transmission of regulatory signals. It is noteworthy that CK2 has substrates throughout the cell. Furthermore, CK2 has been localized with a number of distinct cellular structures including the plasma membrane, endoplasmic reticulum and golgi, nuclear matrix and mitotic spindle and undergoes transport to the nucleus [39–45]. Consequently, it is anticipated that CK2 could interact with a number of distinct proteins that regulate its subcellular distribution and/or control its access to specific substrates in a manner similar to that observed with the AKAPs (A-kinase anchoring proteins) that regulate cAMP-dependent protein kinase [37, 38]. In support of this prediction, it has been demonstrated that CK2 interacts or copurifies with a variety of distinct cellular proteins. However, it appears unlikely that most of these CK2-interacting proteins will have any direct role in regulating the subcellular distribution of CK2 or differentiating between CK2α and CK2α′. In fact, most of the interactions involving CK2 that have been characterized to date appear to reflect interactions between CK2 and potential substrates such as p53, BRCA1, nucleolin, NOPP140, and dishevelled [46–52]. Moreover, many of the interactions involving CK2 and proteins such as A-Raf, p53, c-Mos and CD5 are mediated through the regulatory CK2β subunit [46, 47, 53–56]. Since CK2β does not offer any selectivity towards the catalytic subunits of CK2, it is unlikely that these interactors will offer any isoform specificity. CK2 has also been shown to interact with chaperones such as hsp90 and Cdc37 that are most likely involved in facilitating assembly or stabilization of CK2 during biosynthesis [57, 58]. Overall, it is clear that CK2 is a regulatory participant in a complex array of cellular events, and thus, it is likely that CK2 will be involved in a complex series of protein:protein interactions. Although a number of distinct CK2-interacting proteins have been identified, their identity has provided only limited insights into how the subcellular localization of CK2 is regulated or how CK2α and CK2α′ exhibit functional specialization.

In performing a yeast two-hybrid screen to identify CK2 interacting proteins, we had two main objectives. First, we were interested in identifying proteins that exhibit specific interactions with either CK2α or CK2α′, as a possible mechanism for explaining functional differences between the two CK2 isoforms. Second, we were interested in identifying regulators of CK2 rather than CK2 substrates. With the identification of CKIP-1, it would appear that we have fulfilled the first objective since CKIP-1 interacts with CK2α and not with CK2α′. Furthermore, the properties of CKIP-1 are consistent with a model in which CKIP-1 functions as a non-enzymatic regulator of CK2α. As a non-enzymatic regulator of CK2, CKIP-1 could play a role as a targeting protein that recruits CK2 to specific cellular locations to regulate phosphorylation of specific target proteins. Although we have no data at the present time, it is intriguing to speculate that the apparent regulated localization of CKIP-1 to the nucleus may be related to the import or export of CK2. An alternative manner by which CKIP-1 could regulate CK2α could be through a role as an adapter protein that mediates interactions between CK2α and other signaling molecules to facilitate coordinated transmission of regulatory signals. These possibilities are currently under investigation. Although our current understanding of CKIP-1 remains limited, it may offer a new paradigm for the cellular regulation of CK2 that is similar to that seen with the AKAPs that regulate cAMP-dependent protein kinase. Whether or not CKIP-1 functions in an analogous manner to an AKAP remains to be determined. Furthermore, whether or not there are a large number of distinct CKIPs as there are a large number of AKAPs also remains to be investigated.

Additional considerations

The studies described in this report were initiated when only two isoforms of CK2 (i.e. CK2α and CK2α′) had been definitively identified. With the apparent identification of a novel isoform of CK2 (designated CK2α″ [59]) and the prospect of additional yet to be discovered isoforms to be revealed with the final release of the human genome sequence, we must recognize the possibility that our current approaches must again be refined to ensure that we distinguish between all forms of CK2.

In utilizing the yeast two hybrid system to identify CK2-interacting proteins, there are clearly limitations that must be acknowledged. For example, using a standard two-hybrid system, we would not expect the bait constructs encoding CK2α to be phosphorylated at its mitotic phosphorylation sites. Consequently, phosphorylation-dependent interactions between CK2α and interaction partners such as Pin-1 will not be detected in these assays. The fact that Pin-1 was not isolated in our two hybrid library screens, yet interacts effectively with phosphorylated CK2α in GST-pulldown assays lends support to this suggestion. Furthermore, the presentation of CK2α or CK2α′ as baits in a two-hybrid system may be appropriate for the identification of proteins that interact with isolated catalytic CK2 subunits but might not properly represent tetrameric CK2. With the development of mass spectrometric and proteomic strategies for the identification of proteins that co-purify from cellular extracts with CK2, it is likely that many additional CK2-interacting proteins will be identified. We are willing to speculate that a major mechanism for the regulation of CK2 within cells and for the functional specialization of CK2 isoforms is through interactions with specific binding proteins that regulate precise subsets of CK2-mediated events within cells.

Acknowledgements

The work described in this paper was supported by operating grants from the Canadian Institutes of Health Research and the National Cancer Institute of Canada with funds from the Canadian Cancer Society and the Terry Fox Run. Greg Vilk is the recipient of a studentship from the National Cancer Institute of Canada, David Canton is the recipient of a studentship from the Natural Sciences and Engineering Research Council of Canada and Cunjie Zhang is supported by the Premier's Research Excellence Award program.

References

1. Allende JE, Allende CC: Protein kinase CK2: An enzyme with multiple substrates and a puzzling regulation. FASEB J 9: 313–323, 1995
2. Pinna L, Meggio F: Protein kinase CK2 ('casein kinase-2') and its implication in cell division and proliferation. Prog Cell Cycle Res 3: 77–97, 1997
3. Glover CV: On the physiological role of casein kinase II in *Saccharomyces cerevisiae*. Prog Nucleic Acid Res Mol Biol 59: 95–133, 1998
4. Guerra B, Issinger O-G: Protein kinase CK2 and its role in cellular proliferation, development and pathology. Electrophoresis 20: 391–408, 1999
5. Tawfic S, Faust RA, Gapany M, Ahmed K: Nuclear matrix as an anchor for protein kinase CK2 nuclear signalling. J Cell Biochem 62:165–171, 1996
6. Litchfield DW., Lüscher B: Casein kinase II in signal transduction and cell cycle regulation. Mol Cell Biochem 127–128: 187–200, 1993
7. Litchfield DW, Lozeman FJ, Piening C, Sommercorn J, Takio K, Walsh KA, Krebs EG: Subunit structure of casein kinase II from bovine testis: Demonstration that the α and α' subunits are distinct polypeptides. J Biol Chem 265: 7638–7644, 1990
8. Lozeman FJ, Litchfield DW, Piening C, Takio K, Walsh KA, Krebs EG: Isolation and characterization of human cDNA clones encoding the α and α' subunits of casein kinase II. Biochemistry 29: 8436–8447, 1990
9. Yang-Feng TL, Naiman T, Kopatz I, Eli D, Dafni N, Canaani D: Assignment of the human casein kinase II alpha' gene (CSNK2A1) to chromosome 16p13.2–p12.3. Genomics 19: 173, 1994
10. Wirkner U, Voss H, Lichter P, Ansorge W, Pyerin W: The human gene (CSNK2A1) coding for the casein kinase II subunit alpha is located on chromosome 20 and contains tandemly arranged Alu repeats. Genomics 19: 257–265, 1994
11. Bodenbach L, Fauss J, Robitzki A, Krehan A, Lorenz P, Lozeman FJ, Pyerin W: Recombinant casein kinase II. Eur J Biochem 220: 263–273, 1994
12. Xu X, Toselli PA, Russell LD, Seldin DC: Globozoospermia in mice lacking the casein kinase II alpha' catalytic subunit. Nat Genet 23: 118–121, 1999
13. Maridor G, Park W, Krek W, Nigg EA: Casein kinase II. cDNA sequences, developmental expression and tissue distribution of mRNAs for alpha, alpha' and beta subunits of the chicken enzyme. J Biol Chem 266: 2362–2368, 1991
14 Hanna DE, Rethinaswamy A, Glover CVC: Casein kinase II is required for cell cycle progression during G1 and G2/M in *Saccharomyces cerevisiae*. J Biol Chem 270: 25905–25914, 1995
15. Rethinaswamy A, Birnbaum MJ, Glover CV:Temperature-sensitive mutations of the CKA1 gene reveal a role for casein kinase II in maintenance of cell polarity in *Saccharomyces cerevisiae*. J Biol Chem 273: 5869–5877, 1998
16. Yu IJ, Spector DL, Bae YS, Marshak DR: Immunocytochemical localization of casein kinase II during interphase and mitosis. J Cell Biol 114: 1217–1232, 1991
17. Litchfield DW, Lüscher B, Lozeman FJ, Eisenman RN, Krebs EG: Phosphorylation of casein kinase II by $p34^{cdc2}$ *in vitro* and at mitosis. J Biol Chem 267: 13943–13951, 1992
18. Litchfield DW, Bosc DG, Slominski E: The protein kinase from mitotic human cells that phosphorylates Ser209 on the casein kinase II β subunit is $p34^{cdc2}$. Biochim Biophys Acta 1269: 69–78, 1995
19. Bosc DG, Slominski E, Sichler C, Litchfield DW: Phosphorylation of casein kinase II by $p34^{cdc2}$: Identification of phosphorylation sites using phosphorylation site mutants *in vitro*. J Biol Chem 270: 25872–25878, 1995
20. Bosc DG, Luscher B, Litchfield DW: Expression and regulation of protein kinase CK2 during the cell cycle. Mol Cell Biochem 191: 213–222, 1999
21. Heriche JK, Lebrin F, Rabilloud T, Leroy D, Chambaz EM, Goldberg Y: Regulation of protein phosphatase 2A by direct interaction with casein kinase 2alpha. Science 276: 952–955, 1997
22. Vilk G, Saulnier RB, St Pierre R, Litchfield DW: Inducible expression of protein kinase CK2 in mammalian cells: Evidence for functional specialization of CK2 isoforms. J Biol Chem 274: 14406–14414, 1999
23. Bosc DG, Graham KC, Saulnier RB, Zhang C, Prober D, Gietz RD, Litchfield DW: Identification and characterization of CKIP-1, a novel pleckstrin homology domain containing protein that interacts with protein kinase CK2. J Biol Chem 275: 14296–14306, 2000
24. Litchfield DW, Dobrowolska G, Krebs EG: Regulation of casein kinase II by growth factors: A re-evaluation. Cell Mol Biol Res 40: 373–381, 1994
25. Penner CG, Wang Z, Litchfield DW: Expression and localization of epitope-tagged protein kinase CK2. J Cell Biochem 64: 525–537, 1997
26. Heller-Harrison RA, Czech MP: Enhanced casein kinase II activity in COS-1 cells upon overexpression of either its catalytic or noncatalytic subunit. J Biol Chem 266: 14435–14439, 1991
27. Shen M, Stukenberg PT, Kirschner MW, Lu KP: The essential mitotic peptidyl-prolyl isomerase Pin1 binds and regulates mitosis-specific phosphoproteins. Genes Dev 12: 706–720, 1998
28. Yaffe MB, Schutkowski M, Shen M, Zhou XZ, Stukenberg PT, Rahfeld JU, Xu J, Kuang J, Kirschner MW, Fischer G, Cantley LC, Lu KP: Sequence-specific and phosphorylation-dependent proline isomerization: A potential mitotic regulatory mechanism. Science 278: 1957–1960, 1997
29. Messenger M, Gilchrist A, Saulnier RB, Gorbsky GJ, Litchfield DW: The peptide-prolyl isomerase Pin-1 interacts with protein kinase CK2 in a phosphorylation dependent manner (submitted for publication)
30. Lemmon MA, Falasca M, Ferguson KM, Schlessinger J: Regulatory recruitment of signalling molecules to the cell membrane by pleckstrin-

homology domains. Trends Cell Biol 7: 237–242, 1997
31. Falasca M, Logan SK, Lehto VP, Baccante G, Lemmon MA, Schlessinger J: Activation of phospholipase Cgamma by PI 3-kinase-induced PH domain-mediated membrane targeting. EMBO J 17: 414–422, 1998
32. Lorenz P, Pepperkok R, Ansorge W, Pyerin W: Cell biological studies with monoclonal and polyclonal antibodies against human casein kinase II subunit beta demonstrate participation of the kinase in mitogenic signaling. J Biol Chem 268: 2733–2739, 1993
33. Pepperkok R, Lorenz P, Ansorge W, Pyerin W: Casein kinase II is required for transition of G0/G1, early G1, and G1/S phases of the cell cycle. J Biol Chem 269: 6986, 1994
34. Gerber DA, Souquere-Besse S, Puvion F, Dubois MF, Bensaude O, Cochet C: Heat-induced relocalization of protein kinase CK2. Implication of CK2 in the context of cellular stress. J Biol Chem 275: 23919–23926, 2000
35. Sayed M, Kim SO, Salh BS, Issinger OG, Pelech SL: Stress-induced activation of protein kinase CK2 by direct interaction with p38 mitogen-activated protein kinase. J Biol Chem 275: 16569–16573, 2000
36. Keller DM, Zeng X, Wang Y, Zhang QH, Kapoor M, Shu H, Goodman R, Lozano G, Zhao Y, Lu H: A DNA damage-induced p53 serine 392 kinase complex contains CK2, hSpt16, and SSRP1. Mol Cell 7: 283–292, 2001
37. Pawson T, Scott JD: Signalling through scaffold, anchoring and adaptor proteins. Science 278: 2075–2080, 1997
38. Colledge M, Scott JD: AKAPs: From structure to function. Trends Cell Biol 9: 216–221, 1999
39. Sarrouilhe D, Filhol O, Leroy D, Bonello G, Baudry M, Chambaz E, Cochet C: The tight association of protein kinase CK2 with plasma membrane is mediated by a specific domain of its regulatory β-subunit. Biochim Biophys Acta 1403: 199–210, 1998
40. Wong HN, Ward MA, Bell AW, Chevet E, Bains S, Blackstock WP, Solari R, Thomas DY, Bergeron JJ: Conserved *in vivo* phosphorylation of calnexin at casein kinase II sites as well as a protein kinase C/proline-directed kinase site. J Biol Chem 273: 17227–17235, 1998
41. Mauxion F, Le Borgne R, Munier-Lehmann H, Hoflack B: A casein kinase II phosphorylation site in the cytoplasmic domain of the cation-dependent mannose 6-phosphate receptor determines the high affinity interaction of the AP-1 Golgi assembly proteins with membranes. J Biol Chem 271: 2171–2178, 1996
42. Molloy SS, Thomas L, Kamibayashi C, Mumby MC, Thomas G: Regulation of endosome sorting by a specific PP2A isoform. J Cell Biol 142: 1399–1411, 1998
43. Tawfic S, Davis AT, Faust RA, Gapany M, Ahmed K: Association of protein kinase CK2 with nuclear matrix: Influence of method of preparation of nuclear matrix. J Cell Biochem 64: 499–504, 1997
44. Krek W, Maridor G, Nigg EA: Casein kinase II is a predominantly nuclear enzyme. J Cell Biol 116: 43–55, 1992
45. Filhol O, Cochet C, Chambaz EM: Cytoplasmic and nuclear distribution of casein kinase II: Characterization of the enzyme uptake by bovine adrenocortical nuclear preparation. Biochemistry 29: 9928–9936, 1990
46. Filhol O, Baudier J, Delphin C, Loue-Mackenbach P, Chambaz E, Cochet C: Casein kinase II and the tumor supressor protein p53 associate in a molecular complex that is negatively regulated upon p53 phosphorylation. J Biol Chem 267: 20577–20583, 1992
47. Appel K, Wagner P, Boldyreff B, Issinger O-G, Montenarh M: Mapping of the interaction sites of the growth suppressor protein p53 with the regulatory Beta-subunit of protein kinase CK2. Oncogene 11: 1971–1978, 1995
48. O'Brien KA, Lemke SJ, Cocke KS, Rao RN, Beckmann RP: Casein kinase 2 binds to and phosphorylates BRCA1. Biochem Biophys Res Commun 260: 658–664, 1999
49. Li D, Dobrowolska G, Krebs EG: The physical association of casein kinase 2 with nucleolin. J Biol Chem 271: 15662–15668, 1996
50. Li D, Meier T, Dobrowolska G, Krebs EG: Specific interaction between casein kinase 2 and the nucleolar protein Nopp140. J Biol Chem 272: 3773–3779, 1997
51. Willert K, Brink M, Wodarz A, Varmus H, Nusse R: Casein kinase 2 associates with and phosphorylates dishevelled. EMBO J 16: 3089–3096, 1997
52. Song DH, Sussman DJ, Seldin DC: Endogenous protein kinase CK2 participates in Wnt signaling in mammary epithelial cells. J Biol Chem 275: 23790–23797, 2000
53. Boldyreff B, Issinger O-G: A-Raf kinase is a new interacting partner of protein kinase CK2 beta subunit. FEBS Lett 403: 197–199, 1997
54. Hageman C, Kalmes A, Wixler V, Wixler L, Schuster T, Rapp UR: The regulatory subunit of protein kinase CK2 is a specific A-Raf activator. FEBS Lett 403: 200–202, 1997
55. Chen M, Li D, Krebs EG, Cooper J: The casein kinase II β subunit binds to Mos and inhibits Mos activity. Mol Cell Biol 17: 1904–1912, 1997
56. Raman C, Kuo A, Deshane J, Litchfield DW, Kimberly RP: Regulation of casein kinase 2 by direct interaction with cell surface receptor CD5. J Biol Chem 273: 19183–19189, 1998
57. Miyata Y, Yahara I: Interaction between casein kinase II and the 90-kDa stress protein, HSP90. Biochemistry 34: 8123–8129, 1995
58. Kimura Y, Rutherford SL, Miyata Y, Yahara I, Freeman BC, Yue L, Morimoto RI, Lindquist S: Cdc37 is a molecular chaperone with specific functions in signal transduction. Genes Dev 11: 1775–1785, 1997
59. Shi X, Potvin B, Huang T, Hilgard P, Spray DC, Suadicani SO, Wolkoff AW, Stanley P, Stockert RJ: A novel casein kinase 2 alpha-subunit regulates membrane protein traffic in the human hepatoma cell line HuH-7. J Biol Chem 276: 2075–2082, 2001

Molecular and Cellular Biochemistry **227**: 31–36, 2001.

A surface plasmon resonance study of the interactions between the component subunits of protein kinase CK2 and two protein substrates, casein and calmodulin

María J. Benítez,[1] Claude Cochet[2] and Juan S. Jiménez[1]

[1]*Departamento de Química Física Aplicada, y Centro de Biología Molecular Severo Ochoa (UAM-CSIC), Universidad Autónoma de Madrid, Cantoblanco, Madrid, Spain;* [2]*Laboratoire de Biochimie des Régulations Cellulaires Endocrines, Unité INSERM 244, Département de Biologie Moléculaire et Structurale, CEA Grenoble, Grenoble, France*

Abstract

Surface plasmon resonance has been used to study the interaction between the subunits composing protein kinase CK2 (two catalytic, α-subunits, and two regulatory, β-subunits), as well as the interaction of each subunit with two types of protein substrates, casein, the phosphorylation of which is activated by the regulatory subunit, and calmodulin, which belongs to the kind of substrates on which the catalytic subunit is downregulated by the regulatory subunit. The interaction of casein with the catalytic subunit differs from the interaction with the holoenzyme. Similarly to the interaction with the regulatory subunit, the catalytic subunit interacts with the protein substrate forming a very stable, irreversible complex. The reconstituted holoenzyme, however, binds casein reversibly, displaying a binding mode similar to that displayed by the regulatory subunit. The interaction of calmodulin with the catalytic subunit gives place, like in the case of casein, to an irreversible complex. The interactions with the regulatory subunit and with the holoenzyme were practically negligible, and the interaction with the regulatory subunit disappeared upon increasing the temperature value to close to 30°C. The presence of polylysine induced a high increase in the extent of calmodulin binding to the holoenzyme. The results obtained suggest that CK2β subunit and protein substrates share a common, or at least an overlapping, site of interaction on the catalytic subunit. The interaction between both subunits would prevent substrates from binding irreversibly to α subunit, and, at the same time, it would generate a new and milder site of interaction between the whole holoenzyme and the protein substrate. The main difference between casein and calmodulin would consist in the lower affinity display by the last for the new site generated upon the binding of the regulatory subunit, in the absence of polycations like polylysine. (Mol Cell Biochem **227**: 31–36, 2001)

Key words: plasmon resonance, protein kinase CK2 component subunits

Introduction

Protein kinase CK2 (CK2) is composed of two types of catalytic subunits (α and α'), and two regulatory subunits (β), forming heterotetramers $\alpha_2\beta_2$, $\alpha\alpha'\beta_2$, or $\alpha'_2\beta_2$. The enzyme has been found in the cytoplasm and the nucleus of all eukaryotic organisms investigated. It is probably one of the most pleiotropic kinases, acting on signalling, cell proliferation, and gene expression mechanisms. It catalyses the phosphorylation of serine and threonine residues of a large number of proteins, including enzymes related to the synthesis of nucleic acids, oncogenes products, tumour suppressors, and cytoskeleton and structural proteins (for reviews, see [1–6]).

One of the most singular features of this enzyme is that, in spite of its pleiotropic character and the huge number of protein substrates it has, a mechanism regulating its enzymatic

Address for offprints: J.S. Jiménez, Departamento de Química Física Aplicada, y Centro de Biología Molecular Severo Ochoa (UAM-CSIC), Universidad Autónoma de Madrid, Cantoblanco, 28049 Madrid, Spain (E-mail: juans.jimenez@uam.es)

activity is still lacking. No covalent modification and no second-messenger has been found to modulate the phosphorylating activity of CK2. The inhibitory effect of some negatively charged molecules, and the stimulatory effect of some polycations like polylysine or polyamines remain as the simple mechanisms by which the enzyme activity can be modified. In the last few years, a considerable effort has been made to find some clues to the enzyme regulation in its quaternary structure. Thus, the crystal structure has been solved for both subunits [7, 8], and many structural elements concerning the enzymatic properties, and intersubunit interaction have been finely mapped. The substantial architecture of the intersubunit interaction has been elucidated from different laboratories. β-subunit (CK2β) forms homodimers that bind the α-subunit (CK2α) to form the whole holoenzyme with an overall architecture in which the α-subunits do not contact each other [9–11].

It is well known that both CK2α and CK2β subunits bind each other with high affinity, forming a very stable complex that only becomes dissociated under strong denaturating conditions [12, 13]. The role of CK2β in this complex, however, remains controversial. Some years ago it was apparently clear that CK2β subunit was required for the maximal enzyme activity. However, it has become evident that this regulatory subunit has an inhibitory function when some substrates are used, as is the case with calmodulin [14, 15]. Therefore, the clear role of CK2β in regulating the holoenzyme function remains elusive. In order to contribute to the elucidation of the mode of action of this regulatory subunit on CK2, we have studied with Surface Plasmon Resonance (SPR) the intersubunit interactions to form the whole holoenzyme, as well as the mode of interaction of each subunit with two protein substrates. One of them, casein, represents a case in which the action of CK2β is activatory, and the other one, calmodulin, is inhibited by CK2β in its phosphorylation by the catalytic CK2α subunit.

Materials and methods

The β-subunit (residues 1–193) of protein kinase CK2 and the α-subunit were expressed in *E. coli* as His_6-tagged proteins, and the recombinant subunits purified by a column of Ni-agarose beads. Poly-L-lysine (MW [vis] = 29300), heparin (average MW = 6000), polyadenylic acid, poly(A), polyuridylic acid poly(U), casein, and Hepes were from Sigma. Where indicated, NaOH-solution stands for a solution composed of 0.01 M NaOH and 0.1 M NaCl.

Plasmon resonance was carried out as described originally by Liedberg *et al.* [16], as described elsewhere [17]. A 15 mW helium/neon laser was used as source of p-polarized light of wavelength 632.8 nm. Metal coatings of gold were prepared by evaporation on glass slides. Gold surfaces were then covered by a 3-layer structure composed of poly-L-lysine (polylysine) irreversibly bound to gold, followed by a second layer of heparin, and a third layer of polylysine. This 3-layer structure of polylysine-heparin-polylysine remains irreversibly bound to gold, preventing macromolecules from coming into direct contact with gold. After binding of a biomolecule to the 3-layer structure, the interaction with a second macromolecule can follow. Once this interaction was finished, the full complex was formed by the two interacting macromolecules, together with the second layer of heparin, and the third layer of polylysine can be broken down just by treatment with an alkaline solution having a pH value above the pK value of the amino groups of polylysine. The first polylysine layer remains irreversibly bound to gold, ready to form a new 3-layer structure, and therefore, to support a new macromolecular interaction on the same regenerated surface. Macromolecular adsorption on to the sensor surface was continuously monitored by the increasing reflectance. Once the magnitude of this reflectance had levelled off, denoting saturation, running buffer was always passed through, before injecting the next interacting macromolecule. N-phase complex Fresnel calculation [18] was used to obtain the optical constants of the bare gold surface, as well as the thickness of successive layers of macromolecules bound to the surface. Unless otherwise indicated, all experiments were carried out in a neutral running buffer composed of 10 mM Hepes and 0.1 M NaCl, at pH 6.9.

Results and discussion

Figure 1 shows the binding of CK2α and CK2β to a positively charged monolayer of polylysine and to negatively charged monolayers of heparin and the homopolymer of adenylic acid, poly(A). It is worth noting that CK2α subunit has the capability of binding positively charged monolayers of polylysine as well as negatively charged monolayers of heparin and polyadenylic acid. These interactions were irreversible under the pH and ionic strength conditions used here. CK2β, however, shows a much higher affinity for the positively charged surface of polylysine. Since CK2β is purified as a dimer, and taking into account the hypothesized topology of interunit contacts within the holoenzyme, the interaction between both subunits was made by flowing a solution of β-dimers on to CK2α immobilized on the sensor surface. As can be seen in Fig. 2, when CK2α was immobilized by attaching it to polylysine, an irreversible binding to CK2β could be obtained. In contrast, CK2α immobilized on the negatively charged surface of poly(A) binds CK2β to a much lower extent. This result suggest that some basic residues located on the α-subunit surface, which might be es-

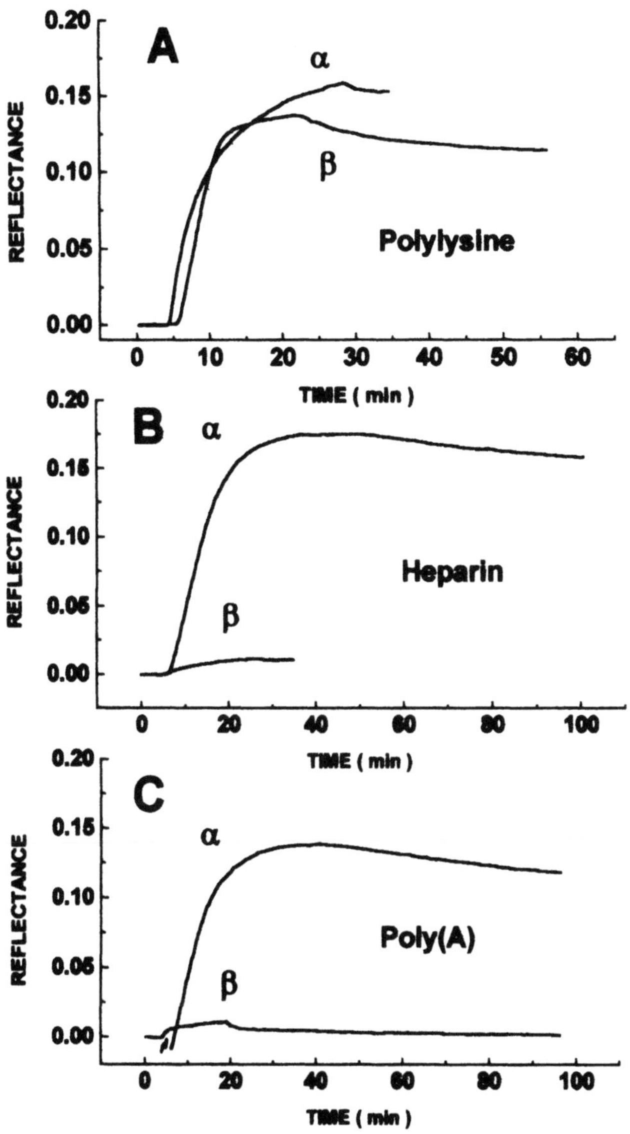

Fig. 1. Immobilization of CK2α and CK2β subunits on to polylysine, heparin, and poly(A). (A) After building a three-layer structure of polylysine-heparin-polylysine, solutions of CK2α (0.024 mg/ml) and CK2β (0.018 mg/ml) in neutral running buffer were flowed through the sensor cell. (B) A two-layer structure of polylysine-heparin was built, and then solutions of CK2α (0.024 mg/ml) and CK2β (0.02 mg/ml) in neutral running buffer were flowed through the sensor cell. (C) After building a three-layer structure of polylysine-heparin-polylysine, a solution of poly(A) (0.04 mg/ml) was flowed through the cell. Once the poly(A) binding on polylysine was complete, solutions of CK2α (0.024 mg/ml) and CK2β (0.012 mg/ml) in neutral running buffer were flowed as described above.

sential for interaction with β-subunit, may be buried by making contact with the poly(A) monolayer. In agreement with this suggestion is that the basic stretch 74–83 on CK2α has been proposed [7, 8, 19, 20] to make contact with the N-terminal region of β-subunit, and to be involved in the inhibition by heparin. Indeed, it would be also plausible that the

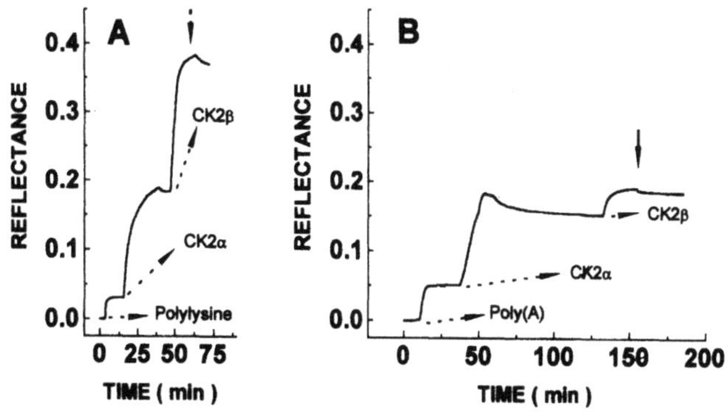

Fig. 2. Interaction between CK2α and CK2β subunits. (A) After immobilization of CK2α subunit on polylysine (A) or poly(A) (B), as described in the legend of Fig. 1, solutions of CK2β (0.027 mg/ml (A), and 0.02 mg/ml (B), in neutral running buffer) were injected into the flow cell. The arrows indicate the times at which running neutral buffer was injected after the CK2β subunit solutions.

site on CK2α for the proposed tight interaction with the C-terminal region of CK2β may not be already accessible to CK2β, as a consequence of the interaction with the negatively charged surface of poly(A).

Beside being involved in making contact with the N-terminal region of CK2β, the cluster of positively charged side chains 74–83, located at the helix αC of CK2α has also been proposed to be involved in substrate recognition [21]. Therefore, it would be expected that protein substrates could interact with CK2α immobilized on polylysine. Figure 3 shows that this is indeed the case. The binding of casein to CK2α immobilized on to polylysine, like the binding of CK2β, was

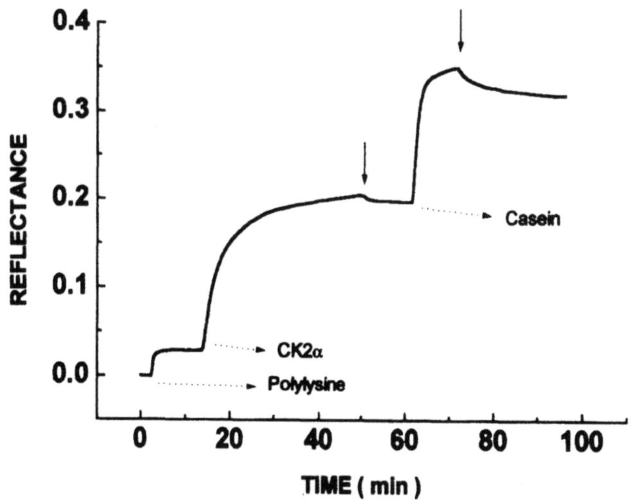

Fig. 3. Interaction between CK2α subunit and casein. A three layer structure of polylysine-heparin-polylysine was built as described in the text. After immobilization of CK2α subunit on the last layer of polylysine, a solution of casein (0.03 mg/ml in neutral running buffer) was injected into the flow cell. The arrows indicate the time at which running buffer was injected.

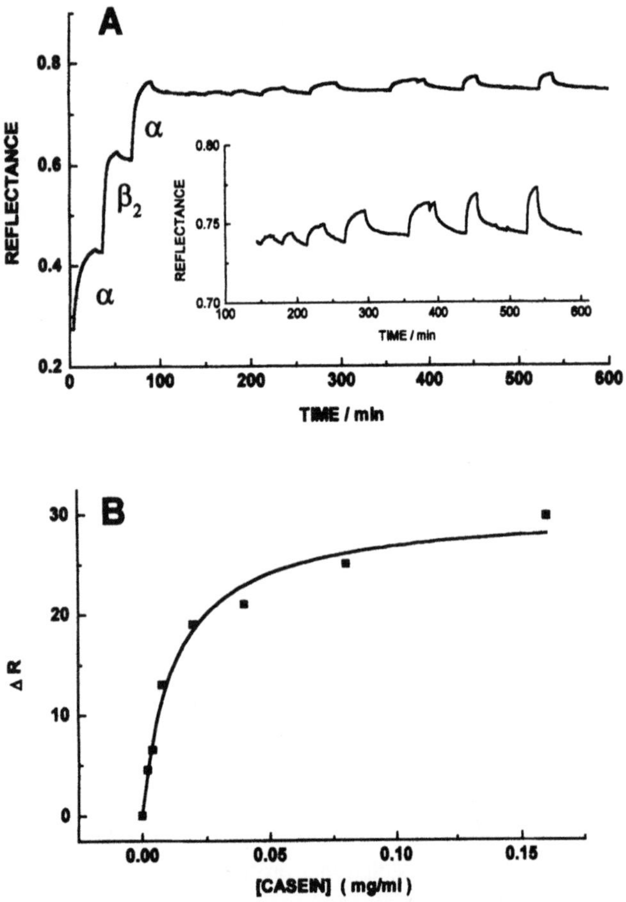

Fig. 4. Interaction of the whole holoenzyme with casein. After building the three-layer structure of polylysine-heparin-polylysine, solutions of α-subunit (0.024 mg/ml in neutral running buffer), and dimers of β-subunit (0.027 mg/ml in the same buffer) were injected into the flow cell at the indicated times. Solutions of increasing concentration of casein in the same buffer (in mg/ml, 0.002, 0.004, 0.008, 0.02, 0.04, 0.08, and 0.16) were then injected into the flow cell. After injection of each casein solution, and once the reflectance had levelled off, denoting saturation, neutral running buffer was injected before the next casein solution injection. (B) Fitting of the reflectance values obtained at equilibrium, for each casein concentration, to a rectangular-hyperbolic equation from which the equilibrium constant was obtained (see the text).

practically irreversible, and no dissociation of the protein substrate could be observed after running neutral buffer through the flow cell.

The results described above, concerning the binding of casein and CK2β to CK2α immobilized on polylysine, are in agreement with the proposed idea that protein substrates and CK2β have a common, or at least overlapping, docking site on CK2α. If this were the case one would expect the interaction between casein and polylysine-immobilized CK2α to be different from the interaction of casein with CK2α forming part of the whole holoenzyme, in which the cluster of positively charged side chains 74–83 must be involved in the interaction with the N-terminal regions of the β-subunit dimer. We carried out then a reconstitution of the heterotetramer $\alpha\beta_2\alpha$ on the flow cell. After immobilizing the α-subunit onto a layer of polylysine, a solution of dimers of β-subunit, β_2, was flowed through the cell, followed by a new solution of α-subunit (Fig. 4A). Both, α-β_2, and ($\alpha\beta_2$)-α interactions were irreversible, in agreement with the known stability of the heterotetramer. The sequence of interactions used: interaction of immobilized α-subunit with running dimers of β-subunit, followed by the interaction of the immobilized α-β_2 complex with running α-subunits, to form the $\alpha\beta_2\alpha$ holoenzyme, is in agreement with the overall architecture of the interunits contacts [9–11]. Assuming that both type of subunits have the same refraction index, and using a complex-Fresnel calculation, values of 51, 86, and 50 Å were obtained for the thickness of each one of the successive layers of α-, β_2-, and α-subunits. These values agree well with the size reported from x-ray studies of both subunits [7, 8], and indicate that three successive monolayers of α, β_2, and α are formed, giving place finally to a monolayer of the complex.

The $\alpha\beta_2\alpha$ complex remained irreversibly bound to the sensor surface by the 3-layer structure of polylysine-heparin-polylysine. Solutions of casein were then flowed through the cell, and as can be seen in the same figure, the protein substrate displayed a reversible binding to the holoenzyme, in contrast with the irreversible monolayer of casein formed when this protein was flowed on to the CK2α alone. From the maximal reflectance reached for each casein concentration, an equilibrium dissociation constant for the casein interaction with the holoenzyme, K = 0.013 mg/ml, could be obtained (Fig. 4B).

The results described above suggest that the binding of the CK2β dimers to the CK2α subunits would involve an interaction between the N-terminal regions of the CK2β dimers and the cluster of positively charged side chains 74–83 on CK2α, therefore preventing casein from coming into direct contact with CK2α to form the irreversible complex shown in Fig. 3. At the same time, CK2β would contribute with the generation of a milder site of interaction between the whole holoenzyme and the protein substrate. In fact, CK2β dimers immobilized on a polylysine-heparin-polylysine structure can bind casein in an almost reversible way, displaying a behaviour pattern for the binding of casein very similar to that shown by the whole holoenzyme (see Fig. 5).

As mentioned above, CK2α subunit alone may catalyse the phosphorylation of calmodulin, while the presence of the regulatory β-subunit produces a downregulation of the CK2α activity. We studied the interaction of calmodulin with CK2α subunit, CK2β subunit, and with the whole holoenzyme reconstituted *in situ* as described above. The interaction of calmodulin with CK2α (see Fig. 6) was quite similar to that of casein. A monolayer of calmodulin seems to be irreversibly formed when the protein substrate was flowed on to the

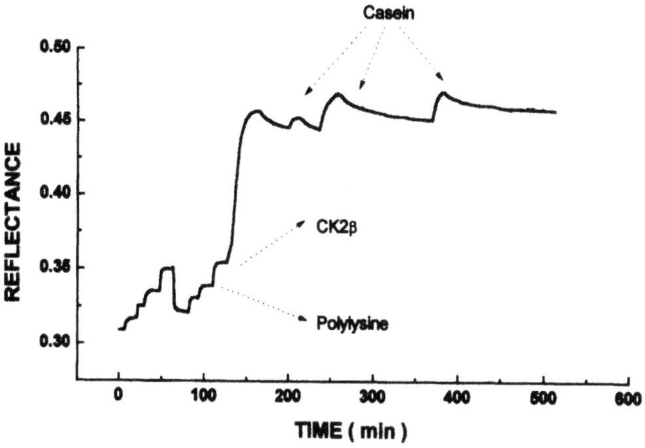

Fig. 5. CK2β subunit interaction with casein. CK2β was immobilized on the last layer of a polylysine-heparin-polylysine structure. Solutions of increasing concentration of casein in neutral running buffer (in mg/ml, 0.01, 0.02, and 0.03) were then injected into the flow cell.

polylysine-immobilized CK2α subunit. On the other hand, polylysine-immobilized CK2β, and the whole holoenzyme displayed a binding behaviour on calmodulin similar to that described above for casein, in that both the regulatory subunit alone and the full complex bind calmodulin reversibly. However, the binding of the protein substrate now is almost negligible (Fig. 6B), and in fact, the rise in temperature to about 30°C gave place to the full absence of binding of calmodulin to the regulatory subunit, while casein kept its interacting capability.

The results described above bear out the downregulation carried out by the regulatory subunit on the calmodulin phosphorylation by CK2α alone. It seems that the binding of the regulatory dimer to the catalytic α-subunit prevents calmodulin, in a similar mode to that suggested for casein, from contacting the binding site present in CK2α alone, but neither the regulatory subunit nor the whole holoenzyme seem to bind calmodulin in a significant extent. The holoenzyme consequently would not show any activity on calmodulin.

Particularly interesting was the result obtained when polylysine was included in the calmodulin solution. Many biomolecules including substrates of CK2 [17, 22, 23], the CK2 subunits, and the holoenzyme, interact with polylysine. Calmodulin has also been shown to interact with polylysine [24], and it binds the last layer of the polylysine 3-layer structure used here (data not shown). As it can be seen in the inset of Fig. 6C, the presence of polylysine induced a clear increasing of the reflectance signal. After flowing two solutions of calmodulin and polylysine, the flowing of a polylysine-free calmodulin solution gave a strong interaction of calmodulin with the holoenzyme. These results, together with the capability of polylysine to bind enzyme and enzyme-substrates, suggest that polylysine may play an anchoring role between substrate and enzyme, and would be in agreement with the reported polylysine dependency of the holoenzyme to catalyse the phosphorylation of calmodulin [5, 15].

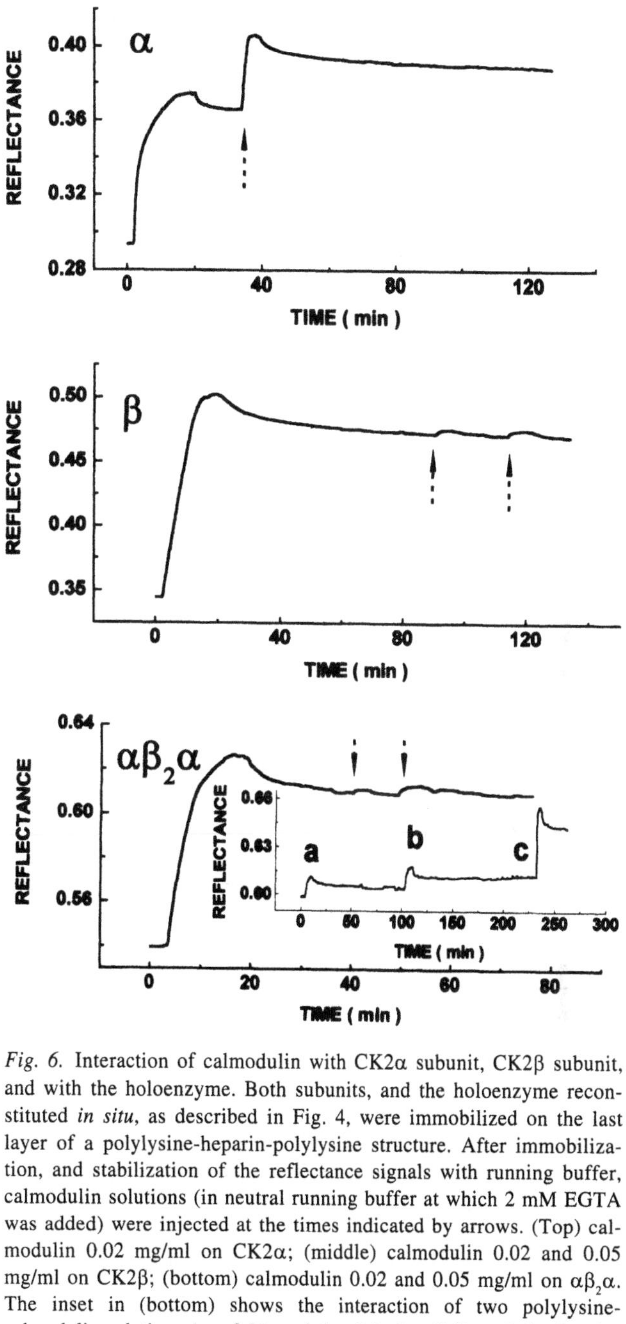

Fig. 6. Interaction of calmodulin with CK2α subunit, CK2β subunit, and with the holoenzyme. Both subunits, and the holoenzyme reconstituted *in situ*, as described in Fig. 4, were immobilized on the last layer of a polylysine-heparin-polylysine structure. After immobilization, and stabilization of the reflectance signals with running buffer, calmodulin solutions (in neutral running buffer at which 2 mM EGTA was added) were injected at the times indicated by arrows. (Top) calmodulin 0.02 mg/ml on CK2α; (middle) calmodulin 0.02 and 0.05 mg/ml on CK2β; (bottom) calmodulin 0.02 and 0.05 mg/ml on $\alpha\beta_2\alpha$. The inset in (bottom) shows the interaction of two polylysine-calmodulin solutions (a = 0.01 mg/ml polylysine, 0.01 mg/ml calmodulin; b = 0.03 mg/ml polylysine, 0.03 mg/ml calmodulin) with the holoenzyme; c = a polylysine-free calmodulin solution at 0.03 mg/ml.

The response of protein kinase CK2 to those substrates represented by casein is different to that observed on those substrates represented by calmodulin, inasmuch as the catalytic activity is considered. In the first case the regulatory β-subunit exerts an activating action, while in the second case, downregulation occurs. However, when the modes of CK2 and CK2-substrate intersubunit interactions are investigated, the response of the enzyme does not seem so distinct. CK2β subunit and protein substrates have a common, or at least an overlapping site of interaction on CK2α subunit. The binding of CK2β subunit to CK2α would prevent substrates from binding CK2α at this site, and at the same time, it would generate a site of a new nature, probably contributed by both subunits. The difference between casein and calmodulin would be that the last protein has practically no affinity for this new site in the absence of polycations like polylysine.

Acknowledgements

Thanks are given to Fernando Moreno for the gold film preparations. We are also grateful to Dr. Jesús Ávila for helpful discussion. This work has been supported by grants from DGICYT of the Spanish Government and from the European Union.

References

1. Edelman AM, Blumenthal DX, Krebs EG: Protein serine-threonine kinases. Annu Rev Biochem 56: 567–613, 1987
2. Pinna LA: Casein kinase 2: An 'eminence grise' in cellular regulation? Biochim Biophys Acta 1054: 267–284, 1990
3. Issinger OG: Casein kinases: Pleiotropic mediators of cellular regulation. Pharmacol Ther 59: 1–30, 1993
4. Allende JE, Allende CC: Protein kinase CK2: An enzyme with multiple substrates and a puzzling regulation. FASEB J 9: 313–323, 1995
5. Pinna LA, Meggio F: Protein kinase CK2 ('casein kinase-2') and its implication in cell division and proliferation. Prog Cell Cycle Res 3: 77–97, 1997
6. Blanquet PR: Casein kinase 2 as a potentially important enzyme in the nervous system. Prog Neurobiol 60: 211–246, 2000
7. Niefind K, Guerra B, Pinna LA, Issinger OG, Schomburg D: Crystal structure of the catalytic subunit of protein kinase CK2 from *Zea mays* at 2.1 Å resolution. EMBO J 17: 2451–2462, 1998
8. Chantalat L, Leroy D, Filhol O, Nueda A, Benítez MJ, Chambaz EM, Cochet C, Dideberg O: Crystal structure of the human protein kinase CK2 regulatory subunit reveals its zinc-finger-mediated dimerization. EMBO J 18: 2930–2940, 1999
9. Gietz D, Graham KC, Litchfield DW: Interactions between the subunits of casein kinase II. J Biol Chem 270: 13017–13021, 1995
10. Kusk M, Bendixen C, Duno M, Westergaard O, Thomsen B: Genetic dissection of intersubunit contacts within human protein kinase CK2. J Mol Biol 253: 702–711, 1995
11. Boldyreff B, Issinger O-G: A-Raf kinase is a new interacting partner of protein kinase CK2β subunit. FEBS Lett 403: 197–199, 1997
12. Cochet C, Chambaz EM: Oligomeric structure and catalytic activity of G type casein kinase. Isolation of the two subunits and renaturation experiments. J Biol Chem 258: 1403–1406, 1983
13. Valero E, De Bonis S, Filhol O, Wade RH, Langowski J, Chambaz EM, Cochet C: Quaternary structure of casein kinase 2. Characterization of multiple oligomeric states and relation with its catalytic activity. J Biol Chem 270: 8345–8352, 1995
14. Meggio F, Boldyreff B, Marin O, Marchiori, F, Perich JW, Issinger OG, Pinna LA: The effect of polylysine on casein kinase-2 activity is influenced by both the structure of the protein/peptide substrates and the subunit composition of the enzyme. Eur J Biochem 205: 939–945, 1992
15. Bidwai AP, Reed JC, Glover CVC: Phosphorylation of calmodulin by the catalytic subunit of casein kinase II is inhibited by the regulatory subunit. Arch Biochem Biophys 300: 265–270, 1993
16. Liedberg B, Nylander C, Lundstrom I: Surface Plasmon Resonance for gas detection and biosensing. Sensors Actuators 4: 299–302, 1983
17. Benítez MJ, Mier G, Briones F, Moreno FJ, Jiménez JS: A surface-plasmon-resonance analysis of polylysine interactions with a peptide substrate of protein kinase CK2 and with the enzyme. Biochem J 324: 987–994, 1997
18. Hansen WN: Electric fields produced by the propagation of plane coherent electromagnetic radiation in a stratified medium. J Opt Soc Am 58: 380–390, 1968
19. Vaglio P, Sarno S, Marin O, Meggio F, Issinger OG, Pinna LA: Mapping the residues of protein kinase CK2α subunit responsible for responsiveness to polyanionic inhibitors. FEBS Lett 380: 25–28, 1996
20. Krehan A, Lorenz P, Plana-Coll M, Pyerin W: Interaction sites between catalytic and regulatory subunits in human protein kinase CK2 holoenzymes as indicated by chemical cross-linking and immunological investigations. Biochemistry 35: 4966–4975, 1996
21. Sarno S, Vaglio P, Meggio F, Issinger, OG, Pinna LA: Protein kinase CK2 mutants defective in substrate recognition. J Biol Chem 271: 10595–10601, 1996
22. Moreno F, Lechuga CG, Collado M, Benítez MJ, Jiménez JS: A polylysine-induced aggregation of substrates accompanies the stimulation of casein kinase II polylysine. Biochem J 289: 631–635, 1993
23. Benítez MJ, Mier G, Briones F, Moreno FJ, Jiménez JS: Binding of polylysine to protein kinase CK2, measured by surface plasmon resonance. Mol Cell Biochem 191: 29–33, 1999
24. Hardy RW, McDonald JM, Remsen EE, d'Avignon DA, Sacks DB: The interaction of calmodulin and polylysine as study by ^{1}H NMR spectroscopy and sedimentation equilibrium centrifugation. Biochem Biophys Res Commun 198: 309–317, 1994

Molecular and Cellular Biochemistry **227**: 37–44, 2001.

The activity of CK2 in the extracts of COS-7 cells transfected with wild type and mutant subunits of protein kinase CK2

Iris Korn, Germaine Jacob, Catherine C. Allende and Jorge E. Allende

Programa de Biología Celular y Molecular, Instituto de Ciencias Biomédicas, Facultad de Medicina, Universidad de Chile, Santiago, Chili

Abstract

Protein kinase CK2 is ubiquitous in eukaryotes and is known to phosphorylate many protein substrates. The enzyme is normally a heterotetramer composed of catalytic (α and α′) and regulatory (β) subunits. The physiological regulation of the enzyme is still unknown but one of the factors that may play an important role in this regulation is the ratio of the catalytic and regulatory subunits present in cells. The possible existence of 'free' CK2 subunits, not forming part of the holoenzyme, may be relevant to the physiological function of the enzyme in substrate selection or in the interaction of the subunits with other partners. The objective of this work was to study in COS-7 cells the effects of transient expression of CK2 subunits and mutants of the catalytic subunit on the CK2 phosphorylating activity of the extracts of these cells. Using pCEFL vectors that introduce hemagglutinin (HA) or a heptapeptide (AU5) tags in the expressed proteins, COS-7 cells were transfected with α and β subunits of *Xenopus* CK2, with the α′ subunit of *D. rerio*, and with *Xl* CK2αA^{156}, which although inactive can bind tightly to CK2β, and with *Xl* CK2α$E^{75}E^{76}$, which is resistant to heparin and polyanion inhibition. The efficiency of transient transfection was of 10–20% of treated cells.

Expression of CK2α or CK2α$E^{75}E^{76}$ in COS-7 cells caused an increase of 5–7-fold of the CK2 activity in the soluble cell extracts. If these catalytic subunits were cotransfected with CK2β, the activity increased further to 15–20-fold of the controls. Transfection of CK2β alone also increase the activity of the extracts about 2-fold. Transfection with the inactive CK2αA^{156} yielded extracts with CK2 activities not significantly different from those transfected with the empty vectors. However, cotransfection of CK2α or CK2α$E^{75}E^{76}$ with CK2αA^{156} caused a 60–70% decrease in the CK2 activity as compared to those of cells transfected with only the active CK2α subunits. These results can be interpreted as meaning that CK2αA^{156} is a dominant negative mutant that can compete with the other catalytic subunits for the CK2β subunit. Addition of recombinant CK2β to the assay system of extracts of cells transfected with catalytic subunits causes a very significant increase in their CK2 activity, demonstrating that CK2β subunit is limiting in the extracts and that an excess of free CK2α has been produced in the transfected cells. Transfection of cells with CK2α$E^{75}E^{76}$ results in a CK2 activity of extracts that is 90% resistant to heparin demonstrating that a very large proportion of the CK2 activity is derived from the expression of the exogenous mutant. In both the *in vivo* and *in vitro* systems, the sensitivity of CK2α$E^{75}E^{76}$ to heparin increases considerably when it forms part of the holoenzyme CK2$\alpha_2\beta_2$. (Mol Cell Biochem **227**: 37–44, 2001)

Key words: casein kinase II, dominant-negative mutant of CK2, heparin-resistant CK2α mutant, CK2α/β subunit ratio

Introduction

Protein kinase CK2 is ubiquitous in eukaryotes and is known to phosphorylate a large number of protein substrates. The holoenzyme is a heterotetramer composed of catalytic (α and α′) and regulatory (β) subunits with a $\alpha_2\beta_2$, $\alpha\alpha'b_2$ and $\alpha'_2\beta_2$ composition observed in most animal cells. Protein kinase CK2 appears to be involved in the regulation of cell division,

Address for offprints: J.E. Allende, Programa de Biología Celular y Molecular, Instituto de Ciencias Biomédicas, Facultad de Medicina, Universidad de Chile, Casilla 70086, Santiago 7, Chile (E-mail: jallende@abello.dic.uchile.cl)

but there are still many important questions concerning the regulation of its own activity [1–3].

One aspect that still is a matter of debate is whether, under *in vivo* conditions, the ratio of the subunits differs from one. This is an important consideration because some substrates of CK2 such as calmodulin and MDM-2 are phosphorylated preferably by the free CK2α subunit and much less or not at all by the holoenzyme which has the bound regulatory subunit [4]. Most other known substrates are much more efficiently phosphorylated by the holoenzyme than by the free catalytic subunit. The possible existence of free catalytic or regulatory subunits is also important because it has been shown that these subunits can interact with and regulate other proteins. Free CK2β, but not the $\alpha_2\beta_2$ holoenzyme, can interact with several other kinases such as A-Raf, c-mos and $p90^{RSK}$, up- or downregulating their activity [5]. This led us to propose that CK2β might be a 'wild-card' subunit that can bind and regulate a number of kinases and not exclusively CK2α [6]. The free catalytic subunit CK2α has also been shown to interact with other proteins such as the protein phosphatase 2A catalytic subunit [7]. The binding of CK2α with PP2Ac is competitive with CK2β and causes an increase in the phosphatase activity.

In addition to these interactions that require the free CK2 subunits, the holozyme has been shown to form stable complexes with many other proteins, most of which are substrates of CK2 [5].

The above considerations prompted us to study the effect of overexpression of the wild-type and mutant subunits of CK2 in transfected cells. This overexpression of one of the subunits would alter the supposed ratio of 1:1 of α/β and would make it possible to study the effect of this variation on the phosphorylation of specific substrates and on overall cell growth.

In the experiments described in this report, two CK2α mutants in addition to the wild-type subunits have been transfected in COS-7 cells. The mutant $CK2\alpha A^{156}$ is a catalytically inactive mutant that still binds CK2β almost as tightly as the wild type CK2α [8], and mutant $CK2\alpha E^{75}E^{76}$ has two lysines replaced by glutamic acids in a basic stretch present in the catalytic subunit. The replacement makes this mutant highly resistant to heparin and other polyanionic inhibitors of CK2 [9].

We provide evidence that overexpression of CK2α and CK2β can increase the CK2 activity of transfected cell extracts. This increment is much larger when both subunits are cotransfected. Cotransfection of wild-type CK2α and $CK2\alpha A^{156}$, on the other hand, causes a drastic lowering of the activity observed when compared to that seen with the transfection of wild-type CK2α alone. This result indicates that $CK2\alpha A^{156}$ can be considered a dominant negative mutant, for at least the exogenously expressed CK2 activity. Supplementing the *in vitro* assay mixture with free recombinant CK2β increases the activity of extracts transfected with CK2α subunits and even the extract of control cells transfected with the empty vector. This result suggests that these extracts contain free CK2α subunits. Extracts of cells transfected with $CK2\alpha E^{75}E^{76}$ have a CK2 activity that is highly resistant to heparin inhibition demonstrating that a large part of the activity measured is due to the expression of the exogenous mutated gene.

Methods and materials

Subcloning CK2 subunits and their mutants into the pCEFL vectors

The cDNAs coding for the subunits wild type *Xenopus laevis* (*Xl*) CK2α, *Xl* $CK2\alpha A^{156}$ and *Xl* $CK2\alpha E^{75}E^{76}$ had been subcloned into Nde I and Sal I sites of pT7-7 H_6 as described [8]. In the three cases the Nde I site was changed to Bgl II and the Hind III site to Not I, using synthetic oligonucleotides. The new recombinant vectors containing the cDNAs for *Xl* CK2α wild type and *Xl* $CK2\alpha A^{156}$ were digested with Bgl II and Not I and the purified inserts coding for the CK2 subunits were ligated to pCEFL vectors containing an AU5 or HA epitope tag and that had also been digested with Blg II and Not I. Competent cells of *E. coli* DH5α were transformed with the recombinant pCEFL plasmids and positive colonies were analyzed by PCR to test for the presence of the 1000 bp insert corresponding to CK2α. The correct ligation was further tested by restriction enzyme analysis.

A similar protocol was used to generate the pCEFL-HA and pCEFL-AU5 recombinant plasmids containing the *Danio rerio* (*Dr*, zebrafish) CK2α' cDNA [10]. In this case the *Dr* CK2α' and the inactive *Dr* $CK2\alpha A^{156}$ were cloned into EcoRI site of pCEFL-HA and pCEFL-AU5.

For the case of *Xl* CK2β, this cDNA cloned in pGEX-2T [11] was cut with EcoRI and first ligated into pCDNA3 vector in its EcoRI site to correct the reading frame. It was then cut with BamH I and Not I and inserted into the Bgl II and Not I of pCEFL-Ha and pCEFL-AU5.

In all cases the correct insertion and structure of the constructs was confirmed by DNA sequencing.

Transfections of COS-7 cells

Transfections of COS-7 cells were performed as described originally by Cullen *et al.* [12] using lipofectamine. Cells were subcultured in p60 plates the day before transfection. Immediately previous to transfection, the Dulbecco's Modified Eagle's Medium (DMEM) containing 10% fetal bovine serum (FBS) was withdrawn by aspiration and replaced with

2 ml of DMEM without serum and without antibiotics. The DNA-lipofectamine mixture prepared in 500 µl of DMEM contained 2–5 µg of the recombinant vector DNA and 12 µl of lipofectamine and was incubated at room temperature for 15 min and then added dropwise to the p60 plates containing the COS-7 cells. The cells were incubated for 5 h at 37°C in the CO_2 incubator. Subsequently 2.5 ml of DMEM with 20% FBS was added to each plate to yield a final concentration of 10% FBS. The following day, this medium which is antibiotic-free, was replaced by complete DMEM (with antibiotics) and 10% FBS. Transfected cells were grown for 48 h after the transfection treatment and then analyzed. The efficiency of the transfection protocol was tested using a pCMVB-gal vector and colormetric staining used to determine the percent of blue colonies. The efficiency of lipofectamine transient transfections was 10–20%.

Preparation of cell lysates

Cells were washed 3 times with PBS. Cell lysis buffer (containing 50 mM Tris-HCl pH7.5, 150 mM NaCl, 1% NP-40, 1 mM phenylmethylsulfonylfluoride, 2 mM orthovanadate and 20 µg/ml of aprotinin, antipain, leupeptin and pepstatin) was added directly to the cell plates (300 µl for p60 plates and 900 µl for p100 dishes). Cells were collected with a cell scraper, incubated 10 min in ice and centrifuged at 14000 rpm for 15 min at 4°C in an Eppendorf microcentrifuge. The soluble phase of the lysate was obtained and protein determined by the Bradford method [13]. Aliquots of lysate were stored at –70°C.

Assays for CK2 activity

Cell lysates containing 2.5–5 µg of protein (linear range) were assayed in a reaction mixture of 30 µl containing 50 mM Hepes pH 7.4, 100 mM NaCl, 10 mM $MgCl_2$, 0.5 mM DTT, 100 µM [^{32}P]γ ATP (1500 CPM/pmol), 1 mM of the specific peptide substrate RRRDDDSDDD. Incubation was for 10 min at 30°C and the reaction was stopped by absorbing the mixture onto 2 cm^2 pieces of p81 phosphocellulose paper. The papers were washed 3 times for 15 min with 75 mM phosphoric acid, dried and counted in a scintillation counter. All assays were performed in duplicate or triplicate samples. Each separate assay included controls in which incubations were run in the absence of the peptide substrate and another run in the presence of 5 µg/ml of heparin. All kinase activity values presented are corrected for the amount of incorporation in the presence of heparin and non-CK2 activity of the extracts. For the *in vitro* assay of the CK2 activity of the purified recombinant CK2 the same system was used. For the assay of activity of immunoprecipitates, cell lysates containing 1 mg of protein were immunoprecipitated with 0.8 µg of mouse monoclonal anti-HA or anti-AU5 antibody or 4 µg antihuman CK2α monoclonal antibody. After overnight incubation at 4°C, G-agarose was added, incubation continued for 45 min at 4°C, and the immunoprecipitate was washed extensively with 1% PBS/NP-40 and finally with kinase buffer as above but with 5 mg/ml of casein instead of the peptide substrate. The reaction was stopped by the addition of gel sample buffer and the results analyzed on 10% SDS-PAGE and autoradiography.

Western blots

After running lysates (20–40 µg total protein) on 10% SDS-PAGE, the proteins were transferred to polyvinylidene fluoride membranes. The membrane was blocked with 4% bovine serum albumin and incubated for 1 h at room temperature with the primary antibody at 1:1000 or 1:500 dilution in a buffer containing 0.4% BSA. The presence of the primary antibody was detected by a secondary antibody conjugated to peroxidase.

Materials

Antibodies were obtained as follows: monoclonal antihuman CK2α, Upstate Biotechnologies; monoclonal anti HA (F-7), Santa Cruz; monoclonal anti AU5, BabCo.

Heparin, protease inhibitors, immobilon membranes and β-casein were purchased from Sigma. Radioisotope [^{32}P]γ ATP were acquired from NEN. Specific synthetic peptide substrates for CK2 was prepared by Oligopeptido, the core facility of the Faculty of Medicine, University of Chile. The same core facility performed the DNA sequence analysis in an automatic ABI 377 DNA Sequencer.

Results

Expression of CK2 subunits in transfected COS-7 cells

Transient transfections of COS-7 cells were carried out using constructs of cDNAs of the CK2 subunits or mutants thereof in pCEFL vectors that generate the proteins with HA or AU5 tag peptides detectable by monoclonal antibodies. The efficiency of transfection varied between 10–20% of the treated cells.

Figure 1 shows Western blots of the lysates of cells transfected with the different CK2 subunits and their mutants. It can be observed that the expression of the transfected subunit genes is very strong. In part C of that figure, in which an

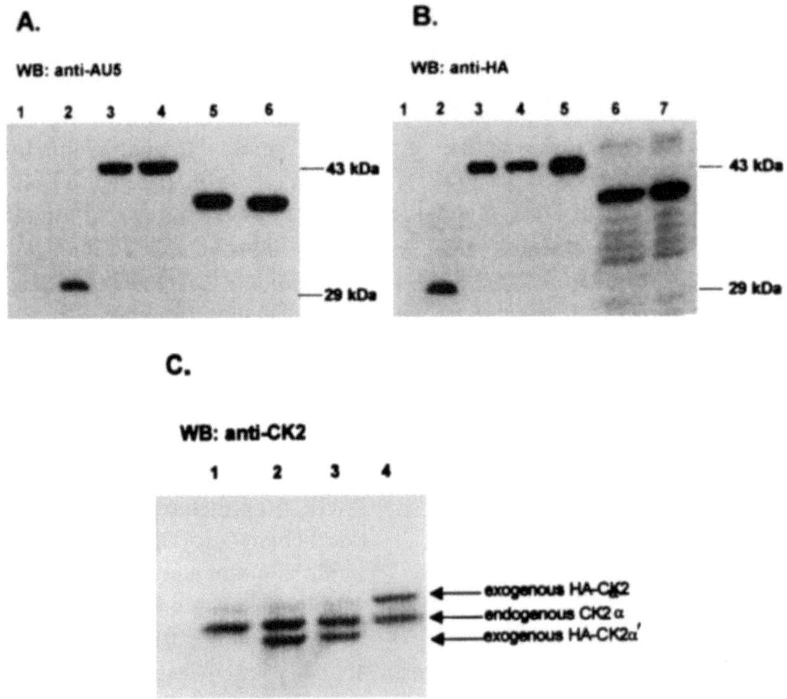

Fig. 1. Expression of CK2 subunits in transfected cells. Cos-7 cells were transfected with the vectors indicated and at 48 h cell lysates were prepared and analyzed by Western blot as given in Materials and methods. (A) Antibody anti-AU5 was used and, in the lanes indicated, transfections performed with vectors as follows: pCEFL-AU5 (1), AU5-CK2β (2), AU5-CK2α (3), AU5-CK2αA^{156} (4), AU5-CK2α′ (5) and AU5-CK2α′A^{156} (6). (B) Antibody anti-HA was used and, in the lanes indicated, transfections performed with vectors pCEFL-HA (1), HA-CK2β (2), HA-CK2α (3), HA-CK2αA^{156} (4), HA-CK2α$E^{75}E^{76}$ (5), HA-CK2α′ (6), and HA-CK2α′A^{156} (7). (C) Antibody anti-human CK2α was used and, in the lanes indicated, transfections performed with vectors pCEFL-HA (1), HA-CK2α′ (2), HA-CK2α′A^{156} (3) and HA-CK2α (4).

antibody against human CKα or CK2α′ is used to detect the protein, the expression of the endogenous CK2α and exogenous α subunit can be compared and found to be quite similar in intensity. Exogenous CK2α migration is slower than the endogenous enzyme because of the inserted tag peptide. Endogenous CK2α′ was not detected in these experiments possibly due to lower expression of this subunit in COS-7 cells.

CK2 activity of lysates of transfected cells

The CK2 activity due to the expression of the exogenous genes was assayed by immunoprecipitation with antibodies against the HA tag and assay of the phosphorylating activity of the immunoprecipitates by incubation with casein and [γ-^{32}P]ATP. The results of two separate experiments can be observed in Fig. 2. Casein substrate phosphorylation catalyzed by immunoprecipitates of cells transfected with CK2α wild-type (lane 1) is much greater than that observed with those of cells transfected with the inactive CK2αA^{156} (lane 2). The casein phosphorylating activity is greatly enhanced by co-transfection of CK2α and CK2β (lane 3), while cotransfection of the regulatory subunit does not significantly stimulate the slight activity of the cells transfected with CK2αA^{156} (lane 4). A detectable increase of phosphorylation, however, is observed when CK2β is transfected alone, suggesting that this subunit can interact and coimmunoprecipitate with endogenous free CK2α.

In an independent experiment (lanes 6–9) the autoradiography obtained was exposed for a longer period. This experiment indicates that cotransfection of wild type CK2α with the inactive CK2αA^{156} (lane 9) causes an important decrease of the casein phosphorylating activity observed in the immunoprecipitate as compared to the activity observed with the CK2α wild type alone (lane 7). This result would indicate that CK2αA^{156} could act as a dominant negative of the exogenously transfected CK2α wild type.

These experiments were complemented with determinations of CK2 activity directly on the cellular lysates using the peptide substrate (RRRDDDSDDD) which is specific for the CK2 kinase. The results shown in Fig. 3 represent the average values of several experiments. The phosphorylation obtained in the presence of heparin or in the case of the heparin-resistant mutant CK2α$E^{75}E^{76}$, in the absence of the specific peptide substrate, has been subtracted from these results. In Fig. 3 we see the activity of CK2 obtained with lysates of cells transfected with CK2α, CK2β, mutants of CK2α and some

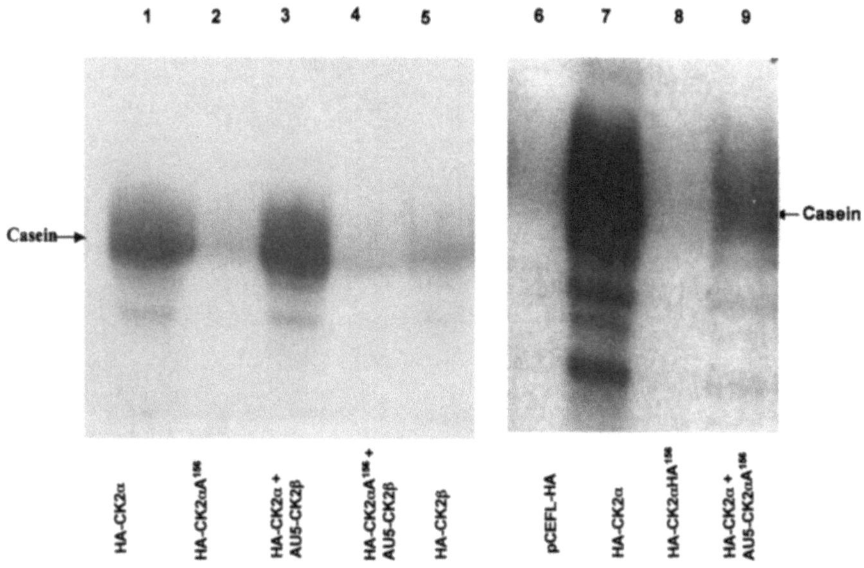

Fig. 2. Autoradiography of CK2 kinase activity present in immunoprecipitates of transfected Cos-7 cells. Cell lysates were prepared 48 h after transfection with the vectors indicated and immunoprecipitated using anti-HA antibody as given in Materials and methods. Immunoprecipitates were assayed by adding substrates casein and [γ^{32}P]ATP using the standard assay. The reaction was stopped by the addition of gel buffer and SDS-gel electrophoresis and autoradiography were performed. Transfections were as follows: pCEFL-HA-CK2α (lane 1), pCEFL-HA-CK2αA^{156} (lane 2), pCEFL-HA-CK2α and pCEFL-AU5-CK2β (lane3), pCEFL-HACK2αA^{156} and pCEFL-AU5-CK2β (lane 4), pCEFL-CK2β (lane 5), empty vector pCEFL-HA (lane 6), pCEFL-HA-CK2α (lane 7), pCEFL-HA-CK2αA^{156} (lane 8) and pCEFL-HA-CK2α and pCEFL-UA5-CK2αA^{156} (lane 9).

combinations of the genes for these subunits. Again, with this more direct measurement, we can observe that transfection with CK2α wild type and the active CK2αE^{75}E^{76} mutant result in large increases of CK2 activity. A smaller but significant increment (2-fold) in the CK2 activity is observed when

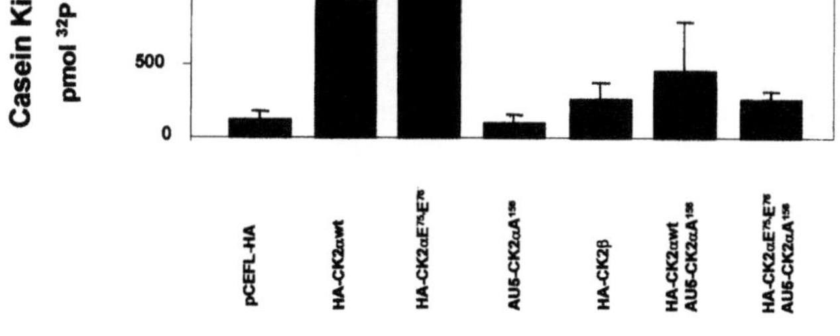

Fig. 3. CK2 kinase activity of cellular lysates from transfected Cos-7 cells using specific peptide substrate. Cells transfected with different vectors as given were assayed 48 h after transfection using conditions given in Material and methods and assayed using 2.5 μg of lysate protein, 1 mM specific peptide RRRDDDSDDD and 100 μM [γ^{32}P]ATP. Statistical analyses of results from independent transfection experiments are presented where n = 4 for pCELF-HA, pCELF-HA-CK2α, pCELF-HA-CK2α^{156}(each alone and combined), pCELF-HA-CK2β, pCELF-HA-CK2α with pCELF-HA-CK2αE^{75}E^{76}, and n = 11 for pCELF-HA-CK2α E^{75}E^{76}.

CK2β is transfected. The activity measured after transfection with the inactive CK2αA^{156} is not significantly different from the activity obtained with cells transfected with the empty vector. The most significant effect is that the cotransfection of wild type CK2α or CK2αE^{75}E^{76} with the inactive CK2αA^{156} caused a very important reduction in the CK2 activity. Similar results were obtained with CK2α′ and CK2α′A^{156} transfections (not shown).

CK2β is limiting in the lysates of cells transfected with CK2α subunits

The most direct way in which CK2αA^{156} could act as a 'dominant-negative' mutant and reduce the CK2 activity of the cells transfected with genes coding for active CK2α would be through competition for CK2β to form inactive or less active tetrameric holoenzyme. This interpretation requires CK2β to be limiting in the extract of these cells. In order to test whether β was truly limiting, assays for CK2 activity were conducted in presence of 25 ng of purified recombinant CK2β. Figure 4 shows the results obtained by assaying the CK2 activity of lysates of cells transfected with different constructs in the presence and absence of added recombinant CK2β. These results demonstrate that the addition of the regulatory β subunit very significantly stimulates the activity of lysates of all the cells tested except those cells that were transfected both with CK2αE^{75}E^{76} and with CK2β.

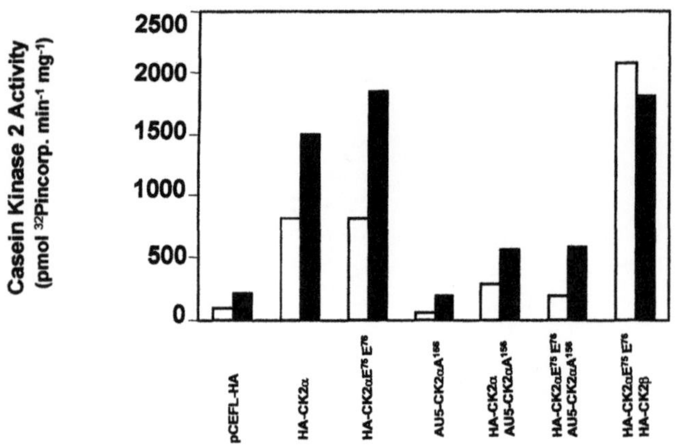

Fig. 4. Effect of the addition of recombinant CK2β on the CK2-specific protein kinase activity present in lysates from CK2-transfected Cos-7 cells. Lysates were prepared from cells transfected with the indicated vectors and assays carried out with specific peptide substrate as given in Fig. 2. Assays were in the presence (black-filled bars) or absence (open bars) of 25 ng recombinant CK2β regulatory subunit.

Interestingly, the addition of CK2β even stimulated the activity in cells transfected with the empty vector indicating that in these cells there probably is some free CK2α and consequently limiting CK2β.

The heparin sensitivity of the CK2 activity of cells transfected with CK2αE^{75}E^{76}

Previous *in vitro* experiments had demonstrated that the mutant CK2αE^{75}E^{76} was very resistant to heparin inhibition [9]. It seemed interesting to test this characteristic in the activity of cells transfected with this mutant, since this property could help to discriminate between the endogenous activity and the expression of the exogenous enzyme.

Figure 5 shows the results obtained when the lysates transfected with CK2α wild type and with CK2αE^{75}E^{76} are compared in relation to inhibition by 5 μg/ml of heparin in the presence or absence of exogenous CK2α added to the assay mixture. In this figure, the value obtained without added substrate peptide is not substracted but presented separately in order to compare how much of the specific CK2 activity is susceptible to heparin. It can be observed that 88% of the activity of cells transfected with CK2α can be inhibited by heparin whether or not CK2β is added to the assay. With transfection of CK2αE^{75}E^{76} and the assay carried out in the absence of added CK2β, close to 85% of the activity remains in the presence of heparin. In the presence of added CK2β, the CK2 activity of this extract is greatly stimulated, as described in Fig. 4. Surprisingly, more than 50% of this activity was inhibited by heparin. This finding prompted us to test *in vitro,* with the purified recombinant proteins, the heparin sensitivity of the mutant and wild type CK2α in the presence and absence of CK2β. As seen in Table 1, the inhibition of CK2αE^{75}E^{76} with 3 μg/ml of heparin is much greater (49%) in the presence of CK2β than in its absence (25%). In the case of the wild-type CK2α, the difference of its susceptibility to heparin caused by the presence of CK2β is much lower.

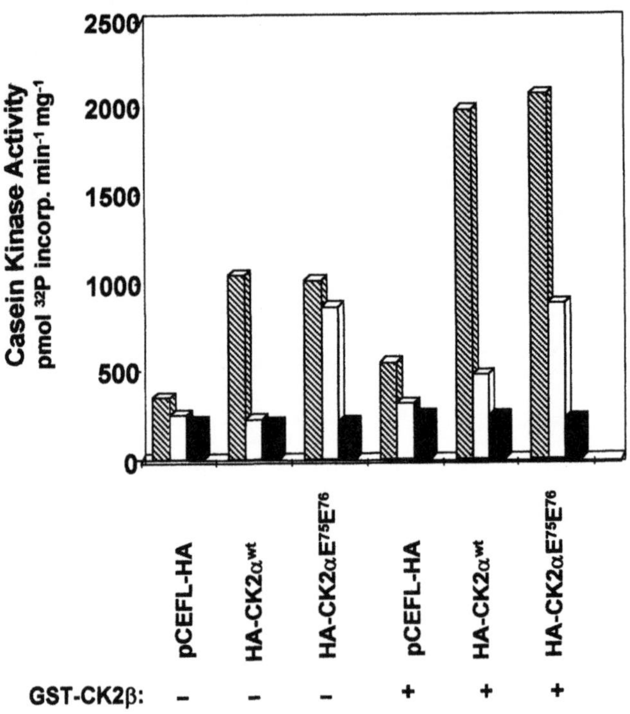

Fig. 5. Effect of heparin on the CK2-specific protein kinase activity present in lysates from CK2-transfected Cos-7 cells. Lysates were prepared from cells tranfected with the indicated vectors and assays performed with the specific peptide substrate (where indicated) as given in Fig. 2. Assays were performed in the absence (cross-hatched bars) or presence of 5 μg/ml heparin (open bars) and in the absence of CK2-specific peptide (black bars). The assays were also performed in the absence or presence of 25 ng recombinant CK2β subunit as indicated below the figure.

Table 1. The differential inhibitory effect of heparin on the activity of recombinant CK2α catalytic subunit and of holozyme *in vitro*

Catalytic subunit	Additions	CK2 activity (pmoles ^{32}P/min/mg)	Inhibition (%)
CK2α wild type	–	1230	0
	heparin*	98	92
	CK2β	6150	0
	CK2β + heparin	1045	83
CK2αE^{75}E^{76}	–	1450	0
	heparin*	1087	25
	CK2β	8410	0
	CK2β + heparin	4290	49

*Heparin concentration = 3 μg/ml.

Discussion

The results obtained by assaying the CK2 activity of lysates of cells transiently transfected with subunits of CK2 and two types of mutants of CK2α provide some interesting information.

Using tags to detect the expression of the transfected subunits it was possible to determine that the vectors were efficient in directing the expression of these subunits in the COS-7 cells. The low efficiency of transfection, however, of only 10–20% of the cells treated, limits the interpretation of the effect of the exogenous mutants on the endogenous CK2. However, using an antibody against human CK2α (Fig. 1C), it was possible to observe that the expression of the exogenous CK2α or CK2α′ was similar in amount to that of endogenous CK2α present in these cells. To account for this observation, one may surmise that the expression of the exogenous enzyme in the few cells that are actually transfected is 5–10 times higher than the amount of endogenous enzyme in those cells.

As discussed by other groups [14–16], transfection of CK2α or of both CK2α and CK2β cause a large increment in the CK2 activity that can be measured in the cell extracts. The increment in the activity is of the order of 5–7-fold with CK2α or CK2α$E^{75}E^{76}$ and close to 20-fold when both CK2α$E^{75}E^{76}$ and CK2β are cotransfected. This result is somewhat surprising because, as we have already mentioned, the Western blots indicate that the amount of endogenous CK2α is similar to the amount of exogenous CK2α ectopically expressed. The experiments carried out with the heparin sensitivity of cells transfected with CK2α$E^{75}E^{76}$ (Fig. 5) confirm the observation that more than 90% of the CK2 activity being measured in the extract is of exogenous origin. No experiments have been carried out to answer this inconsistency, however, a simplistic explanation might be proposed stating that the endogenous enzyme might be inactive, compartmentalized or inhibited in the intact cell while the exogenous enzyme escapes this destiny and is more readily available. A lack of correlation between the intensity of immunoblots for CK2α and the CK2 activity measured in extracts of these cells has been noted previously by Bosc *et al.* [17].

The COS-7 cells transiently transfected with CK2α subunit are limiting in CK2β and therefore contain a significant amount of free CK2α subunit. The limitation of the regulatory subunit was demonstrated by the experiments in Fig. 4 in which the assay was carried out in the presence of added CK2β, resulting in a very important increase of activity. The fact that even the extracts of cells transfected with the empty vector show an increase of CK2 activity when assayed in the presence of added recombinant CK2β would indicate that some free CK2α subunit is present in these cells. This conclusion fits with the observation that transfection of CK2β also causes an increase in the activity of extracts, an observation that agrees with a previous report [14].

The effect of the cotransfection of CK2α or CK2α$E^{75}E^{76}$ together with the inactive CK2αA^{156} is clear and interesting. The coexpression of CK2αA^{156} greatly reduces the CK2 activity of cell extracts that also received and expressed the genes for the active subunits. This inhibitory effect is probably due to the capacity of CK2αA^{156} to trap CK2β in inactive tetrameric complexes. Some free active wild-type or CK2α$E^{75}E^{76}$ subunits are present in the extracts as evidenced by the increment of activity observed when CK2β is added to the phosphorylation assay (Fig. 5). The presence of free CK2αA^{156} subunits, of course, could not be measured in this assay.

Unfortunately, due to the low efficiency of transfection, the possible effect of CK2αA^{156} on the endogenous CK2 activity is not detectable. A maximum of 20% inhibition would be expected if the mutant were to be totally inhibitory to the CK2 activity of the transfected cells. This variation is too low to be determined reliably.

Different approaches will have to be entertained if we are to determine whether CK2αA^{156} can truly act as a dominant negative for the endogenous CK2 activity. One approach may be to cotransfect specific substrates of CK2 whose phosphorylation is stimulated by the presence of CK2β together with CK2αA^{156}. The phosphorylation of these substrates introduced into COS-7 cells should be decreased when the CK2αA^{156} is cotransfected if the mutant can affect the endogenous activity as it does the exogenous active subunits. A second approach might be to use inducible vectors carrying these mutants to prepare stable cell transfectants. In these cells the induction of CK2αA^{156} should affect the endogenous activity measured in extracts.

Transfection of the mutant CK2α$E^{75}E^{76}$ results in cells that have a CK2 activity that is highly resistant to heparin and polyanionic inhibitors. This observation may be useful in determining whether CK2 is physiologically regulated by these types of inhibitors since cells expressing the mutation should be resistant to that inhibition and may have higher phosphorylating activity *in vivo*.

Acknowledgements

This work was supported by FONDECYT-Chile and by the ICGEB (Trieste, Italy). Iris Korn was a recipient of a doctoral fellowship of DAAD.

References

1. Allende JE, Allende CC: Protein kinase CK2: An enzyme with multiple substrates and a puzzling regulations. FASEB J 9: 313–323, 1995
2. Guerra B, Issinger OG: Protein kinase CK2 and its role in cellular proliferation, development and pathology. Electrophoresis 20: 391–408, 1999

3. Pinna LA, Meggio F: Protein kinase CK2 and its implication in cell division and proliferation. Prog Cell Cycle Res 3: 77–97, 1997
4. Meggio F, Boldyreff B, Issinger OG, Pinna, LA: Casein kinase-2 down regulation and activation by polybasic peptides are mediated by acidic residues in the 55–64 region of the β-subunit. A study with calmodulin as a phosphorylatable substrate. Biochemistry 33: 4336–4342, 1994
5. Kusk M, Ahmed R, Thomsen B, Bendixen C, Issinger OG, Boldyreff B: Interactions of protein kinase CK2β subunit within the holoenzyme and with other proteins. Mol Cell Biochem 191: 51–58, 1999
6. Allende CC, Allende JE: Promiscuous subunit interactions: A possible mechanism for the regulation of protein kinase CK2. J Cell Biochem (suppl 30/31): 129–136, 1998
7. Heriché JK, Lebrin F, Rabilloud T, Leroy D, Chambaz EM, Goldberg Y: Regulation of protein phosphatase 2A by direct interaction with casein kinase 2α. Science 276: 952–955, 1998
8. Cosmelli D, Antonelli M, Allende CC, Allende JE: An inactive mutant of the α subunit of protein kinase CK2 that traps the regulatory CK2β subunit. FEBS Lett 410: 391–396, 1997
9. Gatica M, Jedlicki A, Allende CC, Allende JE: Activity of the $E^{75}E^{76}$ mutant of the α subunit of casein kinase II from *Xenopus laevis*. FEBS Lett 339: 93–96, 1994
10. Antonelli M, Daniotti JL, Rojo D, Allende CC, Allende JE: Cloning, expression and properties of the α′ subunit of casein kinase 2 from zebrafish (*Danio rerio*). Eur J Biochem 214: 272–279, 1996
11. Jedlicki A, Hinrichs MV, Allende CC, Allende JE: The cDNAs coding for the α and β subunits of *Xenopus laevis* casein kinase II. FEBS Lett 297: 280–284, 1992
12. Cullen BR: Methods of eukaryotic expression technology in the functional analysis of cloned genes. Meth Enzymol 152: 684–704, 1987
13. Bradford JJ: A rapid and sensitive method for the quantitation of microgram quantities of protein utilizing the principle of protein-dye binding. Anal Biochem 72: 248–254, 1976
14. Heller-Harrison RA, Czech MP: Enhanced casein kinase II activity in Cos-1 cells upon overexpression of either its catalytic or non-catalytic subunit. J Biol Chem 266: 14435–14439, 1991
15. Penner CG, Wang Z, Litchfield DW: Expression and localization of epitope-tagged protein kinase CK2. J Cell Biochem 64: 525–537, 1997
16. Vilk G, Saulnier RB, Pierre RS, Litchfield DW: Inducible expression of protein kinase CK2 in mammalian cells. Evidence for functional specialization of CK2 isoforms. J Biol Chem 274:14406–14414, 1999
17. Bosc DG, Lüscher B, Litchfield DW: Expression and regulation of protein kinase CK2 during cell cycle. Mol Cell Biochem 191: 213–222, 1999

Molecular and Cellular Biochemistry **227**: 45–57, 2001.

Transcriptional coordination of the genes encoding catalytic (CK2α) and regulatory (CK2β) subunits of human protein kinase CK2

Walter Pyerin and Karin Ackermann
Biochemische Zellphysiologie, Deutsches Krebsforschungszentrum, Heidelberg, Germany

Abstract

Little is known of how protein kinase CK2 genes are regulated, and it is unclear whether there are mechanisms of transcriptional coordination. Response elements present in the promoter sequences of the human catalytic (CK2α) and regulatory (CK2β) subunit genes have been examined for the significance in transcriptional control using reporter gene assays, electrophoretic mobility shift assays, site-directed mutagenesis, ectopic protein expressions, and transcript assessments. Most strikingly, in both promoters the regions of highest transcriptional activity contain two adjoining, completely identical and conserved Ets1 response elements, and both the mutation of motifs and the overexpression of Ets1 affect significantly transcriptional activity. Also in common are Sp1 response elements that cooperate with Ets1, and Sp1 is phosphorylatable by CK2 holoenzyme but not by individual CK2α, the phosphorylation negatively affecting DNA binding. CK2α and CK2β transcript levels and stoichiometries of mRNA species turned out quite constant in cultured cells despite progressing through various stages of proliferation and differentiation. The data seem to indicate transcriptional coordination of the human genes encoding CK2α and CK2β based on an Ets1 double motif common to both genes cooperating with Sp1 motifs and amenable to negative feedback control by the gene products which, following complexation into CK2 holoenzyme, could phosphorylate Sp1 (and Ets1?) and thus downregulate transcription and contribute to the observed constant cellular CK2α and CK2β transcripts situation. (Mol Cell Biochem **227**: 45–57, 2001)

Key words: protein kinase CK2 genes, transcriptional control, Ets1, Sp1, gene transcription coordination

Introduction

Evidence has been provided by many laboratories that protein kinase CK2 represents a tetrameric holoenzyme composed of two catalytically active subunits (CK2α and isoforms), and two regulatory subunits (CK2β and isoforms) occurring ubiquitously among eukaryotes and representing a vital cellular entity. Both the sequences of CK2 subunits and the tetrameric structure of holoenzymes are highly conserved. Deviations from stoichiometry and deviations in cellular levels have been linked to diseased states (reviewed in [1–11]). Despite the resulting importance of CK2 expression control, little is known of how the transcription of the genes encoding CK2 subunits are regulated, and it is unclear whether there are mechanisms of transcriptional coordination of the genes encoding catalytically active and regulatory CK2 subunits.

The human genome contains four CK2 gene loci (reviewed in [9]). Two of these, at chromosomes 11p15 and 20p13, have been characterized as a silent CK2α pseudogene and a transcriptionally active CK2α gene, respectively [12–15]. The CK2α′ gene has been mapped to chromosome 16p13.2–13.3 [16], and the CK2β gene to chromosome 6p21.3 [17, 18]. So far, the structures of both of the CK2α genes and that of the CK2β gene have been unraveled, including adjacently upstream sequences, while respective informations on the CK2α′ gene are lacking. Thus, comparisons of genes coding for catalytically active CK2 subunits and regulatory subunits are currently limited to the active CK2α and the CK2β genes. The genes vary considerably in length, the active CK2α gene and the CK2β gene spanning roughly 70 kb and 4.2 kb, respectively, but share a number of structural features substantial for transcriptional control. Both have more than one

Address for offprints: W. Pyerin, Biochemische Zellphysiologie (B0200), Deutsches Krebsforschungszentrum, Im Neuenheimer Feld 280, 69120 Heidelberg, Germany (E-mail: w.pyerin@dkfz-heidelberg.de)

transcription start site (two and three start sites in the active CK2α gene and the CK2β gene, respectively), both genes lack a TATA box, possess several GC boxes including non-standard positioned CAAT-related motifs, and have a CpG island around Exon 1 [14, 15]. These features identify the genes as so-called housekeeping genes implying permanent expression with little regulation. However, highly tissue-specific regulation has been shown to occur with this type of promoter (e.g. [19, 20]). In addition to the shared *cis*-acting elements, each of the two CK2 gene promoters contains a gene-specific set of transcription factor binding motifs and characteristics. For instance, a NF-κB motif (two overlapping sites) with significant effects on transcriptional activity is present in the CK2α gene promoter but not in the CK2β gene promoter, and, on the other hand, the CK2β gene but not the CK2α gene has CAAT-related motifs with significant effects on transcriptional activity within an upstream enhancer [15, 21–23]. Such motifs might, therefore, have a meaning in tissue-specific tuning of expression levels and stoichiometries of CK2 subunits.

The present report examines comparatively the role played by *cis*-acting response elements in the promoters of the human CK2α and CK2β genes focusing on elements that are common to both promoters. A 12 base-pair homology region turned out to be present in areas of maximum transcriptional activity of both gene promoters, representing two adjoining consensus motifs of transcription factor Ets1. In addition, Sp1 motif clusters occur in both of the promoters. Mutational analysis seems to indicate that these motifs are cooperating with each other and are significant for promoter activity regulation and thus for transcriptional coordination of the CK2α and CK2β genes. This regulation is hypothesized to possibly include a negative feedback control by the gene products which could, following their complexation into CK2 holoenzyme, phosphorylate Sp1 (and possibly Ets1) and negatively affect DNA binding and transcription factor cooperation. Such a control mechanism would help explain the occurrence of constant levels of CK2α and CK2β gene transcripts in human cultured cells despite their transition through different stages of proliferation and differentiation.

Materials and methods

Materials

pGL3-vector system, luciferase- and β-galactosidase assay kits, and recombinant transcription factors were purchased from Promega. Stratagene was provider of the *pfu*-turbo polymerase and Quick Change™ site directed mutagenesis kit; pSVK3-vector system and enzymes were from Amersham Pharmacia Biotech, Stratagene and MBI Fermentas. Chromatography material, Hybond-C membrane and radioactively labeled nucleotides were from Amersham Pharmacia Biotech, Microcon filtration units from Amicon, polyvinylidene difluoride (PVDF) membrane was from Millipore, Superfect from Qiagen, antibodies from Santa Cruz Biotechnology. Plasmid kits, fetal calf serum, lipofectamine and media were from Gibco. All other chemicals originated from Roth, Sigma and Merck.

Methods

Construction of luciferase reporter gene vectors
Oligonucleotides derived from sense and antisense strand of CK2α or CK2β promoter regions were synthesized (oligonucleotide synthesis group, DKFZ, Heidelberg, Germany) with additional terminal adapter sequences of XmaI, BglII or KpnI/HindIII restriction sites and cloned into the multiple cloning sites of pGL3 basic (pGL3BV), enhancer (pGL3EV), or promoter (pGL3PM) vectors (Promega). Correct insertion and orientation was confirmed by sequencing. Promoter regions exceeding 80 basepair in length were PCR amplified from the respective CK2α or CK2β genomic fragment with primers containing terminal KpnI and Hind III restriction sites.

Cloning of transcription factor c-Ets1
HeLa cDNA was prepared by reverse transcription. Using adequate PCR-primer with a terminal NdeI site (5′ primer; GTC ATA TGA AGG CGG CCG TCG ATC TCA A) or ApaI site (3′ primer; GTG GAT CCT CAC TCG TCG GCA TCT G) cDNA was PCR amplified with *pfu* polymerase, cloned into a Flag-tag containing pSVK3-vector and sequenced.

Mutational modification of transcription factor consensus motifs
Consensus motifs for transcription factor binding in CK2α and CK2β promoters were identified by computer search with HUSAR program (DKFZ Heidelberg). The 'QuickChange™ site directed mutagenesis' method (Stratagene) was used for modifying defined nucleotides of promoter sequences and for deletion mutagenesis. Mutational exchange was confirmed by sequencing. Risk of random mutations in the vector sequence was eliminated by recloning of modified promoter inserts into pGL3 vectors. For mutated sequences see Results.

Cell culture, transfection and reporter gene assays
HeLa cells grown in modified Eagle's minimal essential medium (MEM) supplemented with 10% fetal calf serum (FCS) at 37°C, 95% humidity, and 5% CO_2 in 6 well plates to reach 60–80% confluency at the time of transfection. Prior to transfection, Lipofectamine (4 μg/well) or Superfect (10 μl/well) was incubated with 4 μg of the respective luciferase vector

and 1 μg of a β-gal/Renilla reporter gene vector for 35/10 min in a vol. of 50 μl Opti-MEM and added to each well. Medium was adjusted to 10% FCS after 5 h followed by incubation for another 17 h. Cells were harvested in 120 μl of cell lysis buffer (Promega) and an ensuing 1-min centrifugation step (20000 × g) yielded a luciferase-containing supernatant. In both cases aliquots of 20 μl supernatant were tested for luciferase activity and for β-gal activity to adjust for transfection efficiency. For co-transfection experiments 2 μg of expression vector and the respective reporter gene vector were applied. Medium was changed after 3 h incubation (10% FCS in MEM) and cells were harvested in 120 μl of lysis buffer. A 30-sec centrifugation step (20000 × g) yielded a luciferase-containing supernatant. In both cases aliquots of 20 μl supernatant were tested for luciferase activity of the reporter gene vector and the Renilla vector as internal control to adjust for transfection efficiency (Dual Luciferase Reporter Assay System, Promega). JEG-3 choriocarcinoma cells, obtained from American Type Culture Collection (HTB36), were grown in Eagle's MEM supplemented with 10% FCS (Eurobio) at 37°C in a 95% air and 5% CO_2 atmosphere. In cases of serum-free propagations, cells were cultured in the presence of serum until reaching subconfluency. Then the medium was changed to serum-free Eagle's MEM and incubation continued. Cells were harvested by scraping off in phosphate-buffered saline (PBS) and sedimented at 800 × g. If not analyzed immediately, cell pellets were stored at –80°C until use.

Gels and Western blots

Samples were applied on 12% SDS-PAGE (sodium dodecyl-polyacrylamide gel electrophoresis) and either run at 200 V (const.) for 47 min or at 11 mA (const.) for 16–18 h. Gels were either dried and autoradiographed directly or used for semi-dry blotting on PVDF-membrane for immunological detection of separated proteins [24]. In the latter case the membranes were either blocked with a solution containing 1 μg polyvinylalcohol/ml in PBS or overnight with Tris-buffered saline containing 5% milk powder. Membranes were incubated for 2 h at room temperature with specific antibodies (diluted 1:4000 for anti-α329–343; 1:2000 for anti α′ and 1:10,000 for anti-β171–186; anti-Ets1 and anti-Sp1 antibodies were diluted 1:1500) treated with Protein A-biotin (1:10,000) for 45 min at room temperature and followed by an 1 h incubation with peroxidase-coupled streptavidin. All steps were followed by 4 washings for 5 min [22, 23]. Detection was achieved using 4 chloronaphthol as substrate [25].

Northern blot

Cytoplasmic RNA was purified according to the method of Gough with some modifications as described in detail previ-

Table 1. Mutation of transcription factor Ets1 and Sp1 consensus motifs in human CK2α and CK2β gene promoters and effect on transcriptional activity

Gene promoter	Transcription factor	Mutation: Naming	Mutation: Sequence	Mutation: Position	Transcription activity (%)
CK2α	Ets1	wt.	–^{24}TTTCCGCTTCCGG–		100
(–256/144)		mutEts1	–^{24}TGTACGCGTACGG–	24, 26, 29, 31	46
	Sp1	wt.	–$^{-25}$GGGGCG–		100
		mutSp1-1	–$^{-25}$GTGACA–	–24, –22, –20	25
		wt.	–$^{-5}$CTCGCCCC–		100
		mutSp1-2	–$^{-5}$CGCACACC–	–4, –2, +1	110
	Sp1 + Sp1	mutSp1-1 + mutSp1-2			54
	Etsl + Sp1	mutEtsl + mut Sp1-1			n.d.
		mutEts1 + mut Sp1-2			19
		mutEts1 + mutSp1-1 + Sp1-2			3
CK2β	Ets1	wt.	–$^{-14}$TTCCGCTTCCGG–		100
(–167/122)		mutEts1	–$^{-14}$GTACGCGTACGG–	–14, –12, –8, –6	2
	Sp1	wt.	–$^{-96}$CCCC//CCTC^{-77}–		100
		mutSp1-1	– deleted –	–96/–77	92
		wt.	–^{14}CCCC//CTCC^{37}		100
		mutSp1-2	– deleted –	14/37	68
	Sp1 + Sp1	mutSp1 + mutSp1-2			78
	Ets1 + Sp1	mutEts1 + mutSp1-1			58
		mutEts1 + mutSp1-2			64
		mutEts1 + mutSp1–1 + Sp1-2			n.d.

Ets1 and Sp1 consensus motifs in CK2α and CK2β gene promoters (genomic fragments –256/144 and –167/122, respectively), were point mutated at underlined positions or deleted as indicated. Mutations were confirmed by sequencing. Risk of random mutations in the vector sequence was eliminated by recloning. Given are luciferase activities of indicator gene constructs carrying the given mutations (mutEts1; mutSp1) relative to the respective wild type constructs (wt; set 100%). Given are mean values of three separate assays each conducted in triplicate. Standard deviation of triplicates and of separate assays was below 15%. wt = wild type sequence; mut = mutation in consensus motif named adjacently; n.d. = not determined.

ously [26]. RNA (10 µg) was denatured and electrophoretically separated followed by capillary transfer and fixation to nylon membranes, hybridization, and autoradiography. For Aromatase mRNA detection, Arom II/V was used as a hybridization probe, a 495 bp Aromatase cDNA *BamHI/KpnI* fragment (position 125–619) comprising part of exon II plus exons III and IV and part of exon V. CK2 hybridization probes were cDNAs comprising the complete coding sequences of CK2α or CK2β, respectively [27, 28]. For control of applied RNA amounts, the filters were re-probed with a 1.7 kb *EcoRI* cDNA fragment of GAPDH (glyceraldehydephosphate dehydrogenase).

Electrophoretic mobility shift assay (EMSA)

Oligonucleotides representing segments of CK2β or CK2α promoters (as given in Table 1) were synthesized in sense and antisense orientation, annealed and end-labeled with T4-polynucleotide kinase (MBI Fermentas). Surplus radioactivity was removed by Amersham Pharmacia Biotech G25 Microspin columns. Gelshift assays were performed using 40,000 cpm of ^{32}P end-labeled oligonucleotides and 5 µg of HeLa S3 nuclear extracts per assay in a final volume of 20 µl. Presence of poly(dIdC) prevented non-specific protein-DNA binding. All other components were taken from Amersham Pharmacia Biotech Gel-shift kit. An incubation for 20 min preceded electrophoretic separation on a native 6% polyacrylamide gel (100 V const.) at 4°C. Gels were dried and autoradiographed. For competition reactions, unlabeled oligonucleotides were added in 25-fold surplus (comp. DNA), for Ets 1, GATCTCGAGCAGGAAGTTCGA, and for Sp1, AATCGATCGGGGCGGGGCGAGC.

Phosphorylation assay

2 fpu of recombinant Sp1 (Promega) was phosphorylated in the presence of 100 mM MOPS pH 6.9, 80 µM γ^{32}P-ATP (0.4 µCi) and 10 mM $MgCl_2$ for 10 min at 30°C using 0.02 pmol of either CK2α or CK2 holoenzyme in a final volume of 10 µl. The reaction was stopped by adding an equal volume of 2 × SDS-PAGE sample buffer and heating at 95°C for 5 min. Samples were separated on a 12% SDS-PAGE. After Coomassie staining the gel was dried and autoradiographed.

Results

Defining transcriptionally active regions of the human CK2α and CKβ gene promoters

Stepwise deletions from the 3′ and the 5′ ends of genomic fragments containing either the promoter of the CK2α gene [14, 15] or that of the CKβ gene [21] were carried out. The resulting fragments were cloned into pGL3 luciferase reporter gene vectors and tested for transcriptional activity in human cultured cells (HeLa). Deletions at the 3′ end of the CK2α gene fragment comprising positions –256 to 144 and including the two transcription start sites (TS1 and TS2 at positions +1 and 50, respectively) had little effect on indicator gene transcription up to position 65 (fragment –256/65), but significantly affected transcriptional activity when further deleted, resulting in a complete loss of activity upon TS1 and TS2 elimination (fragment –256/–40) (Fig. 1A). Deleting at the 5′ end at position –69 had moderately negative effect on transcriptional activity that was not changed upon further deletion down to position –26 (fragments –69/144, –39/144, –26/144), whereas deletions near to or beyond TS1 had sig-

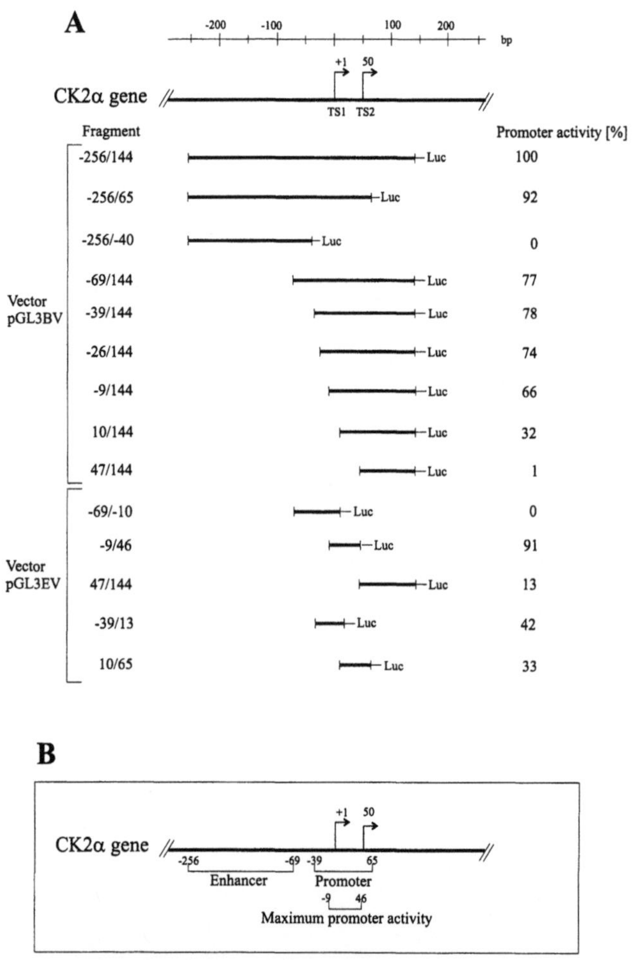

Fig. 1. Determination of transcriptionally active regions at the human CK2α gene promoter. (A) Locating transcriptional activities. The given 3′ and 5′ deletions and segments of the genomic CK2α fragment –256/144 were cloned into luciferase reporter vectors (pGL3BV; pGL3EV) and transfected into cultured human cells (HeLa). Cell homogenates were assayed for luciferase activity, and activity was calculated relative to that of fragment –256/144 (set 100). Given are mean values of three separate assays each conducted in triplicates. Standard deviation of triplicates, and of separate assays was below 15%. TS = transcription start site; Luc = luciferase indicator gene. (B) Summarizing scheme of human CK2α gene promoter situation. Numbers provide nucleotide positions.

nificantly negative effects (one-third and two-third activity losses with fragments –9/144 and 10/144, respectively), and deletion near TS2 resulted in complete inactivation (fragments 45/144 and 47/144). Sub-dividing the CK2α gene region –69/144 into short segments and cloning into reporter gene vector endowed with an enhancer element (pGL3 EV) in order to detect even minor transcriptional activities, the by far highest activity was obtained with segment –9/46, while the adjacent upstream segment (–69/–10) and the adjacent downstream segment (47/144) were inactive and only weakly active, respectively. Activity was also found with segments containing either TS1 (–39/13) or TS2 (10/65), but activities were significantly below that of segment –9/46. Taken together, the data indicate that the promoter of the CK2α gene spans roughly positions –39 to 65, has an enhancer element(s) upstream position –69, possesses maximum promoter activity at positions –9 to 46, and has functional links between the regions surrounding TS1 and TS2 (Fig. 1B).

Deletions at the 3′ end of the CK2β gene promoter fragment comprising positions –682 to 122 and including all three

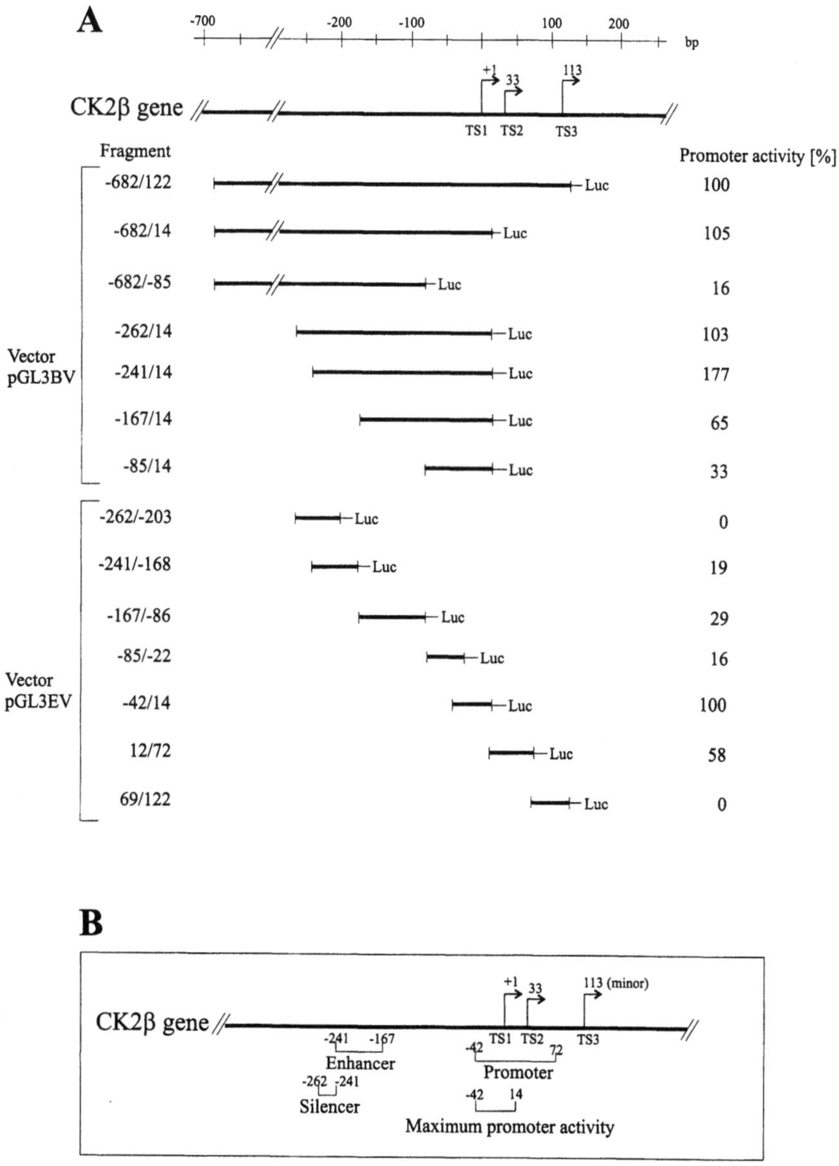

Fig. 2. Determination of transcriptionally active regions at the human CK2β gene promoter. (A) Locating transcriptional activities. The given 3′ and 5′ deletions and segments of the genomic CK2β fragment –682/122 were cloned into luciferase reporter vectors (pGL3BV; pGL3EV) and transfected into cultured human cells (HeLa). Cell homogenates were assayed for luciferase activity, and activity was calculated relative to that of fragment –682/122 (set 100; vector pGL3BV) or fragment –42/14 (set 100; vector pGL3EV). Given are mean values of three separate assays each conducted in triplicates. Standard deviation of triplicates, and of separate assays was below 15%. TS = transcription start site; Luc = luciferase indicator gene. (B) Summarizing scheme of human CK2β gene promoter situation. Numbers provide nucleotide positions.

transcription start sites (TS1, TS2, and TS3 at positions 1, 33, and 113, respectively) had little effect on indicator gene transcription activity up to position 14 (fragment –682/14) although eliminating TS2 and TS3, whereas elimination of all three sites caused a dramatic decrease (fragment –682/–85) (Fig. 2A). Stepwise deletions at the 5′ end down to position –262 had little effect (fragment –262/14), but provoked a significant increase (up to 177%) when set only 21bp further downstream (fragment –241/14), obviously unmasking a silencer. Further deletions resulted in further activity decreases (by one-third and two-third with fragments –167/14 and –85/14, respectively). Subdividing CK2β gene region –262/122 and cloning into pGL3 EV vector (see above), maximum transcriptional activity was found with segment –42/14 and significant activity with segment 12/72, the segments containing TS1 and TS2, respectively, whereas segment 69/122 containing TS3 was inactive. No activity was seen with segment –262/–203, and some activity with segments –241/–168, –167/–86, and –85/–22. Taken together, the data indicate that the promoter of the CK2β gene spans roughly positions –42 to 72, has enhancer and silencer elements upstream at positions –241 to –167 and –262 to –241, respectively, has maximum promoter activity at positions –42 to 14, and, in good agreement with our previous primer extension analysis [21], starts transcription predominately at TS1 and TS2 (Fig. 2B).

Transcription-affecting features common to CK2α and CKβ gene promoters

The promoter regions of the human CK2α gene and the CKβ gene share not only structural features such as mentioned above (see Introduction), but also a number of consensus motifs for transcription factor binding. Within the transcriptionally active regions, these include several clusters of Sp1 motifs and, of particular concern, a double motif for Ets1. The CK2α gene promoter contains two clusters of overlapping Sp1 motifs somewhat upstream of TS1 (Sp1-1, position –25) and right at TS1 (Sp1-2, position –5), respectively, and the CKβ gene promoter has two such clusters upstream of TS1 (Sp1-1, position –96) and between TS1 and TS2 (Sp1-2, position 14). The promoters of both the CK2α gene and the CKβ gene possess an identical Ets1 double motif comprising 12 bp at positions 24 and –14, respectively. Because binding of individual transcription factors to consensus motifs is sufficiently specific to be affected by mutations, mutational analysis was applied to examine the significance of these motifs for transcriptional activity.

As shown in Table 1, when the Ets1 double motif of the CK2α gene promoter was mutated by exchanging two bases per motif, more than half of indicator gene transcriptional activity was lost. The same mutation set in the Ets1 double motif of the CK2β gene promoter decreased activity to background level (the same mutation set in larger CK2β gene fragments decreased activity less strongly; for instance, when set in a promoter-active 2.8 kb genomic fragment, activity decreased to 51%; [23]). Identical mutations of the Ets1 double motif in smaller fragments of both gene promoters led in all cases to significant decreases or even complete loss of activity [22, 23]. Mutations of Sp1 motifs had varying effects. In the CK2α gene promoter, Sp1 motif mutations had practically no effect (mutSp1-1) or led to significant decreases (mutSp1-2; mutSp1-1 and mutSp1-2 simultaneously), and in the CK2β gene promoter had either a small (mutSp1-1) or a moderately decreasing effect (mutSp1-2; mutSp1-1 and mutSp1-2 simultaneously). Particularly interesting were the effects observed with simultaneous mutations of Ets1 and Sp1 motifs. In the CK2α gene promoter, these strongly decreased (simultaneous mutations of Ets1 and Sp1-2) or nearly silenced transcriptional activity (simultaneous mutations of Ets1, Sp1-1 and Sp1-2). Also the CK2β gene promoter was negatively affected by simultaneous Ets1 and Sp1 mutations (simultaneous mutations of Ets1 and Sp1-1, and of Ets1 and Sp1-2), but did rather not exceed the silencing effect seen with Ets1 motif mutations alone.

Together, the data support the implications that (i) Ets1 and Sp1 are significant transcription-affecting motifs common to both the CK2α gene promoter and the CK2β gene promoter, and (ii) that Ets1 and Sp1 motifs are contributing not only individually to transcriptional control but via an intimate cooperation with each other.

Table 2. Occurrence of Ets1 homology regions in promoters of CK2 genes of various organisms

Species	Ets1 homology region(s)		References
	CK2α gene	CK2β gene	
Human	^{24}TTCCGCTTCCGG	$^{-14}$TTCCGCTTCCGG	14, 21
Mouse	n.a.	$^{-33}$TTCCGCTTCCGG	73
Worm (*C. elegans*)	$^{-128}$TTCATGTTCTGG	^{78}TTCCACTTCGGT $^{-129}$TTCTGTTTCGAC	74, 75
Frog (*X. leavis*)	TTCCGCTTCCGG	n.a.	76

Compared are sequences given by the cited references. Numbers indicate positions. n.a. = sequence region not available.

Availability of Ets1 and Sp1 proteins, overexpression and modification by phosphorylation

Mediators of the roles played by Ets1 and Sp1 motifs are the respective proteins, Ets1 and Sp1. Evidence for their availability in the nucleus of cells used in the present study was obtained previously using various methods, including Western blotting of nuclear extracts (NEs) and immunodetection by antibodies specific to Ets1 and Sp1; electrophoretic mobility shift assays (EMSAs) of NEs using oligonucleotides synthesized according to sequences in question (binding motifs plus some flanking regions) of the CK2α and CK2β gene promoters, including oligonucleotide competition (oligonucleotides with and without consensus motif mutations) as well as up-shifts with specific antibodies; UV-cross-linking studies of DNA fragments incubated with NEs; affinity chromatography using columns prepared by immobilizing respective DNA fragments to solid supports, applying various measures against non-specific binding, and Western blotting of eluate fractions [22, 23].

The Ets1 double motif's striking conservation (see Table 2) and strong negative effects on transcriptional activity when mutated, seems to indicate a role in transcriptional control of both the CK2α and the CK2β gene. If so, a simple increase in cellular Ets1 level was thought to be possibly already enough to affect transcriptional control to an extent allowing detection by indicator gene assays. To proof this idea, Ets1 was ectopically expressed in human cultured cells and these cells were used to carry out indicator gene assays with the promoter-containing segments –256/144 and –682/122 of the CK2α and the CK2β gene, respectively. As a result, in cells overexpressing Ets1 a moderately, but significantly, enhanced transcriptional activity with both of the promoters was detected (Fig. 3A). The data support the notion that Ets1 may be involved in the control of the promoters of both CK2 genes, thus identifying Ets1 as a factor involved in their transcriptional coordination.

CK2 holoenzyme but not individual CK2α has been shown to phosphorylate transcription factors such as UBF, CREB and c-Jun [29]. Under similar conditions and using identical molar amounts of kinase, phosphorylation of Sp1 was examined. As a result, phosphorylation of Sp1 was clearly seen with CK2 holoenzyme but not with individual CK2α (Fig. 3B). The phosphorylation affected DNA binding; EMSAs carried out with Sp1 motif-containing segments such as –7/11 of the CK2α gene promoter revealed that the interaction of DNA with Sp1 was significantly decreased due to the phosphorylation (Fig. 3B). This result opened the possibility that the gene products, CK2α and CK2β, might, following their complexation into CK2 holoenzyme, negatively regulate back onto the promoters of their genes via phosphorylation of factors such as Sp1. Because containing CK2 consensus motifs (see below), phosphorylation might, although not proved so far, occur also with Ets1.

Transcript levels of the CK2α and CK2β genes are kept constant

While the cellular stoichiometry of catalytic and regulatory subunits are known to be kept quite constant at protein level [9, 30], the situation at transcript level is less clear. Therefore, presence of the transcripts of the CK2α and CK2β genes were determined in cultured human cells at various stages of proliferation and differentiation. For this purpose, a cell line (JEG-3) with a known differentiation marker (aromatase gene expression) was analyzed over time for transcripts of CK2 genes, differentiation marker gene (aromatase) and standard control genes (GAPDH). Probes were taken at seeding of very dilute cell suspensions and at various times post-seeding until dense confluent cell layers were formed, a process running from a predominantly proliferating cell population over various stages of differentiation to a contact-inhibited differentiated cell population. As a result, the transcripts of the CK2 genes (three CK2α mRNA and one CK2β mRNA bands, respectively; see also [27, 28]) were found at same level throughout this process as compared to the GAPDH controls and, moreover, without any changes in the stoichiometry of mRNA species (Fig. 4). By contrast, the aromatase gene transcripts (three bands; see also [26]) were low at beginning, i.e. in predominantly proliferating cell populations, increasing dramatically at later stages, i.e. in predominantly differentiated cell populations. The data seem to indicate that cells tend to adjust a relatively constant transcript situation of CK2 genes independent of the respective cellular status. The common features seen in the transcription machinery of the CK2α and CK2β genes and the possibility of a negative feedback on transcription by the gene products are perfectly compatible with this result.

Discussion

According to the shown transcript analysis, cells obviously tend under normal conditions to adjust a relatively constant transcript situation for CK2α and CK2β independent of their respective proliferation or differentiation status. Mechanistically, this may, among others, find explanation by a coordinate control of gene transcription. If so, common features are expected to exist in the transcription machinery of the CK2α and CK2β genes. This seems indeed to be the case. The genes have transcription activity-determining response elements in common that are cooperating with each other and, moreover, are recognized by transcription factors which are demonstrated (or potential) substrates of CK2 holoenzyme and thus possibly are amenable to a feedback control via phosphorylation by the gene products following their complexation into holoenzyme.

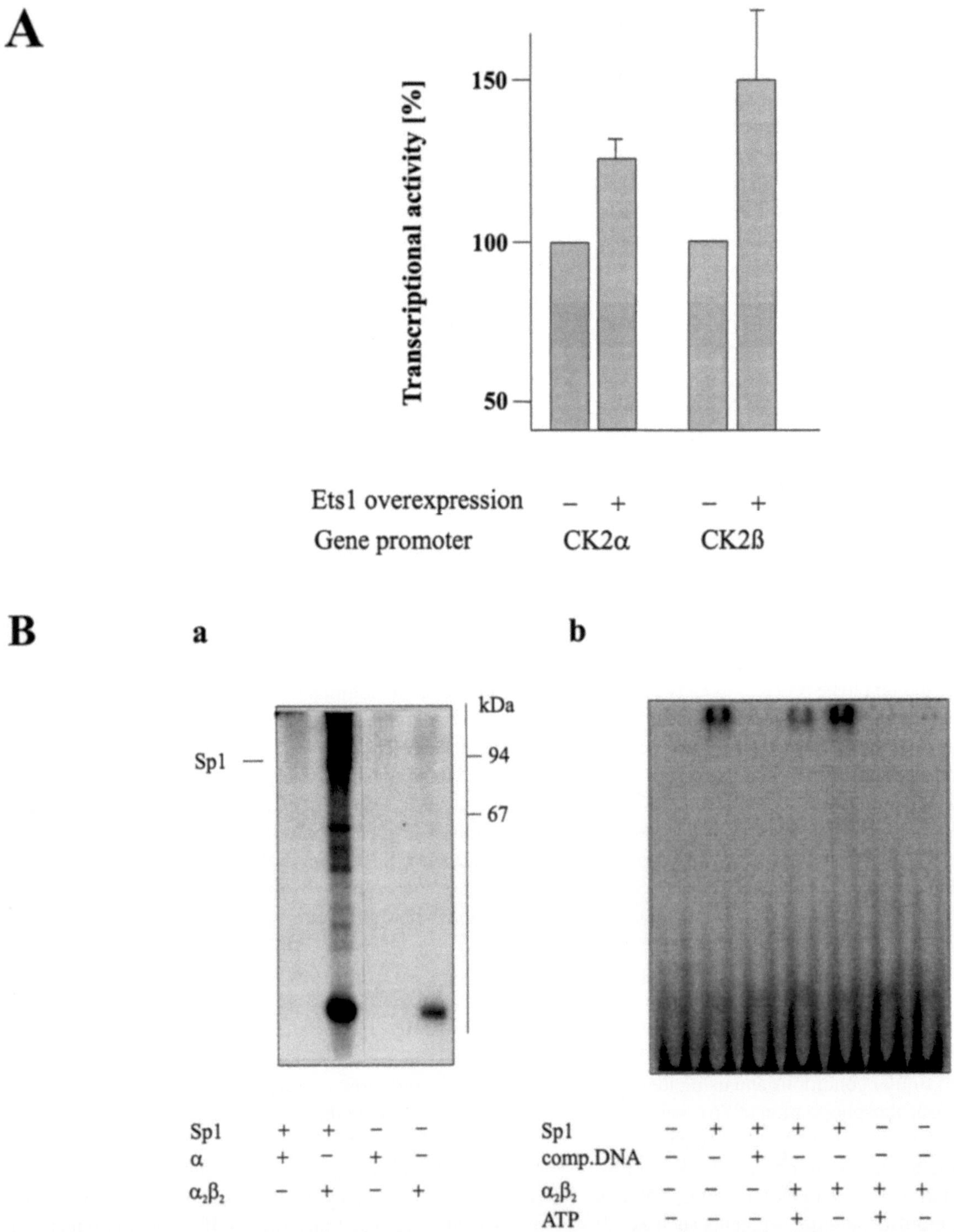

Fig. 3. Effect of Ets1 overexpression on transcriptional activity and of Sp1 phosphorylation on DNA binding. (A) Effect of ectopic expression of Ets1 on transcriptional activity of CK2α and CK2β gene promoters. Ets1 was overexpressed in HeLa cells (Ets1+) and the effect tested on activity of indicator gene (luciferase) constructs of CK2α gene promoter (fragment –256/144) and CK2β gene promoter (fragment –682/122). Controls received indicator gene constructs and expression vector devoid of the Ets1 sequence. Given are means of three independent assays with standard deviations as error-bars. (B) Phosphorylation of Sp1 and effect on DNA binding. (a) phosphorylation of Sp1. Sp1 was assayed for phosphorylation by CK2α (α) and CK2 holoenzyme ($\alpha_2\beta_2$), phosphorylated samples were separated on 12% SDS-PAGE and autoradiographed. Assay compositions as indicated at bottom (+ = presence of indicated component; – = absence of indicated component), molecular weight markers at right. (b) effect of phosphorylation on DNA binding by EMSA. Sp1 was tested for binding to a labeled DNA fragment containing Sp1 response elements (CK2α promoter fragment –7/11) without and with phosphorylation by CK2 holoenzyme ($\alpha_2\beta_2$). Component compositions in assays were as indicated at bottom of picture (+ = presence of indicated component; – = absence of indicated component), including competitor DNA (comp.DNA) to test for specificity of Sp1-DNA interaction.

Of particular importance seems to be an Ets1 response element. The transcriptionally most active regions of the promoters of both the CK2α gene and the CK2β gene contain an identical 12βp region representing an Ets1 double motif. This motif is highly conserved and found also in all the promoters of CK2 genes available so far, including human, mouse, nematode worm, and frog (Table 2). Alignment of the human and mouse gene upstream sequences indicates 100% identity of the Ets1 motifs at only a 55% identity of the surrounding sequences. Further, the Ets1 motifs have similar relative distances to respective transcription start sites. The pairwise occurrence of Ets1 motifs has frequently been observed and seems to relate to a relatively low Ets1 binding affinity of individual sites depicted, aside others, by DNaseI footprints

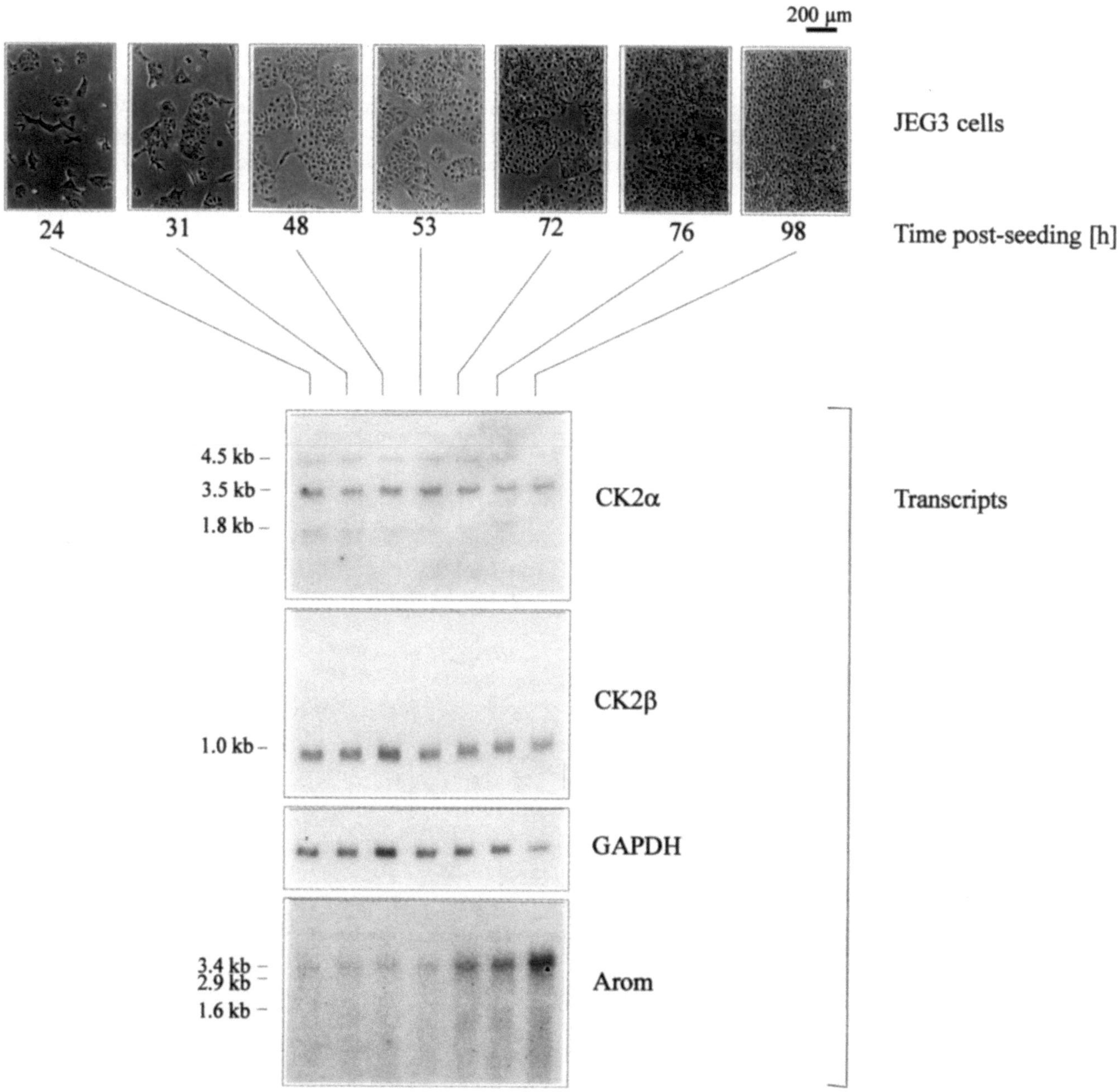

Fig. 4. Transcript situation of protein kinase CK2 genes in human cultured cells varying in differentiation/proliferation status. JEG-3 cells were seeded on petri dishes (scale given by bar), harvested 24, 31, 48, 53, 72, 76 and 98 h post-seeding, and analyzed by Northern blotting (10 µg of total RNA; specific probes for respective gene transcripts) and densitometric scanning of autoradiograms for the presence of CK2α and CK2β gene transcripts as well as of aromatase (Arom) and GAPDH (GAPDH) gene transcripts.

exceeding significantly the Ets-1 motif sequence [31, 32]. In TATA-less promoters, Ets motifs are usually located closely to transcription initiation sites, and may participate in the formation of transcription initiation complexes (e.g. [33–41]). Further, some of the Ets-family binding sites have been demonstrated to function as initiator elements for transcription in TATA-less genes [34, 35, 42]. Mutating the Ets1 double motif results in significant decreases of transcriptional activity of CK2α and CK2β genes independent of the lengths of the DNA fragments tested in indicator gene assays (lengths ranging from several kb to some twenty bp). Appropriately, overexpression of Ets1, whose presence in the employed cells and motif-dependent binding to DNA has been documented (see also [22, 23]), causes moderate but significant increases of transcriptional activity. The impact of Ets1 on CK2 gene expression has also been supported by the antisense approach (CK2α gene expression; HeLa cells; ets1-antisense oligonucleotides; see [22]). Therefore, it is tempting to assume that Ets1 may represent a feature common to and important for the transcription machineries of both the CK2α gene and the CK2β gene, and thus playing a role in coordinating their expression.

Ets1 belongs to a family of transcriptionally active proteins functioning frequently in combination with other factors, including Sp1 [43–45]. Our analysis seems to show that Sp1 response elements contribute also significantly to the transcriptional control of the CK2α and CK2β genes. Mutation of individual Sp1 motifs, which are present in the promoters in form of clusters (some three to eight overlapping motifs), reduces basal transcriptional activity in both CK2 gene promoters, and, particularly marked with the CK2α gene promoter, simultaneous mutation of Sp1 motifs (clusters) may enhance effect indicating cooperation of motifs. In addition, Sp1 elements have been shown to also compensate for each other, so that complete deletions are usually required in order to abolish effects provoked by ectopically expressed Sp1 [43]. Sp1 is a reported GC-box binding activator of transcription found in many promoters and enhancers [44–49] that is

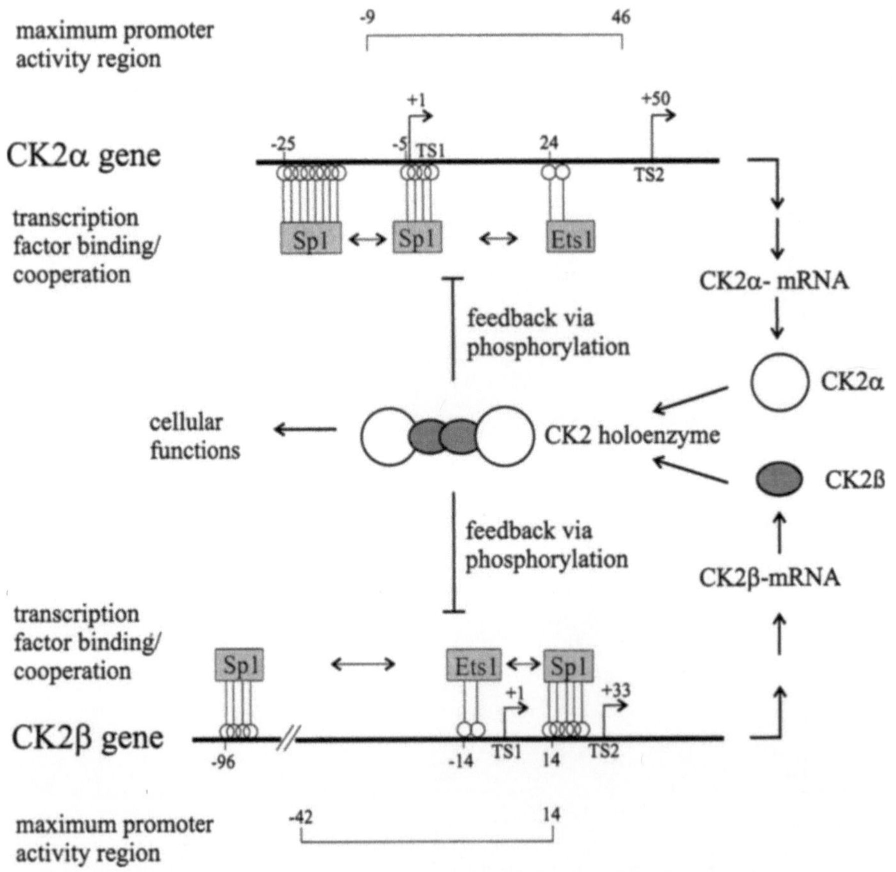

Fig. 5. Schematic representation of transcriptional coordination hypothesis of the human CK2α and CK2β genes. Ets1 response element (double motif; circles symbolizing individual Ets1 elements; not to scale) common to transcriptionally active regions of both the human CK2α (region –9 to 46) and CK2β (region –42 to 14) gene promoters might activate in cooperation with Sp1 response elements (overlapping motif clusters, circles symbolizing individual Sp1 elements; not to scale) the transcription of genes. The generated mRNAs (CK2α-mRNA; CK2β-mRNA) could be translated into CK2 subunit proteins (CK2α = open circle; CK2β = shaded oval) which may readily complex into tetrameric CK2 holoenzyme. The holoenzyme might, aside catalyzing various cellular functions, feedback to gene transcription via phosphorylation of Sp1 (and Ets1?) resulting in a down-regulation of transcription.

capable of recruiting factors such as TFIID required also by TATA-less promoters in order to activate RNA polymerase II [50], and that has a known ability to link distant transcription control elements, i.e. to mediate cooperation between transcriptionally active proteins [33, 46]. Although Ets1 and Sp1 may function separately to initiate transcription, more likely seem to be concerted actions as demonstrated by a practical inactivation of the CK2α gene promoter by simultaneous mutations of Ets1 and Sp1 motifs, and as demonstrated for various genes, including virus-mediated Ets1-Sp1 complex formation and transactivation [43, 44, 51, 52].

The Ets1-Sp1 cooperation seems to provide (part of) a mechanistic feature helpful in explaining how the adjustment of a relatively constant transcript situation for CK2α and CK2β might be achieved. Like a number of other transcription factors [29], Sp1 turned out phosphorylatable by CK2, and Ets1 to possess several minimal CK2 phosphorylation consensus sites (no direct evidence yet for a phosphorylation by CK2). Two features of the phosphorylation are of importance. It occurs with the CK2 holoenzyme but not with individual CK2α, and the phosphorylation decreases the binding capacity to DNA, a result supported by earlier data that identified phosphorylation at positions such as T579 [53]. The phosphorylation of transcription factors by CK2 holoenzyme but not individual CK2α provides a basis for a working hypothesis of how the expression control might occur in order to achieve quasi constant cellular CK2 transcript levels. As schematically outlined in Fig. 5, it is tempting to hypothesize that activation of the genes would occur under participation of Ets1 and Sp1 by their promoter binding and cross-talk. Activation would continue until a certain cellular level of newly generated CK2α and CK2β would be reached which, due to their high mutual affinity [54], would readily complex to CK2 holoenzyme tetramers. The holoenzyme, but not the individual CK2α, would then phosphorylate Sp1 (and Ets1?) loosening promoter binding and/or disturbing cross-talks between transcription factors. As a consequence, transcriptional activation would decrease and expression of CK2α and CK2β genes decline.

Such a negative feedback control of CK2α and CK2β onto their own genes appears to provide an appropriate mechanistic device for cells in helping to keep a certain transcript level and stoichiometry constant (or at least to contribute to this process), and thus to avoid harmful deviations such as described for various diseased states, including *Theileriosis* and tumorigenesis [55–58]. In fact, various diseased states with deviating CK2 situations seem to link to Ets1. For instance, the Ets1 function is modulated by mitogenic signals transmitted via the Ras/Raf-kinase pathway [59], and activated Raf kinase has been described to phosphorylate a conserved threonine (T38) causing activation of Ets1 and enhanced transcription of genes related to early mitogenic events [60], including CK2 genes [59]. This may possibly help explain earlier reports on mitogen-induced increases in CK2 expression levels in cell cultures [61, 62], or on high expression levels of CK2 in proliferative tissues and in tumors [63–65], matching the fact that Ras and Ets1 are both proto-oncogene products, and that Ets1 represents in addition a well-documented regulator of other proto-oncogenes and genes correlated with metastasis [66–72].

Acknowledgements

We would like to acknowledge the engaged work of all those former group members who participated in the gene expression program, in particular doctors Helenia Ansuini, Ludger Bodenbach, Oliver Böcher, Swen Grein, Andreas Krehan, Ute Wirkner, and Hartmut Voss. Further, we acknowledge the technical assistance of Andrea Waxmann and Michael Emmenlauer. This work was supported by the Commission of the European Communities (Biomed2; BMH4-CT96-0047) and the Deutsche Forschungsgemeinschaft (Py2/2-2).

References

1. Hathaway GM, Traugh JA: Casein kinases – multipotential protein kinases. Curr Topics Cell Regulation 21: 101–127, 1982
2. Pinna LA: Casein kinase 2: An 'eminence grise' in cellular regulation? Biochem Biophys Acta 1054: 267–284, 1990
3. Tuazon PT, Traugh JA: Casein kinase I and II – multipotential serine protein kinases: Structure, function, and regulation. Adv Sec Mess Phosphoprot Res 23: 123–164, 1991
4. Litchfield DW, Lüscher B: Casein kinase II in signal transduction and cell cycle regulation. Mol Cell Biochem 127/128: 187–199, 1993
5. Issinger OG: Casein kinases: Pleiotropic mediators of cellular regulation. Pharmac Ther 59: 1–30, 1993
6. Ahmed K: Significance of the casein kinase system in cell growth and proliferation with emphasis on studies of the androgenic regulation of the prostate. Cell Mol Biol Res 40: 1–11, 1994
7. Pyerin W: Human casein kinase II: Structures, genes, expression and requirement in cell growth stimulation. Adv Enzyme Regul 34: 225–246, 1994
8. Allende JE, Allende CC: Protein kinases. 4. Protein kinase CK2: An enzyme with multiple substrates and a puzzling regulation. FASEB J 9: 313–323, 1995
9. Pyerin W, Ackermann K, Lorenz P: Casein kinases. In: F. Marks (ed). Protein Phosphorylation. Verlag Chemie, Weinheim, 1996, pp 117–174
10. Pinna LA, Meggio F: Protein kinase CK2 ('casein kinase-2') and its implication in cell division and proliferation. Prog Cell Cyc Res 3: 77–97, 1997
11. Glover CVC: On the physiological role of casein kinase II in *Saccharomyces cerevisiae*. Prog Nucleic Acid Res Mol Biol 59: 95–133, 1998
12. Wirkner U, Voss H, Lichter P, Weitz S, Ansorge W, Pyerin W: Human casein kinase II subunit alpha: Sequence of a processed (pseudo)gene and its localization on chromosome 11. Biochim Biophys Acta 1131: 220–222, 1992
13. Wirkner U, Voss H, Lichter P, Ansorge W, Pyerin W: The human gene (CSNK2A1) coding for the casein kinase II subunit α is located on chromosome 20 and contains tandemly arranged *Alu* repeats. Genomics 19: 257–326, 1994

14. Wirkner U, Voss H, Ansorge W, Pyerin W: Genomic organization and promoter identification of the human protein kinase CK2 catalytic subunit alpha (CSNK2A1). Genomics 48: 71–78, 1998
15. Wirkner U, Pyerin W: CK2alpha loci in the human genome: Structure and transcriptional activity. Mol Cell Biochem 191: 59–64, 1999
16. Yang-Feng TL, Naiman T, Kopatz I, Eli D, Dafni N, Canaani D: Assignment of the human casein kinase II alpha' subunit gene (CSNK2A1) to chromosome 16p13.2–p13.3. Genomics 19: 173, 1994
17. Wirkner U, Voss H, Lichter P, Pyerin W: Human protein kinase CK2 genes. Cell Mol Biol Res 40: 489–499, 1994
18. Albertella MR, Jones H, Thomson W, Olavesen MG, Campbell RD: Localization of eight additional genes in the human major histocompatibility complex, including the gene encoding the casein kinase II β subunit (CSNK2B). Genomics 36: 240–251, 1996
19. Boehm SK, Gum JR, Erickson RH, Hicks JW, Kim YS: Human dipeptidyl peptidase IV gene promoter: Tissue-specific regulation from a TATA-less GC-rich sequence characteristic of a housekeeping gene promoter. Biochem J 311: 835–843, 1995
20. Gumina RJ, Kirschbaum NE, Piotrowski K, Newman PJ: Characterization of the human platelet/endothelial cell adhesion molecule-1 promoter: Identification of a GATA-2 binding element. Blood 89: 1260–1269, 1997
21. Voss H, Wirkner U, Jakobi R, Hewitt NA, Schwager C, Zimmermann J, Ansorge W, Pyerin W: Structure of the gene encoding human casein kinase II subunit beta. J Biol Chem 266: 13706–13711, 1991
22. Krehan A, Ansuini H, Böcher O, Grein S, Wirkner U, Pyerin W: Transcription factors ets1, NF-kappa B, and Sp1 are major determinants of the promoter activity of the human protein kinase CK2alpha gene. J Biol Chem 275: 18327–18336, 2000
23. Krehan A, Schmalzbauer R, Böcher O, Ackermann K, Wirkner U, Brouwers S, Pyerin W: Ets1 is a common element in directing transcription of the α and the β gene of human protein kinase CK2. Eur J Biochem 268: 3243–3252, 2001
24. Kyhse-Andersen J: Electroblotting of multiple gels: A simple apparatus without buffer tank for rapid transfer of proteins from polyacrylamide to nitrocellulose. J Biochem Biophys Meth 10: 203–209, 1984
25. Kobayashi R, Tashima Y: Visualization of antigen on nitrocellulose membrane by the oxidative coupling reaction of N,N'-dimethyl-p-phenylenediamine and 4-chloro-1-naphthol. Anal Biochem 183: 9–12, 1989
26. Ackermann K, Fauss J, Pyerin W: Inhibition of cyclic AMP-triggered aromatase gene expression in human choriocarcinoma cells by antisense oligodeoxynucleotide. Cancer Res 54: 4940–4946, 1994
27. Jakobi R, Voss H, Pyerin W: Human phosvitin/casein kinase type II. Molecular cloning and sequencing of full-length cDNA encoding subunit beta. Eur J Biochem 183: 227–233, 1989
28. Lozeman FJ, Litchfield DW, Piening C, Takio A, Walsh KA, Krebs EG: Isolation and characterization of human cDNA clones encoding the alpha and the alpha' subunits of casein kinase II Interaction sites between catalytic and regulatory subunits in human protein kinase CK2 holoenzymes as indicated by chemical cross-linking and immunological investigations. Biochemistry 29: 8436–8447, 1990
29. Bodenbach L, Fauss J, Robitzki A, Krehan A, Lorenz P, Lozeman FJ, Pyerin W: Recombinant human casein kinase II. A study with the complete set of subunits (alpha, alpha' and beta), site-directed autophosphorylation mutants and a bicistronically expressed holoenzyme. Eur J Biochem 220: 263–273, 1994
30. Lüscher B, Litchfield DW: Biosynthesis of casein kinase II in lymphoid cell lines. Eur J Biochem 220: 521–526, 1994
31. Ho IC, Bhat NK, Gottschalk LR, Lindsten T, Thompson CB, Papas TS, Leiden JM: Sequence-specific binding of human Ets-1 to the T cell receptor alpha gene enhancer. Science 250: 814–818, 1990
32. Seth A, Hodge DR, Thomson DM, Robinson L, Panayiotakis A, Watson DK, Papas TS: Aids Res Hum Retroviruses 9: 1017–1023, 1993
33. Gill G, Pascal E, Tseng ZH, Tijan R: A glutamine-rich hydrophobic patch in transcription factor Sp1 contacts the dTAFII110 component of the Drosophila TFIID complex and mediates transcriptional activation. Proc Natl Acad Sci USA 91: 192–196, 1994
34. Carter RS, Bhat NK, Basu A, Avadhani NG: The basal promoter elements of murine cytochrome c oxidase subunit IV gene consist of tandemly duplicated ets motifs that bind to GABP-related transcription factors. J Biol Chem 267: 23418–23426, 1992
35. Carter RS, Avadhani NG: Cooperative binding of GA-binding protein transcription factors to duplicated transcription initiation region repeats of the cytochrome c oxidase subunit IV gene. J Biol Chem 269: 4381–4387, 1994
36. Wasylyk B, Hahn SL, Giovane A: The Ets family of transcription factors. Eur J Biochem 211: 7–18, 1993
37. Pearson BE, Nasheuer H-P, Wang TS: Human DNA polymerase alpha gene: Sequences controlling expression in cycling and serum-stimulated cells. Mol Cell Biol 11: 2081–2095, 1991
38. Yamaguchi M, Hayashi Y, Matsukage A: Mouse DNA polymerase beta gene promoter: Fine mapping and involvement of Sp1-like mouse transcription factor in its function. Nucl Acids Res 16: 8773–8787, 1988
39. Jolliff K, Li Y, Johnson LF: Multiple protein-DNA interactions in the TATAA-less mouse thymidylate synthase promoter. Nucl Acid Res 19: 2267–2274, 1991
40. Clevers HC, Dunlap S, Wileman TE, Terhorst C: Human CD3-epsilon gene contains three miniexons and is transcribed from a non-TATA promoter. Proc Natl Acad Sci 85: 8156–8160, 1988
41. Siu G, Wurster A, Lipsick JS, Hedrick SM: Expression of the CD4 gene requires a Myb transcription factor. Mol Cell Biol 12: 1592–1604, 1992
42. Sucharov C, Basu A, Carter RS, Avadhani NH: A novel transcriptional initiator activity of the GABP factor binding ets sequence repeat from the murine cytochrome c oxidase Vb gene. Gene Express 5: 93–111, 1995
43. Crepieux P, Coll J, Stehelin D: The Ets family of proteins: Weak modulators of gene expression in quest for transcriptional partners. Crit Rev Oncogen 5: 615–638, 1994
44. Dittmer J, Pise-Masison CA, Clemens KE, Choi K-S, Brady JN: Interaction of human T-cell lymphotropic virus type I Tax, Ets1, and Sp1 in transactivation of the PTHrP P2 promoter. J Biol Chem 272: 4953–4958, 1997
45. Ackermann K, Pyerin W: Protein kinase CK2alpha may induce gene expression but unlikely acts directly as a DNA-binding transcription-activating factor. Mol Cell Biochem 191: 129–134, 1999
46. Gunther M, Frebourg T, Laithier M, Fossar N, Bouziane-Ouartini M, Lavialle C, Brison O: An Sp1 binding site and the minimal promoter contribute to overexpression of the cytokeratin 18 gene in tumorigenic clones relative to that in nontumorigenic clones of a human carcinoma cell line. Mol Cell Biol 15: 2490–2499, 1995
47. VanDijk MA, Holthuizen PE, Sussenbach JS: Elements required for activation of the major promoter of the human insulin-like growth factor II gene. Mol Cell Biol 88: 175–185, 1992
48. Wu RL, Chen T-T, Sun T-T: Functional importance of an Sp1- and an NFκB-related nuclear protein in a keratinocyte-specific promoter of rabbit K3 keratin gene. J Biol Chem 269: 28450–28459, 1994
49. Biggs JR, Kudlow JE, Kraft AS: The role of the transcription factor Sp1 in regulating the expression of the WAF1/CIP1 gene in U937 leukemic cells. Biol Chem 271: 901–906, 1996
50. Kollmar R, Sukow KA, Spongale SK, Farnham PJ: Start site selection at the TATA-less carbamoyl-phosphate synthase (glutamine-hydrolyzing)/aspartate carbamoyltransferase/dihydroorotase promoter. J Biol Chem 269: 2252–2257, 1994
51. Block KL, Shou Y, Poncz M: An Ets/Sp1 interaction in the 5'-flanking region of the megakaryocyte-specific alpha IIb gene appears to sta-

bilize Sp1 binding and is essential for expression of this TATA-less gene. Blood 88: 2071–2080, 1996
52. Sun SC, Elwood J, Beraud C, Greene WC: Human T-cell leukemia virus type I Tax activation of NF-kappa B/Rel involves phosphorylation and degradation of I kappa B alpha and RelA (p65)-mediated induction of the c-rel gene. Mol Cell Biol 14: 7377–7384, 1994
53. Armstrong SA, Barry DA, Leggett RW, Mueller CR: Casein kinase II-mediated phosphorylation of the C terminus of Sp1 decreases its DNA binding activity. J Biol Chem 272: 13489–13495, 1997
54. Cochet C, Chambaz EM: Oligomeric structure and catalytic activity of G type casein kinase. Isolation of the two subunits and renaturation experiments. J Biol Chem 258: 1403–1406, 1983
55. Ole-MoiYoi OK: Casein kinase II in theileriosis. Science 267: 834–836, 1995
56. Seldin DC, Leder P: Casein kinase II alpha transgene-induced murine lymphoma: Relation to theileriosis in cattle. Science 267: 894–897, 1995
57. Kelliher MA, Seldin DC, Leder P: Casein kinase II alpha transgene-induced murine lymphoma: Relation to theileriosis in cattle. EMBO J 15: 5160–5166, 1996
58. Landesman-Bollag E, Channavajhala PL, Cardiff RD, Seldin DC: p53 deficiency and misexpression of protein kinase CK2alpha collaborate in the development of thymic lymphomas in mice. Oncogene 16: 2965–2974, 1998
59. Orlandini M, Semplici F, Ferruzzi R, Meggio F, Pinna LA, Oliviero S: Protein kinase CK2alpha′ is induced by serum as a delayed early gene and cooperates with Ha-ras in fibroblast transformation. J Biol Chem 273: 21291–21297, 1998
60. Yang B-S, Hauser CA, Henkel G, Colman MS, VanBeveren C, Stacey KJ, Hume DA, Maki RA, Ostrowski MC: Ras-mediated phosphorylation of a conserved threonine residue enhances the transactivation activities of c-Ets1 and c-Ets2. Mol Cell Biol 16: 538–547, 1996
61. Litchfield DW, Dobrowolska G, Krebs EG: Regulation of casein kinase II by growth factors: A reevaluation. Cell Mol Biol Res 40: 373–381, 1994
62. Ackerman P, Osheroff N: Regulation of casein kinase II activity by epidermal growth factor in human A-431 carcinoma cells. J Biol Chem 264: 11958–11963, 1989
63. Faust RA, Gapany M, Tristani P, Davis A, Adams GL, Ahmed K: Elevated protein kinase CK2 activity in chromatin of head and neck tumors: Association with malignant transformation. Cancer Lett 101: 31–35, 1996
64. Münstermann U, Fritz G, Seitz G, Yiping L, Schneider HR, Issinger O-G: Casein kinase II is elevated in solid human tumours and rapidly proliferating non-neoplastic tissue. Eur J Biochem 189: 251–257, 1990
65. Daya-Makin M, Sanghera JS, Mogentale TL, Lipp M, Parchomchuk J, Hogg JC, Pelech SL: Activation of a tumor-associated protein kinase (p40TAK) and casein kinase 2 in human squamous cell carcinomas and adenocarcinomas of the lung. Cancer Res 54: 2262–2268, 1994
66. Bhat NK, Fischinger PJ, Seth A, Watson DK, Papas T: Isolation and characterization of a novel gene expressed in multiple cancers. Int J Oncol 8: 841–846, 1996
67. Sato M, Morii E, Komori T, Kawahata H, Sugimoto M, Terai K, Shimizu H, Yasui T, Ogihara H, Yasui N, Ochi T, Kitamura Y, Ito Y, Nomura S: Transcriptional regulation of osteopontin gene *in vivo* by PEBP2-alphaA/CBFA1 and ETS1 in the skeletal tissues. Oncogene 17: 1517–25, 1998
68. Gambarotta G, Boccaccio C, Giordano S, Ando M, Stella MC, Comoglio PM: Ets up-regulates MET transcription. Oncogene 13: 1911–1917, 1996
69. McCracken S, Leung S, Bosselut R, Ghysdael J, Miyamoto NG: Myb and Ets related transcription factors are required for activity of the human lck type I promoter. Oncogene 12: 3609–3615, 1994
70. Majérus M-A, Bibollet-Ruche F, Telliez J-B, Wasylyk B, Bailleul B: Serum, AP-1 and Ets-1 stimulate the human ets-1 promoter. Nucl Acids Res 20: 2699–2703, 1992
71. Bassuk AG, Leiden JM: A direct physical association between ETS and AP-1 transcription factors in normal human T cells. Immunity 3: 223–237, 1995
72. McKinlay LH, Tymms MJ, Thomas RS, Seth A, Hasthorpe S, Hertzog PJ, Kola I: The role of Ets-1 in mast cell granulocyte-macrophage colony-stimulating factor expression and activation. J Immunol 161: 4098–5105, 1998
73. Boldyreff B, Issinger O-G: Structure of the gene encoding the murine protein kinase CK2β subunit. Genomics 29: 253–256, 1995
74. Hu E, Rubin CS: Casein kinase II from *Caenorhabditis elegans*. J Biol Chem 265: 5072–5080, 1990
75. Hu E, Rubin CS: Casein kinase II from *Caenorhabditis elegans*. J Biol Chem 266: 19796–19802, 1991
76. Allende CC, Neckelman G, Wilhelm J, Brown J, Allende JE: The promoter region of the CK2α subunit gene of *Xenopus laevis*. 3rd Int. Symposium on Protein Kinase CK2: From Structure to Function, San Esteban, Chile, Poster Abstracts, January 8–10, 2001

Molecular and Cellular Biochemistry **227**: 59–66, 2001.

Genes targeted by protein kinase CK2: A genome-wide expression array analysis in yeast

Karin Ackermann,[1] Andrea Waxmann,[1] Claiborne V.C. Glover[2] and Walter Pyerin[1]

[1]*Biochemische Zellphysiologie (B0200), Deutsches Krebsforschungszentrum, Heidelberg, Germany;* [2]*Biochemistry and Molecular Biology, University of Georgia, Athens, GA, USA*

Abstract

Protein kinase CK2, a tetramer composed of two catalytically active (CK2α isoforms) and two regulatory (CK2β isoforms) subunits, is suspected to have, among others, a role in gene transcription. To identify the genes targeted by CK2, the transcriptional effect of silencing the CK2 subunit genes in *Saccharomyces cerevisiae* (CK2α isoform genes: CKA1 and CKA2; CK2β isoform genes: CKB1 and CKB2) was examined using genome-wide expression array analysis (oligonucleotide array chips). Silencing did not influence the overwhelming majority (5801) of the over six thousand open reading frames composing the yeast genome. Cells knocked-out for both CKA1 and CKA2 and plasmid-rescued by Cka1 affected specifically at 2-fold discrimination level the transcription of 57 genes, and when rescued by Cka2, the transcription of 118 genes. In CKB1/CKB2 double knock-outs, transcription of 54 genes was specifically altered. Interestingly, aside overlaps between the gene spectra affected by CKA1 and CKA2 silencing, there were overlaps also between those influenced by CK2α and CK2β isoform silencing. The data indicate a distinct role of CK2 in gene transcription control, identify specific functional differences between the two catalytic subunits in gene targeting, and reveal independent effects by the regulatory subunits. (Mol Cell Biochem **227**: 59–66, 2001)

Key words: transcriptional control, CK2 isoform typing, DNAchips, gene expression profiling, protein kinase CK2 (casein kinase II), functional genomics, gene targeting

Introduction

Protein kinase CK2 represents a conserved tetrameric holoenzyme composed of two catalytically active subunits (CK2α and isoforms), and two regulatory subunits (CK2β and isoforms) occurring ubiquitously among eukaryotes and representing a vital cellular entity. Conserved are not only the sequences of CK2 subunits and the tetrameric structure of holoenzymes, but even the structures of the genes encoding subunits. Deviations from stoichiometry and deviations in cellular levels have been linked to diseased states [1–11]. CK2 seems to be involved in various cellular processes, including proliferation signaling and gene transcription, the involvement occuring probably via both phosphorylation and non-phosphorylation interactions [12–14]. The substrate spectrum of CK2 is broad and includes transcription regulation components such as transcription factors and subunits of all three classes of RNA polymerases [11]. Although highly similar, there are indications that CK2 isoforms α and α′ may differ functionally in mammalians and be regulated independently [15, 16]. In yeasts, defects in cell cycle progression and alterations in cell polarity have been linked to isoforms CK2α (Cka1) and CK2α′ (Cka2), respectively [17], and specific functional differences between catalytic subunits were already indicated by a first set of data on isoform-specific targeting of genes [18]. Despite of high cell physiological significance, published work on CK2 isoform-specific processes is rare.

The involvement of CK2 in transcription control let us to perform a genome-wide investigation for genes targeted by CK2 in the budding yeast *Saccharomyces cerevisiae*. This organism has extensively been characterized genetically and the genome has completely been sequenced so that each of

Address for offprints: K. Ackermann, Biochemische Zellphysiologie (B0200), Deutsches Krebsforschungszentrum, Im Neuenheimer Feld 280, 69120 Heidelberg, Germany (E-mail: k.ackermann@dkfz–heidelberg.de)

the over six thousand open reading frames (ORFs) composing the genome are known [19]. In addition, it has the ability to proliferate in haploid state allowing for gene mutations without the complication by a second, non-mutated copy of the gene(s) in question as in diploid cells. The *S. cerevisiae* genome contains four CK2 gene loci. Two of these encode catalytic subunits and are located at chromosomes IX (CKA1) and XV (CKA2), providing the gene products Cka1 and Cka2, the homologs of mammalian CK2α and CK2α′, respectively. The two other genes encode regulatory subunits and are located at chromosomes VII (CKB1) and XV (CKB2), providing the gene products Ckb1 and Ckb2, respectively. While single knock-outs (either *cka1* or *cka2*) are viable and more or less indistinguishable phenotypically from each other (both showing some Na^+ and Li^+ sensitivity compared to wild type), double knock-outs are lethal but may be rescued by transfection of CK2-carrying plasmids even of other organisms such as CK2α of *D. melanogaster* [11]. We have been examining the effects of functional deletions of catalytic and regulatory subunits of CK2 on global gene transcription monitored by oligonucleotide array chips. The results obtained seem to indicate that the overwhelming majority of genes is not affected by these deletions, but that there are distinct sets of genes targeted in a CK2 subunit-specific manner, and that this is not restricted to the catalytic subunits but to concern also the regulatory subunits.

Materials and methods

Yeast strains and cultivation

The *Saccharomyces cerevisiae* strain YPH250 (*MAT*a *CKA1 CKA2 CKB1 CKB2*) was genetically tractated as described elsewhere [11, 17, 20] to provide strains YAR109 (*MAT*a *cka1-Δ1::HIS3 cka2-Δ1::TRP1 CKB1 CKB2;* Cka1-Δ2^{ts}) and YDH13 (*MAT*a *cka1-Δ1::HIS3 cka2-Δ1::TRP1 CKB1 CKB2*; Cka2-13^{ts}). These were grown in YPD medium at 28°C, i.e. at permissive temperature, to allow for the expression of either Cka1 but not Cka2 (YAR109; 'α cells'), or of Cka2 but not Cka1 (YDH13; 'α′ cells'). The yeast strain YPH499 (*MAT*a *CKA1 CKA2 CKB1 CKB2*) was the basis to provide strain YAPB10-2c (*MAT*a *CKA1 CKA2 ckb1-Δ1::HIS3 ckb2-Δ1:: LEU2*; [21]). Growth was in YPD medium at 28°C to allow for expression of Cka1 and Cka2 but not Ckb1 and Ckb2 ('β$^-$ cells'). When reaching mid-log phase (0.46–0.47 optical density at 600 nm), cells were harvested by centrifugation at room temperature at 2100 × g for 5 min, and cell pellets were flash frozen in liquid nitrogen. Frozen pellets were stored at –80°C. Yeast cells were phenotyped by differential interference contrast microscopy according to Nomarski at a laser scanning microscope (LSM 410 invert; Carl Zeiss).

Isolation of total RNA and mRNA

Total RNA was extracted by adding acidic phenol and TES (10 mM Tris-Cl, pH 7.5; 10 mM EDTA; 0.5% SDS) to frozen yeast pellets. The suspension was incubated at 65°C for 20 min with periodic vortexing, and, after freezing at –80°C for 30 min, applied to oakridge tubes and centrifuged at 20000 × g for 10 min. Extraction was repeated with one more time phenol, phenol/chloroform/isoamylalcohol and chloroform each. RNA was precipitated with 3 vol. ethanol and 1/10 vol. of 3 M sodium acetate, pH 5.2. For mRNA isolation 1 g of oligo(dT)-cellulose (Ambion) was washed 3 times with 1 × NETS (500 mM NaCl; 10 mM EDTA; 10 mM Tris, pH 7.4; 0.2% SDS) and finally resuspended in 10 ml of 2 × NETS. 1 ml of resuspended oligo(dT)-cellulose was incubated with 1 mg of total RNA for 2 h on a rotation device at room temperature, and washed in Biospin Columns (Bio-Rad) with 5 times 0.7 ml of 1 × NETS. Poly(A)$^+$mRNA was eluated with 5 times 0.7 ml of 70°C-prewarmed 10 mM Tris pH 7.4. The eluate was extracted with chloroform, and poly(A)$^+$mRNA was precipitated with 1 vol. of isopropanol and 1/10 vol. of 3 M sodium acetate, pH 5.2. After quantitation, the quality of total RNA and poly(A)$^+$mRNA was checked on a 1% agarose gel.

RT-PCR

Reverse transcription of 5 µg total RNA was performed according to the instructions of the SuperScript™ first strand synthesis system for RT-PCR (Gibco BRL). 50 pmol of cDNA-specific primers each for CKA1 (sense: 5′-GAAATGCAGGGTATGGTCAG-3′; antisense: 5′-TGTCCCATCGCTTCCTTAG-3′), CKA2 (sense: 5′-ATGCCATTACCTCCGTCAAC-3′; antisense: 5′-CAACCACTTCAGGAACAGC-3′), CKB1 (sense: 5′-ACTATTCACGTACAGGGTCCTC; antisense: 5′-GTCTCGAACTCGCATTTGG-3′) and CKB2 (sense: 5′-ATGGGCAGTAGATCGGAGA; antisense: 5′-GTTCTTGCCACCTTGTCAGT-3′), 1 µl of template cDNA, 0.2 mM dNTPs, 2 mM $MgCl_2$ and 2.5 U Taq Polymerase (Gibco BRL) in 10 × PCR buffer (200 mM Tris-Cl, pH 8.4; 500 mM KCl) were applied to PCR (94°C, 3 min [94°C, 30 sec; 62°C, 30 sec.; 72°C, 30 sec] 30 cycles; 72°C, 3 min). PCR products were separated on a 1% agarose gel to give approximate lengths of 1 kb for CKA1, 900 bp for CKA2, 800 bp for CKB1 and 700 bp for CKB2, respectively.

cDNA synthesis and preparation of biotin-labeled cRNA

First and second strand cDNA synthesis was performed using SuperScript™Choice System for cDNA synthesis (Gibco BRL). Shortly, for first strand synthesis 5 µg poly(A)$^+$mRNA was reverse transcribed using 100 pmol T7-oligo(dT)-Primer (5′-GGCCAGTGAATTGTAATACGACTCAC TATAGGG-AGGCGG-$(dT)_{24}$-3′), 200 U/µg SuperScriptII RT, 10 mM DTT and 10 pmol dNTPs in 5 × first strand buffer (250 mM Tris-Cl, pH 8.3; 375 mM KCl; 15 mM $MgCl_2$) at 37°C for 1 h. For second strand synthesis 30 pmol dNTPs, 10 U *E. coli* DNA Ligase, 40 U *E. coli* DNA Polymerase I and 2 U *E. coli* RNase H in 5 × second strand buffer (100 mM Tris-Cl, pH 6.9; 450 mM KCl; 23 mM $MgCl_2$; 0.75 mM β-NAD^+; 50 mM $(NH_4)_2SO_4$) were added and incubated at 16°C for 2 h. Finally, the reaction was incubated for further 5 min with 10 U of T4 DNA Polymerase and stopped by adding 10 µl of 0.5 M EDTA. ds cDNA was cleaned up using Phase Lock Gels (PLG heavy; Eppendorf) and precipitated with 2.5 vol. of ethanol and 0.5 vol. of 7.5 M NH_4Ac.

For production of large amounts of hybridizable biotin-labeled RNA targets by *in vitro* transcription the ENZO BioArray™ High Yield™ RNA transcript labeling kit (Enzo Diagnostics) was used. 1 µg of ds cDNA was incubated with 10 × High Yield Reaction buffer, 10 × Biotin-labeled ribonucleotides (ATP, GTP, CTP, UTP with bio-UTP and bio-CTP), 10 × DTT, 10 × RNase Inhibitor Mix and 20 × T7 RNA Polymerase following instructions of the kit manual. *In vitro* transcribed cRNA was purified using the RNeasy Mini Protocol (Qiagen). 50 µg of cRNA was fragmented by adding 5 × RNA Fragmentation buffer (200 mM Tris-acetate, pH 8.1; 500 mM KOAc; 150 mM MgOAc) in a final volume of 40 µl and incubated at 94°C for 35 min. 1 µg of fragmented cRNA was analyzed on a 1% agarose gel to give RNA fragment sizes from approximately 35–300 bases.

Hybridization, staining and scanning of arrays

Using the GeneChip Eukaryotic Hybridization Control Kit (Affymetrix) 15 µg of fragmented cRNA was mixed with 50 pM control oligo B2, various concentrations (1–100 pM) of control cRNA cocktail, 0.1 mg/ml herring sperm DNA, 0.5 mg/ml acetylated BSA in 100 mM MES, 1 M [Na^+], 20 mM EDTA, 0.01% Tween 20 and denatured at 99°C for 5 min. Hybridization was performed at 45°C with 60 rpm for 16 h using Yeast Genome S98 Arrays (Affymetrix). Using the GeneChip Fluidics Station 400 (Affymetrix) arrays were washed and stained according to the instructions of the expression analysis technical manual (Affymetrix) with protocol EukGE-WS2.v3. Shortly, arrays were washed with non-stringent washing buffer (6 × SSPE; 0.01% Tween 20; 0.005% Antifoam) at 25°C and stringent washing buffer (100 mM MES; 0.1 M [Na^+]; 0.01% Tween 20) at 50°C. Staining was performed with R-Phycoerythrin Streptavidin conjugate (Molecular Probes) for a final concentration of 10 µg/ml. For amplification of signal biotinylated anti-streptavidin antibody (Vector Laboratories) was applied at a final concentration of 3 µg/ml. For scanning arrays the HP GeneArray™ scanner (Affymetrix) was used, applying an argon ion laser as the excitation source, with the emission detected by a photomultiplier tube through a 560 nm longpass filter [22, 23].

Data analysis

The resulting image data were analyzed using the GeneChip Expression Analysis software (Microarray Suite vs. 4.0; Affymetrix). After automatically placing a grid over the data files using the corner control regions as markers, the average intensity of each probe cell is calculated based on the intensities of the pixels contained in cell. The software then applies the selected probe array algorithm to determine expression levels for each gene. For further analysis the expression analysis software GeneSpring vs. 3.2.12 (Silicon Genetics) was applied.

Results and discussion

Yeast strain characterization

For examination of the genes targeted by Cka1 and Cka2, the *S. cerevisiae* strain YPH250 (*MAT***a** *CKA1 CKA2 CKB1 CKB2*) was functionally deleted in both the CKA1 gene and the CKA2 gene to provide for a null background of both gene products (*MAT***a** *cka1-Δ1::HIS3 cka2-Δ1::TRP1 CKB1 CKB2*). The cells were rescued by transfection with a plasmid expressing in a temperature-sensitive manner either Cka1 (*cka1*^{ts}; *CEN6/ARSH4 LEU2 cka1-Δ2*^{ts}; strain YAR109; α cells) or Cka2 (*cka2*^{ts}; *CEN6/ARSH4 LEU2 cka2-13*^{ts}; strain YDH13; α′ cells). Strains expressing only Cka1 or only Cka2 had no overt phenotype on normal media (see also [11]), had comparable growth rates, and seemed to have a significant tendency to flocculate (Fig. 1A). For Ckb1 and Ckb2, the strain YPH499 (*MAT***a** *CKA1 CKA2 CKB1 CKB2*; differing from YPH250 only in the *trp1* allele, which is *trp1-Δ1* in YPH250 and trp1-Δ63 in YPH499) was deleted in both the CKB1 and the CKB2 genes to provide for cells devoid in regulatory proteins (*MAT***a** *CKA1 CKA2 ckb1-Δ1::HIS3 ckb2-Δ1::LEU2*; strain YAPB10-2c; $β^-$ cells). Strains devoid in both these genes have shown to be phenotypically indistinguishable from strains carrying deletion mutations either in CKB1 or CKB2 [11].

Expression and lack of expression, respectively, was checked by RT-PCR using CK2 subunit isoform-specific primers that were designed in a way as to generate products differing in length (Fig. 1B). The results unambiguously demonstrated the expression of all four CK2 genes in the wild type strains YPH250 and YPH499, the expression of CKA1 but not CKA2

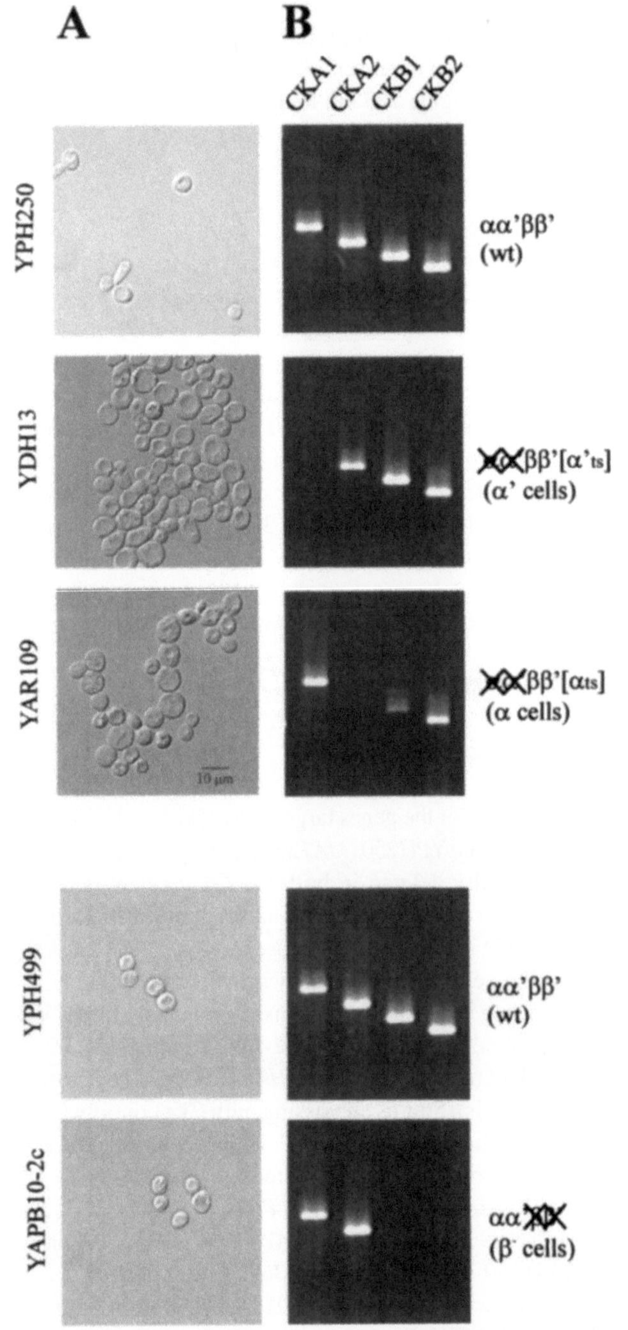

Fig. 1. Yeast strain chracteristics. (A) Morphology. Wildtype strains YPH250 and YPH499 as well as functional CK2 knock-out strains YAR109, YDH13 and YAPB10-2c have been characterized by differential interference contrast microscopy (scale given by bar). (B) Expression of CK2 isoform genes. CK2 isoform-specific RT-PCR was performed employing specific primer pairs for the transcripts of CKA1, CKA2, CKB1 and CKB2, designed to get different lengths of PCR products. Range of CK2 subunits present in cells and cell naming provided at right.

in YAR109 cells (α cells), the expression of CKA2 but not CKA1 in YDH13 cells (α′ cells), and the lack in expression of both CKB1 and CKB2 in YAPB10-2c cells (β^- cells).

Expression profiling

For expression profiling, the diverse yeast cells were grown under identical conditions to comparable optical densities (OD at 600 nm: 0.46–0.47) in order to avoid deviations in gene expression profiles due to metabolites, partial oxygen deficiency, etc. [24]. As schematically outlined in Fig. 2A, total RNA (RNA) was prepared from the various yeast cells harvested at room temperature to avoid expression of temperature-shock genes. From RNA, poly(A)$^+$RNA (mRNA) was prepared and, following quality check, further processed by reverse transcription. The resulting cDNA populations were *in vitro* transcribed in presence of biotinylated nucleotides to provide cRNA populations detectable by the streptavidin system. These cRNA populations were hybridized to DNA chips consisting of a comprehensive set of gene-specific oligonucleotides representing virtually the whole yeast genome and synthesized onto a glass-supported surface. The hybridized chips were scanned by a laser-supported chip reading device, and the resulting signals transformed via computer programs into expression profiles. Individual genes are represented on the chips not by a single oligonucleotide spot but rather by a series of unique oligonucleotide pairs representing a perfect match and a mismatch (negative control) of different parts of the genes. Expressed genes, therefore, appear as colored tiny lanes composed of a row of squares as exemplified here by chips that were hybridized to transcripts of wild type and α′ cells (Fig. 2B).

Expression profile comparisons

Comparing the gene expression profiles of wild type cells and α cells, the overwhelming majority of the genes expressed under the conditions used (over nine tenth of genes) showed similar expression levels, i.e. were within the 2-fold deviation boundary lines used throughout this paper (Fig. 3A). Aside these, abundances of transcripts in α cells deviated significantly from that in wild type cells for a number of genes, both by increases and by decreases of transcript levels. This clearly indicates that despite lack of overt phenotype, cells living with only Cka1 instead of Cka1 plus Cka2 significantly deviate in the expression of various genes. Similarly, comparison of wild type cells and α′ cells shows that an overwhelming majority of genes are expressed to comparable levels, but that there is despite lacking overt phenotype a range of genes with deviating transcript abundances, either increased or decreased (Fig. 3B).

Are the deviatingly expressed genes in α cells and α′ cells identical? A direct comparison makes clear that this is not the case (Fig. 3C). There are obviously genes scattered beyond deviation boundary lines that, therefore, are specifi-

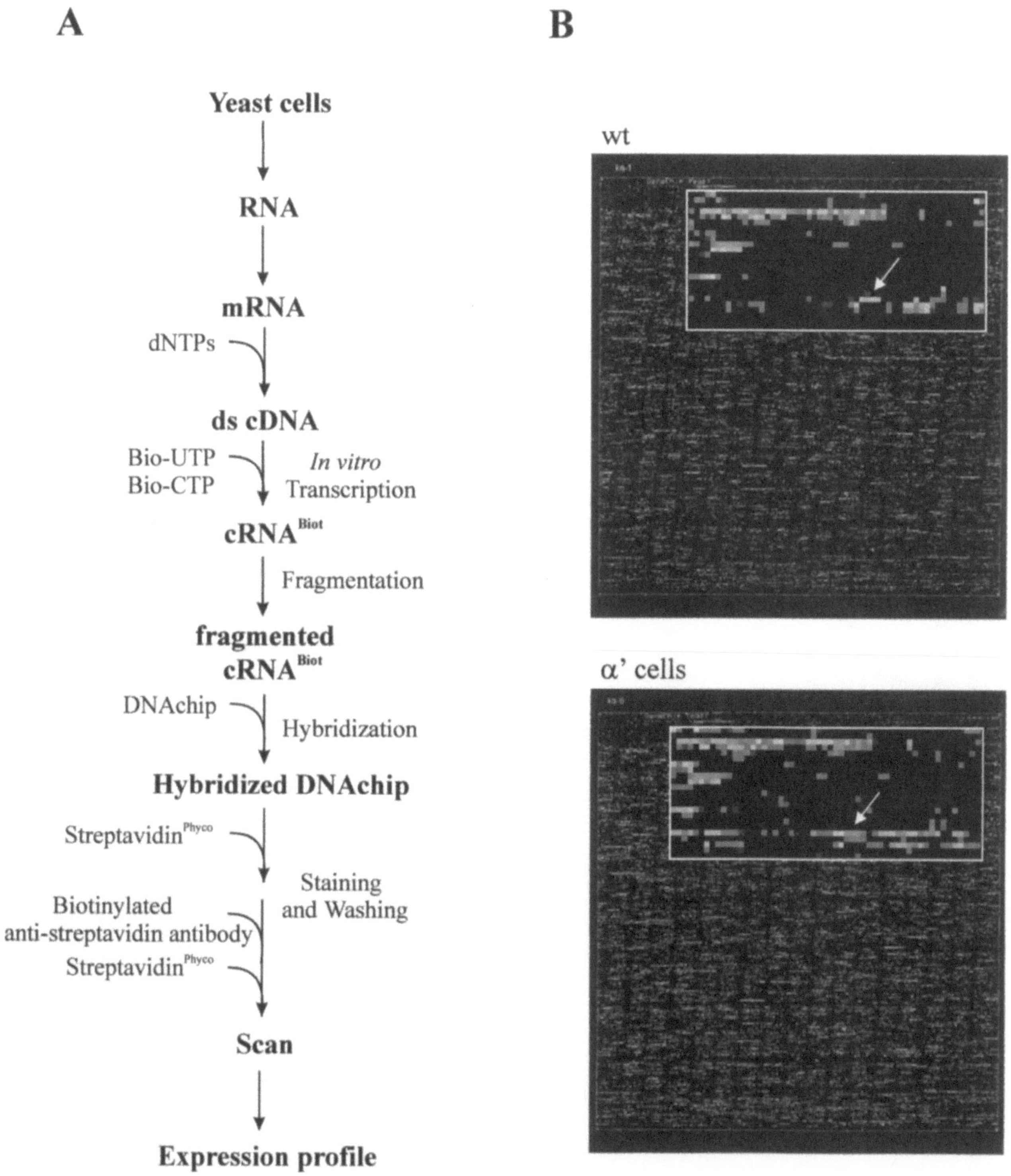

Fig. 2. Expression profiling. (A) Schematic overview of experimental procedure applied in expression monitoring. For details see text. Fragmentation of labeled cRNAs was by heat in presence of magnesium ions, hybridization to chips (oligonucleotide arrays) carried out in flow cells, and scanning occurred in a laser-supported device. (B) Fluorescent array hybridization images. Shown are the gene transcription images of YPH250 cells (wt) and YDH13 cells (α' cells). Boxes represent enlarged details of arrays, demonstrating that ORFs are respresented by rows of approximately 20 gene-specific oligonucleotides each; colours indicate expression levels (blue, very low; gradient increase via green, yellow to red, very high). An example of gene expression deviation is given by arrows (low and high expression levels in wt and α' cells, respectively).

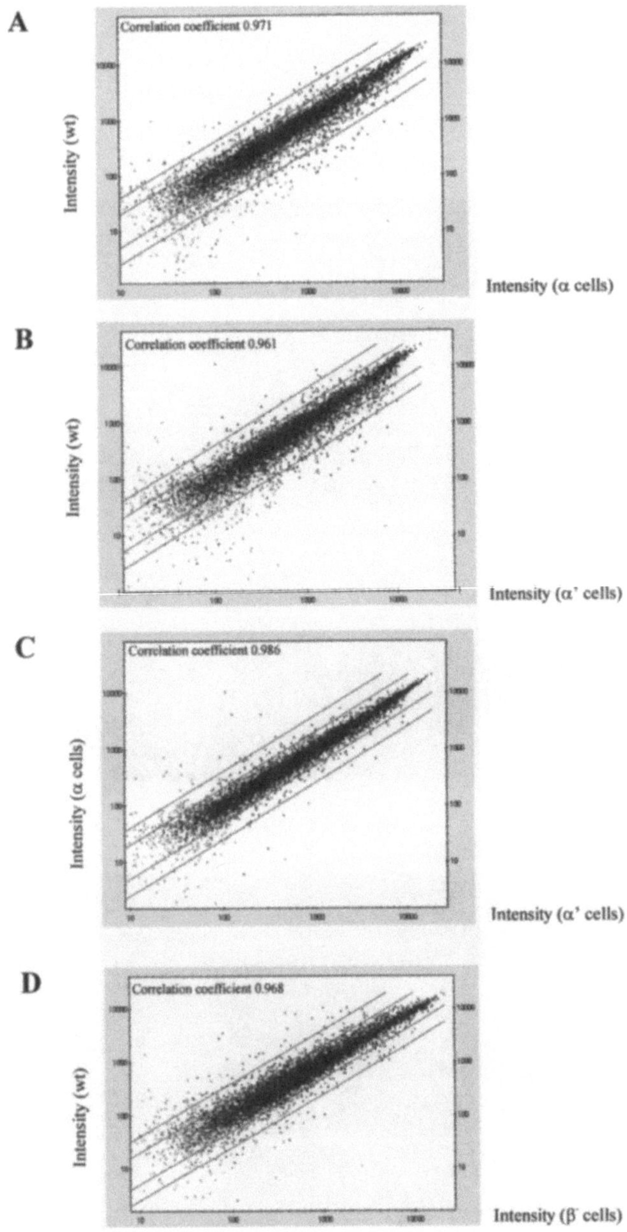

Fig. 3. Correlation analysis of gene expression profiles. Wildtype profiles are compared to those of the CK2 knock-out strains as indicated (A, B, D) or knock-out profiles to each other (C). For cell naming see Fig. 1. Correlation coefficients are provided for each plot. Individual genes are represented by dots; deviation boundary lines given for 2-fold (inner lines) and 4-fold deviations.

cally linked in their expression to either Cka1 or to Cka2. However, the number of deviating genes in this direct comparison is clearly smaller than seen with the above comparisons to wild type (correlation coefficients 0.986 to 0.971 and 0.961, respectively).

Comparing the gene expression profiles of wild type cells and β^- cells, the overwhelming majority of the genes again showed similar expression levels, i.e. were within the deviation boundary lines (Fig. 3D). Aside these, however, abundances of transcripts in β^- cells deviated significantly from that in wild type cells for a number of genes, both enhightened and reduced in levels. Thus, despite lack in overt phenotype, loss of β and/or β′ affects the expression of genes. Functionally, this might occur either by a direct interaction of individual β and β′ proteins with factors of (or linked to) the transcription machinery, or via complexation to Cka1 and/or Cka2.

Together, the expression profiles clearly indicate that the mRNA populations of yeast cells remain unchanged when CK2 genes are functionally inactivated, but that a significant set of gene transcripts deviates in relative abundancy.

On a more quantitative basis, aside 5801 genes unaltered in transcription, the deviations determined in relative abundances at a 2-fold discrimination boundary seems to concern 279 genes in α′ cells, i.e. cells lacking Cka1 and thus living on Cka2 only, 216 genes in α cells, i.e. cells lacking Cka2 and thus living on Cka1 only, and 76 genes in β^- cells, i.e. cells living without any Ckb1 or Ckb2 subunit (Fig. 4). These gene spectra show a number of interesting overlaps. A contingent of 144 genes seems to be common to both of the spectra of genes changed in expression by either Cka1 or Cka2 inactivation. This means that the transcription of 118 genes remains as specifically affected by Cka2 and 57 specifically affected by Cka1. Together with an already provided first set of data obtained with another expression array system [18] this result clearly indicates functional differences between the catalytic subunits of CK2 in gene targeting. A number of examples of functional differences have been found recently also in mammalian cells [15, 16]. Not considering ORFs with so far unknown functions, the nature of the genes deviating in expression fall into relatively few categories. Examples include multidrug resistance genes and structural genes specifically linked to Cka1, and mating genes and flocculation genes to Cka2. The specific differences between isoforms CK2α and CK2α′ in gene targeting may also be useful for isoform typing.

Surprisingly, overlaps are not just seen between the gene expression spectra linked to catalytic subunits. Rather these spectra seem to contain genes that appear also in the gene expression spectrum linked to regulatory subunits. Some 7 genes seem to be common to β^- cells and α′ cells, 5 genes to β^- cells and α cells, and 10 genes to β^- cells and both α′ and α cells.

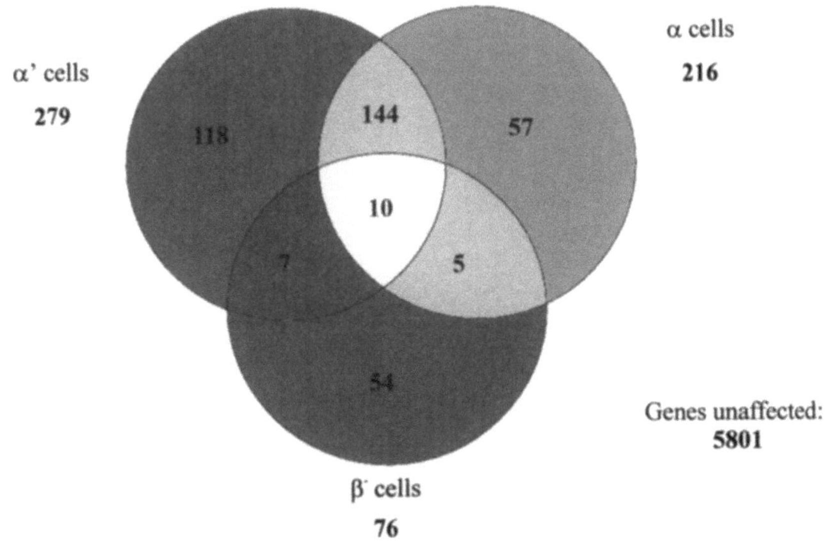

Fig. 4. Genes affected in expression due to manipulation of CK2 subunit genes (Venn diagram). Shown are the respective numbers of genes in α′ cells, α cells and β^- cells (for definition see Fig. 1) altered with significantly expression levels (at least 2-fold) and of unaffected genes. Numbers in circles and circle overlaps represent gene expressions linked specifically to indicated CK2 subunits situation.

Acknowledgements

We thank Dr. M. Hergenhahn for assistance in the Affymetrix GeneChip system handling and Dr. H. Spring for help in laser scanning microscopy. We also thank the members of the Biochemical Cell Physiology group for permanent fruitful discussions. The work was supported by HGF-Strategiefonds III (A1/H161 and B5/H166; 2000–2003) to W.P. and by NIH Grant GM54626 to CVC. G.

I would like to dedicate this paper to my dad, Manfred Ackermann, who died during the course of this work.

References

1. Hathaway GM, Traugh JA: Casein kinases – multipotential protein kinases. Curr Topics Cell Regul 21: 101–127, 1982
2. Pinna LA: Casein kinase 2: An 'eminence grise' in cellular regulation? Biochem Biophys Acta 1054: 267–284, 1990
3. Tuazon PT, Traugh JA: Casein kinase I and II – multipotential serine protein kinases: Structure, function, and regulation. Adv Sec Mess Phosphoprot Res 23: 123–164, 1991
4. Litchfield DW, Lüscher B: Casein kinase II in signal transduction and cell cycle regulation. Mol Cell Biochem 127/128: 187–199, 1993
5. Issinger OG: Casein kinases: Pleiotropic mediators of cellular regulation. Pharmacother 59: 1–30, 1993
6. Ahmed K: Significance of the casein kinase system in cell growth and proliferation with emphasis on studies of the androgenic regulation of the prostate. Cell Mol Biol Res 40: 1–11, 1994
7. Pyerin W: Human casein kinase II: Structures, genes, expression and requirement in cell growth stimulation. Adv Enzyme Regul 34: 225–246, 1994
8. Allende JE, Allende CC: Protein kinases. 4. Protein kinase CK2: An enzyme with multiple substrates and a puzzling regulation. FASEB J 9: 313–323, 1995
9. Pyerin W, Ackermann K, Lorenz P: Casein kinases. In: F. Marks (ed). Protein Phosphorylation. Verlag Chemie, Weinheim, 1996, pp 117–174
10. Pinna LA, Meggio F: Protein kinase CK2 ('casein kinase-2') and its implication in cell division and proliferation. Prog Cell Cyc Res 3: 77–97, 1997
11. Glover CVC: On the physiological role of casein kinase II in *Saccharomyces cerevisiae.* Prog Nucleic Acid Res Mol Biol 59: 95–133, 1998
12. Pepperkok R, Herr S, Lorenz P, Pyerin W, Ansorge W: System for quantitation of gene expression in single cells by computerized microimaging: Application to c-fos expression after microinjection of anti-casein kinase II antibody. Exp Cell Res 204: 278–285, 1993
13. Pepperkok R, Lorenz P, Ansorge W, Pyerin W: Casein kinase II is required for transition of G0/G1, early G1, and G1/S phases of the cell cycle. J Biol Chem 269: 6986–6991, 1994
14. Lorenz P, Ackermann K, Simoes-Wuest P, Pyerin W: Serum-stimulated cell cycle entry of fibroblasts requires undisturbed phosphorylation and non-phosphorylation interactions of the catalytic subunits of protein kinase CK2. FEBS Lett 448: 283–288, 1999
15. Vilk G, Saulnier RB, St Pierre R, Litchfield DW: Inducible expression of protein kinase CK2 in mammalian cells. Evidence for functional specialization of CK2 isoforms. J Biol Chem 274: 14406–14414, 1999
16. Xu X, Toselli PA, Russell LD, Seldin DC: Globozoospermia in mice lacking the casein kinase II alpha' catalytic subunit. Nat Genet 23: 118–121, 1999
17. Rethinaswamy A, Birnbaum MJ, Glover CVC: Temperature-sensitive mutations of the CKA1 gene reveal a role for casein kinase II in maintenance of cell polarity in *Saccharomyces cerevisiae.* J Biol Chem 273: 5869–5877, 1998
18. Ackermann K, DeRisi J, Pyerin W: The genome-wide analysis of gene expression in *Saccharomyces cerevisiae* demonstrates specific func-

tional differences between isoforms α and α′ of protein kinase CK2. Curr Genetics 35: 176, 1999

19. Goffeau A, Barrell BG, Bussey H, Davis RW, Dujon B, Feldmann H, Galibert F, Hoheisel JD, Jacq C, Johnston M, Louis EJ, Mewes HW, Murakami Y, Philippsen P, Tettelin H, Oliver SG: Life with 6000 genes. Science 274: 563–567, 1996
20. Hanna DE, Rethinaswamy A, Glover CVC: Casein kinase II is required for cell cycle progression during G1 and G2/M in *Saccharomyces cerevisiae.* J Biol Chem 270: 25905–25914, 1995
21. Bidwai AP, Reed JC, Glover CVC: Cloning and disruption of CKB1, the gene encoding the 38-kDa β subunit of *Saccharomyces cerevisiae* casein kinase II (CKII). J Biol Chem 270: 10395–10404, 1995
22. Lockhart DJ, Dong H, Byrne MC, Follettie MT, Gallo MV, Chee MS, Mittmann M, Wang C, Kobayashi M, Horton H, Brown EL: Expression monitoring by hybridization to high-density oligonucleotide arrays. Nat Biotech 14: 1675–1680, 1996
23. Wodicka L, Dong H, Mittmann M, Ho M, Lockhart DJ: Genome-wide expression monitoring in *Saccharomyces cerevisiae.* Nat Biotech 15: 1359–1367, 1997
24. DeRisi JL, Iyer VR, Brown PO: Exploring the metabolic and genetic control of gene expression on a genomic scale. Science 278: 680–686, 1997

Molecular and Cellular Biochemistry **227**: 67–71, 2001.

Consequences of CK2 signaling to the nuclear matrix

Shihui Yu, Huamin Wang, Alan Davis and Khalil Ahmed

Cellular and Molecular Biochemistry Research Laboratory (151), Department of Laboratory Medicine and Pathology and University of Minnesota Cancer Center, University of Minnesota and the Department of Veterans Affairs Medical Center, Minneapolis, MN, USA

Abstract

Protein kinase CK2 is recognized as one of the key cellular signals for cell growth and proliferation. Its nuclear targeting appears to be critical to its role in these functions. In the nucleus, nuclear matrix (NM) which plays a major role in growth-related activities is a primary locus for CK2 signaling. A variety of growth stimuli evoke a rapid translocation of the CK2 to the NM whereas removal of these factors has the opposite effect. These studies, employing various experimental models of cell growth (involving different growth-stimulatory factors), have suggested that rapid shuttling of CK2 to the NM is a key feature of early growth control. By contrast, removal of growth-stimulatory factors leading to the loss of cell viability is associated with early loss of CK2 from the NM (and chromatin). This indicates that absence of CK2 from the nuclear compartment is contributory to induction of cell death via apoptosis, implying a protective role for CK2 against cell death. Here, we review the evidence that suggests that CK2 signaling in the NM is not only involved in cell growth but also in cell survival. (Mol Cell Biochem **227**: 67–71, 2001)

Key words: protein kinase CK2, nuclear matrix, cell growth, apoptosis, cell survival, adaptive response

Introduction

Protein kinase CK2 (also known as casein kinase 2 or II), a highly conserved and ubiquitous protein serine/threonine kinase, is intimately involved in the control of cell growth and proliferation (for reviews see e.g. [1–5]). The kinase consists of two catalytic subunits (α and/or α') complexed with two β subunits, existing as $\alpha_2\beta_2$, or $\alpha\alpha'\beta_2$, or $\alpha'_2\beta_2$ heterotetramer. Various lines of evidence have demonstrated that spatio-temporal signaling to the nucleus, where nuclear matrix (NM) and chromatin appear to be its preferential targets, represents a major mode of its regulation in relation to the control of cell growth (see e.g. [2–5]). In these studies, growth stimulatory agents have been found to evoke a shuttling of the kinase to the nuclear compartments such as the NM and chromatin. On the other hand, removal of the growth promoting factors which results in loss of cell growth and induction of cell death via apoptosis is associated with a rapid early loss of CK2 from the nuclear compartment. An interpretation of these observations is that the presence of the CK2 in the nuclear compartment (especially in association with the NM and chromatin) may be essential for cell survival. Recent studies have shown that induction of apoptosis in cells (through the use of apoptosis-inducing chemicals) results in a significant enhancement in the nuclear association of CK2 which appears to be an adaptive protective response of the cell [6]. Here, we provide a brief review of our data that link (a) the elevation of CK2 in the nucleus (and especially the NM) to cell growth, survival, and resistance to apoptosis; and (b) the loss of CK2 from the NM to the cessation of cell growth and induction of apoptosis.

Address for offprints: K. Ahmed, Cellular and Molecular Biochemistry Research Laboratory (151), Veterans Affairs Medical Center, One Veterans Drive, Minneapolis, MN 55417, USA (E-mail: ahmedk@tc.umn.edu)

Materials and methods

Chemicals

Synthetic dodecapeptide substrate (RRRADDSDDDDD) for assay of CK2 activity [6] was purchased from Peptide Technologies (Gaithersburg, MD, USA) or from Sigma-Genosys (The Woodlands, TX, USA). Etoposide and diethylstilbestrol (DES) were purchased from Sigma (St. Louis, MO, USA). Cell proliferation reagent WST-1 was supplied by Roche Molecular Biochemicals (Indianapolis, IN, USA). Anti-CK2α antibody was purchased from Transduction Laboratories (Lexington, KY, USA). All reagents were of the highest purity available.

Animals

Male Sprague–Dawley rats weighing 290–320 g (Harlan Sprague–Dawley, Indianapolis, IN, USA) were maintained as described previously [7–9]. Their androgenic status was manipulated via orchiectomy and treatment with androgen (5α-DHT dissolved in sesame oil), as described previously [7–9].

Cell culture

The cell lines employed in these studies were LNCaP, PC-3 and ALVA-41 derived from human prostate adenocarcinoma. LNCaP and PC-3 cells were purchased from American Type Culture Collection (Manassas, VA, USA), whereas ALVA-41 cell line was obtained from Dr. Richard C. Ostenson (University of Washington, Seattle, WA, USA). The LNCaP and PC-3 cells were maintained in RPMI 1640 (Life Technologies, Grand Island, NY, USA) supplemented with 2 mM L-glutamine, 25 mM HEPES, and 10% defined FBS (Hyclone Laboratories, Logan, UT, USA) in an atmosphere containing 5% CO_2. Culture medium for LNCaP cells also included 10^{-10} M 5α-dihydrotestosterone. The ALVA-41 cells were maintained in RPMI 1640 supplemented with 2 mM L-glutamine, 25 mM HEPES, and 6% defined FBS in an atmosphere containing 5% CO_2.

Cell viability and proliferation assay

The cell proliferation reagent WST-1 (a tetrazolium salt that is cleaved by mitochondrial dehydrogenases in viable cells) was employed to determine cell viability and proliferation in cells treated with various chemicals. An aliquot of 100 μl of cell suspension (containing 0.5–2 × 10^4 cells) was plated in each well of 96-well plates, and cells were allowed to reattach over a period of 24 h. Following the treatment of cells with various agents for the indicated periods of time, WST-1 (10 μl) was added to each well and incubation was carried out at 37°C for an additional 30 min. The OD_{450} was then measured by employing an automated plate reader.

Transfection with expression plasmids

The procedures for the preparation of the expression plasmids pCI-CK2α, pCI-CK2β, and bicistronic pCI-CK2αβ, and transfection of cells with these plasmids have been described previously [10].

Measurement of CK2 activity

CK2 activity in the nuclear matrix (NM) fraction isolated from prostate tissue or cells following various treatments was measured as described previously [6, 10, 11].

Results

We have employed androgenic regulation of rat ventral prostate growth as an experimental model to delineate the dynamics of CK2 in the nuclear compartment (see e.g. [2–5]). These studies have demonstrated a rapid loss of CK2 from the NM that appeared to precede the morphological evidence of apoptosis in the epithelial cells. As shown in Table 1, the CK2 activity in the NM from the ventral prostate gland of animals

Table 1. Relation of growth-stimulus mediated modulation in NM-associated CK2 and apoptosis in rat ventral prostate epithelial cells

Treatment of rats*	CK2 activity in the NM	% cells in apoptosis[1]
C_0	100 ± 2.5	1.2 ± 0.1
C_{24}	20 ± 3.0	3.0 ± 0.2
C_{48}	8 ± 1.3	21.4 ± 1.9
C_{144} plus DHT_0	100 ± 9.4	
C_{144} plus DHT_1	211 ± 15.5	
C_{144} plus DHT_4	329 ± 18.0	

*Rats were subjected to castration for 0–144 h (indicated by C_0, C_{144}, etc.) and treated with 5α-DHT for the periods of 0, 1, or 4 h as indicated. The latter reflect the growth phase of the prostate in response to androgenic stimulus; however, the time periods indicated correspond to the pre-replicative phase only. NM from the prostatic tissue was isolated and CK2 activity associated with this fraction was determined as described under Materials and methods. The results are expressed as percent of the control (C_0 or C_{144}) as indicated. The actual CK2 activity in C_0 control and C_{144} controls respectively was 32.2 ± 1.6 and 1.9 ± 0.4 nmol ^{32}P incorporated in the CK2-specific peptide/mg of NM protein/h (data are adapted in part from [7 and 8]).
[1]Data adapted from [12–14].

subjected to orchiectomy was lost by nearly 80% within 24 h and was essentially completely lost within 48 h following androgen deprivation. Under similar conditions, a progressive increase in apoptosis of prostatic epithelial cells has been well documented [12–14]. Thus, it is reasonable to assume that there may be a link between the loss of CK2 from the NM and apoptosis in the cells resulting from ablation of the androgenic growth stimulus. When castrated rats are treated with androgen which would induce cell growth and proliferation in the epithelial cells of the prostate gland (and reduction of apoptosis to normal levels) [12–14], there is a rapid and extensive increase in the NM-associated CK2 (Table 1). Remarkably, this increase becomes apparent at the pre-replicative stage of epithelial cell growth (seen as early as 1 h following androgen treatment) suggesting its key role in promotion of cell growth and eventual proliferation. Presumably, there would be minimal apoptotic activity under these conditions. Continued treatment of the animals with androgen results in the attainment of the normal size and the level of CK2 in the NM as for normal prostate (data not shown). The change in the NM associated CK2 under these various conditions is due to the shuttling of the kinase between the nuclear and cytoplasmic compartments [7, 15]. Thus, the shuttling of CK2 to the NM may be associated with cell growth and proliferation as well as reduction in the apoptotic activity to a minimal level.

To further explore the link between apoptosis and loss of CK2 from the NM, we determined the response of the latter to treatment of cells with chemical agents know to induce apoptosis. The results in Table 2 show the effects of etoposide and DES at different concentrations on the NM-associated CK2 activity in three prostate cancer cell lines (ALVA-41, PC-3, and LNCaP). Both of these agents have been shown to induce apoptosis in these cells as determined by DNA ladder analysis [6]. The treatment of the three cell lines with these agents for a period of 48 h induced a concentration-dependent elevation (rather than a decrease) in the CK2 activity in the NM fraction isolated from these cells (Table 2). An interpretation of this observation is that induction of apoptosis by chemical agents (i.e. stress) evokes an adaptive response in the cells involving the shuttling of CK2 to the NM. This is further indicated on examination of the effects of etoposide and DES over time following treatment of ALVA-41 cells with a single dose of these agents (Table 3). For example, treatment with etoposide resulted in a progressive loss of viable cells due to apoptosis and was accompanied by an increase in the NM-associated CK2 at the corresponding times. Similar results were observed for cells treated with DES.

To test the hypothesis that enhanced CK2 in the NM may serve as a protective response in cells against apoptosis, experiments were undertaken to overexpress CK2 in cells prior to their exposure to the chemical mediators of apoptosis. ALVA-41 and PC-3 cells were transfected with expression plasmids of the α, β and αβ cDNA. These manipulations resulted in enhanced expression of the CK2 that was apparent in the NM fractions isolated from these cells [6]. Response of these cells to etoposide or DES clearly indicated the protection afforded by the prior overexpression of the α or αβ subunits of CK2; interestingly, the overexpression of the β subunit alone appeared to be without such an effect (Fig. 1). These results provide additional evidence supporting the notion that the increase in CK2 in the NM in response to toxic agents that induce apoptosis is an adaptive survival response of the cells.

Table 2. Response of CK2 signaling to the NM upon chemical induction of apoptosis

Chemical agents*	CK2 activity associated with the NM (% control) in		
Etoposide (mM)	ALVA-41	PC-3	LNCaP
0	100 ± 0	100 ± 6	100 ± 10
10	121 ± 12	103 ± 7	107 ± 9
30	218 ± 21	132 ± 9	211 ± 32
100	242 ± 23	252 ± 20	360 ± 22
DES (μM)			
0	100 ± 6	100 ± 8	100 ± 5
10	128 ± 7	N.D.	N.D.
30	N.D.	131 ± 9	201 ± 21
100	221 ± 12	244 ± 16	N.D.

*Cells were treated with etoposide or DES at the concentrations indicated for a period of 48 h. At this time, NM was prepared from these cells and CK2 activity associated with it was determined as described under Materials and methods. Results are presented as % of control (no drug) and represent mean ± S.E.M. (N.D. = not determined).

Table 3. Comparison of the CK2 signaling to the NM and viability of ALVA-41 cells upon chemical induction of apoptosis

Etoposide (30 μM) at time (h)*	Viable ALVA-41 cells (% of control)	NM-associated CK2 (% of control)
0	100 ± 11	100 ± 10
24	72 ± 7	142 ± 12
48	33 ± 3	261 ± 20
96	9 ± 1	300 ± 22
DES (30 μM) at time (h)*		
0	100 ± 4	100 ± 5
24	88 ± 3	92 ± 6
48	30 ± 3	160 ± 3
96	14 ± 4	181 ± 5

*ALVA-41 cells were treated with etoposide or DES at the concentration indicated for the periods of time indicated. At these times, cell viability and CK2 activity associated with the NM fractions isolated from these cells were determined as described under Materials and methods. Values are compared with the untreated control at each time point. The results are expressed as the mean ± S.E.M.

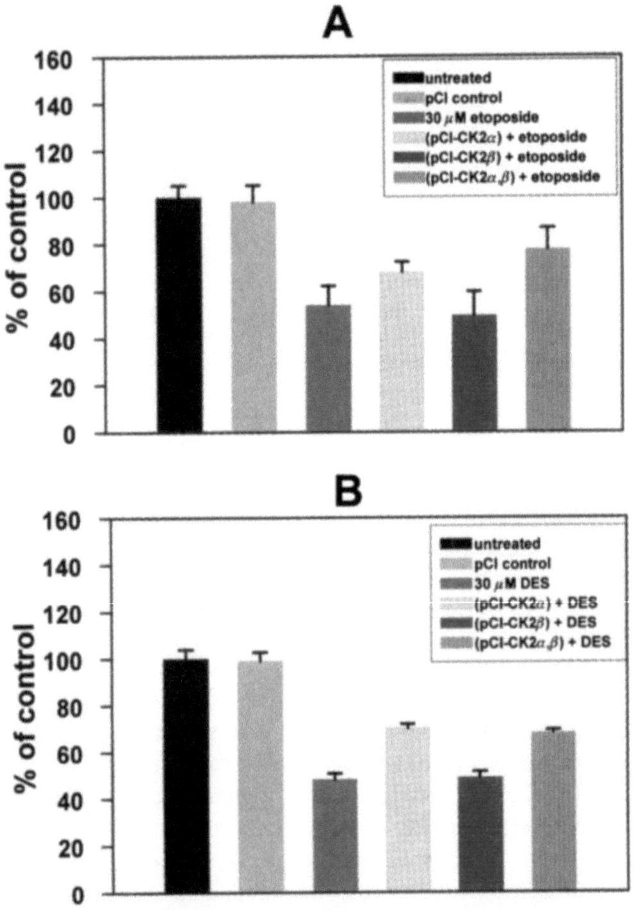

Fig. 1. Protection against cell death on overexpression of CK2. The period of transfection of the cells with various plasmids was 24 h with an estimated transfection efficiency of 50–60%. Following the transfection, PC-3 cells (Panel A) were treated with 30 μM etoposide, and ALVA-41 cells (Panel B) with 30 μM DES. After an additional period of 48 h, cell viability was determined as described under Materials and methods. The control cells were transfected with the pCI vector alone. The results are expressed as percent of control ± S.E.M.

Discussion

The evidence summarized above underscores the role of NM shuttling of CK2 in inducing cell growth and proliferation. In addition, the results also provide a strong support for a novel functional role of the NM-associated CK2, namely that presence of CK2 in the NM serves to act as a survival signal protecting cells against cell death (via apoptosis). This is evident from the experiments involving the removal of cell survival signals that lead to loss of CK2 from the nuclear compartment which is associated with appearance of apoptosis. Provision of the survival or growth signal (hormone or growth factors for prostate cells) reverses these events. Likewise, the shuttling of the CK2 to the NM in response to the chemical toxic agents that are known to induce apoptosis can now be interpreted in terms of a protective response instituted by the cells. This is reinforced from the experiments in which CK2 was overexpressed prior to treatment with the toxic agents. Thus, despite the apparently diverse results, the dynamics of CK2 in all these cases can be interpreted in terms of the growth, proliferative, and protective role of CK2 association with the NM whereas the loss of CK2 from this fraction would be concordant with induction of apoptosis. These views are schematically summarized in Scheme 1.

Scheme 1. Functional consequences of CK2 dynamics in the nuclear matrix.

CK2 dynamics		Associated cellular response
High CK2 in NM	- - - - -	• Stimulation of cell growth • Protection against apoptosis
Low CK2 in NM	- - - - -	• Cessation of cell growth • Induction of apoptosis

In summary, our studies lead to the proposal that NM signaling of CK2 has diverse functional consequences. The shuttling of CK2 to the NM is associated with its role in cell growth and proliferation. Additionally, high CK2 in the NM is associated with resistance of cells to apoptosis. In regard to the latter, it is significant that cells evoke a shuttling of the CK2 to the nuclear compartment when confronted with stress [6, 16]. Conversely, the loss of CK2 from the NM is linked to cessation of cell growth and induction of apoptosis. Studies are now under way to delineate the site(s) of CK2-mediated protective signaling in the apoptotic pathways. Identification of this key new feature of the biological role of CK2 in the cell as an inhibitor of apoptosis is of considerable significance in relation to its role in neoplasia [5, 6]. Tumor cells generally demonstrate altered apoptotic activity. Since CK2 has been found to be invariably elevated in these cells, it is likely that it provides a dual functional role under these conditions, namely stimulation of cell growth and proliferation as well as prevention of apoptosis.

Acknowledgments

This work was supported in part by United States Public Health Service Research Grant CA-15062 awarded by the National Cancer Institute, Department of Health and Human Services, and in part by the Medical Research Fund of the United States Department of Veterans Affairs.

References

1. Guerra B, Issinger O-G: Protein kinase CK2 and its role in cellular proliferation, development and pathology. Electrophoresis 20: 391–408, 1999

2. Ahmed K: Significance of the casein kinase system in cell growth and proliferation with emphasis on studies of the androgenic regulation of the prostate. Cell Mol Biol Res 40: 1–11, 1994
3. Ahmed K: Nuclear matrix and protein kinase CK2 signaling. Crit Rev Eukary Gene Exp 9: 329–336, 1999
4. Ahmed K, Davis AT, Wang H, Faust RA, Yu S, Tawfic S: Significance of protein kinase CK2 nuclear signaling in neoplasia. J Cell Biochem (suppl 35): 130–135, 2000
5. Tawfic S, Yu S, Wang H, Faust R, Davis A, Ahmed K: Protein kinase CK2 signal in neoplasia. Histol Histopathol 16: 573–582, 2001
6. Guo C, Yu S, Davis AT, Wang H, Green JE, Ahmed K: A potential role of nuclear matrix-associated protein kinase CK2 in protection against drug-induced apoptosis in cancer cells. J Biol Chem 276: 5992–5999, 2001
7. Tawfic S, Ahmed K: Growth-stimulus mediated differential translocation of casein kinase 2 to the nuclear matrix: Evidence based on androgen action in the prostate gland. J Biol Chem 269: 24615–24620, 1994
8. Tawfic S, Davis AT, Faust RA, Gapany M, Ahmed K: Association of protein kinase CK2 with nuclear matrix: Influence of method of preparation of nuclear matrix. J Cell Biochem 64: 499–504, 1997
9. Guo C, Davis AT, Yu S, Tawfic S, Ahmed K: Role of protein kinase CK2 in phosphorylation of nucleosomal proteins in relation to transcriptional activity. Mol Cell Biochem 191: 135–142, 1999
10. Yu S, Davis AT, Guo C, Green JE, Ahmed K: Differential targeting of protein kinase CK2 to the nuclear matrix upon transient overexpression of its subunits. J Cell Biochem 74: 127–134, 1998
11. Guo C, Yu S, Davis AT, Ahmed K: Nuclear matrix targeting of the protein kinase CK2 signal as a common downstream response to androgen or growth factor stimulation of prostate cancer cells. Cancer Res 59: 1146–1151, 1999
12. English HF, Santen RJ, Isaacs JT: Response of glandular versus basal rat ventral prostatic epithelial cells to androgen withdrawal and replacement. Prostate 11: 229–242, 1987
13. Berges RR, Furuya Y, Remington L, English HF, Jacks T, Isaacs JT: Cell proliferation, DNA repair, and p53 function are not required for programmed cell death of prostatic glandular cells induced by androgen ablation. Proc Natl Acad Sci USA 90: 8910–8914, 1993
14. Furuya Y, Lin XS, Walsh JC, Nelson WG, Isaacs JT: Androgen ablation-induced programmed death of prostatic glandular cells does not involve recruitment into a defective cell cycle or p53 induction. Endocrinology 136: 1898–1906, 1995
15. Ahmed K, Yenice S, Davis A, Goueli S: Association of casein kinase 2 with nuclear chromatin in relation to androgenic regulation of rat prostate. Proc Natl Acad Sci USA 90: 4426–4430, 1993
16. Gerber DA, Souquere-Besse S, Puvion F, Dubois M-F, Bensaude O, Cochet C: Heat-inducible relocalization of protein kinase CK2. J Biol Chem 275: 23919–23926, 2000

Molecular and Cellular Biochemistry **227**: 73–80, 2001.

Localization of individual subunits of protein kinase CK2 to the endoplasmic reticulum and to the Golgi apparatus

Michael Faust, Martin Jung, Jürgen Günther, Richard Zimmermann and Mathias Montenarh

Medical Biochemistry and Molecular Biology, University of the Saarland, Homburg/Saar, Germany

Abstract

The protein kinase CK2 is composed of two catalytic α- or α′- and two regulatory β-subunits. In mammalian cells there is ample evidence for the presence of individual CK2 subunits beside the holoenzyme. By immunofluorescence studies using peptide antibodies which allow us to detect the CK2α-, α′- and β-subunits we found all three subunits to be co-localized with a 58 KDa Golgi protein which is specific for the Golgi complex. Subfractionation studies using dog pancreas cells revealed the presence of all three subunits of CK2 at the smooth endoplasmic reticulum (sER)/Golgi fraction whereas the rough endoplasmic reticulum (rER) harboured only the catalytic α- and α′-subunits. We found that the microsomal preparation from dog pancreas cells contained CK2 which phosphorylated a CK2 specific synthetic peptide and which was heparin sensitive. Furthermore, we could immunoprecipitate the CK2α-subunit that exhibited a kinase activity which phosphorylated a CK2 specific substrate and which was heparin sensitive. Protease digestion experiments revealed that the CK2 subunits were located on the cytosolic side of the rER and the sER/Golgi complex. Thus, we could demonstrate an asymmetric distribution of the CK2 subunits at the rER and sER/Golgi complex. Since the CK2α- and α′-subunits exhibit a substrate specificity which is different from the CK2 holoenzyme one might speculate that the asymmetric distribution of the CK2 holoenzyme and the CK2 catalytic subunits may have regulatory functions. (Mol Cell Biochem **227**: 73–80, 2001)

Key words: protein kinase, phosphorylation, protein kinase CK2, endoplasmic reticulum, Golgi apparatus

Introduction

Protein kinase CK2 is a ubiquitously expressed serine and threonine kinase. In many organisms the main form of protein kinase CK2 is a heterotetramer composed of two catalytic α- or α′-subunits and two regulatory β-subunits (for review see [1, 2]). In the holoenzyme the β-subunit of CK2 seems to play at least a trifunctional role, (i) it confers stability to the holoenzyme [3], (ii) it increases the enzyme activity [4] and (iii) it determines substrate specificity [3, 5]. The catalytic subunits harbour the ATP and GTP binding sites [6]. The catalytic activity is not only regulated by the β-subunit but also by polyanions like heparin [7, 8]. There is now accumulating evidence for individual subunits of CK2 free from the other subunits. In mouse brain and testicles a CK2β-subunit is found which is not complexed to the α- or α′-subunit [9]. Also the catalytic α- and α′-subunits were found free of the regulatory β-subunit [10]. The functions of the subunits beside their roles in the holoenzyme are not well understood. The β-subunit of CK2 inhibits the DNA binding activity [11] and the transactivation functions [12] of the growth suppressor p53.

Protein kinase CK2 was found almost everywhere in the eukaryotic cell (for review see [13]). In some cases the subcellular location was determined by analysing subfractions of the cell for CK2 activity, in other cases also an immunological analysis was performed and the distribution was determined by immunofluorescence. In most of the early studies the authors used antibodies directed against the holoenzyme

Address for offprints: M. Montenarh, Medical Biochemistry and Molecular Biology, University of the Saarland, Building 44, D-66424 Homburg/Saar, Germany (E-mail: tm13mm@rz.uni-sb.de)

or against the α-subunit. Since the α- and the α′-subunits share nearly all of the amino acids with the exception of some small areas in the C-terminus of both proteins these early studies could not discriminate between α- and α′-subunits. Furthermore, the subcellular localization seems to be dynamic since with progress in the cell cycle CK2 becomes enriched in the nucleus [14].

Protein kinase CK2 has also been identified as an endoplasmic reticulum associated protein kinase [15] by virtue of the phosphorylation of its substrate calnexin, which is associated with the endoplasmic reticulum [16]. Furthermore, it was shown that the calnexin cytosolic domain could be phosphorylated in an *in vitro* kinase reaction by the CK2 holoenzyme as well as by the CK2α-subunit alone [17]. By using anti peptide antibodies which react with the unique regions of the CK2α- or CK2α′- or CK2β-subunit [10] we now attempted to analyse the subcellular location of the individual subunits of CK2. We found all three subunits co-localized with the Golgi complex. We could also demonstrate biochemically the association of all three subunits of CK2 with the smooth endoplasmic reticulum/Golgi complex. In contrast, the rough endoplasmic reticulum but not the ribosomes contained only the catalytic α- and α′-subunits of CK2. Thus, this study is a step forward in demonstrating an asymmetric distribution of individual subunits of CK2 within mammalian cells.

Material and methods

Cell culture

HeLa cells and cos-1 cells were maintained in Dulbecco's modified Eagle's medium (DMEM) supplemented with 10% foetal calf serum (FCS). Cells were grown to 75% subconfluence in 10 cm dishes in a 5% CO_2 atmosphere.

Preparation of cell extracts

Cells were harvested, washed 3 times with phosphate buffered saline (PBS), pH 7.4 and resuspended in lysis buffer (100 mM Tris-HCl, pH 7.0, 100 mM NaCl, 0.5% (v/v) NP40, 1% Trasylol). Proteins were extracted for 1 h on ice. Cell debris was eliminated by centrifugation (4°C, 30 min, 13,000 × g).

Cell fractionation

Dog pancreas microsomes were prepared and treated with nuclease and EDTA as described [18]. The absorbance at 280 nm of the final microsomal suspension was 50, as measured in 2% SDS, which corresponds to a protein concentration of 15 mg/ml. The ribosome free sER/Golgi fraction was isolated from the top of a 1.5/1.7/2.1 M sucrose step gradient during the purification of the rER fractions. Ribosomes were isolated by sedimentation (150,000 × g) on a step gradient (1.5/1.7/2.1 M sucrose) from a precleared cell disruption (14,500 × g supernatant) of dog pancreas. Fractions were tested for cell compartment specific proteins by immunoblotting (Ribosomes: L4; Golgi: p58 KDa; rER: SRP-receptor α-subunit, Sec61 α-subunit).

Protease protection assay

Rough microsomes and sER/Golgi apparatus preparations were treated with sucrose, proteinase K (final concentration 170 μg/ml) or proteinase K and 0.01% Triton X-100 at 0°C for 1 h. The reaction was stopped with 7.5 mM phenylmethylsulfonylfluoride (PMSF) and the proteins were separated by SDS-polyacrylamide gel electrophoresis.

Preparation of microsomal extracts

Dog pancreatic rough microsomes were sedimented by centrifugation. The membranes were resuspended at a concentration of 1 eq/μl (according to [19]) in an extraction buffer containing 20 mM HEPES-KOH (pH 7.8), 5 mM magnesium acetate, 5 mM DTT, 10 μg/ml leupeptin, 10 μg/ml antipain, 5 μg/ml chymostatin, 2 μg/ml pepstatin and 15% glycerol and 0.4% digitonin. After solubilization the mixture was sedimented for 40 min at 425,000 × g in a tabletop ultracentrifuge. The supernatant was removed and the pellet resuspended in a buffer containing 20 mM HEPES-KOH (pH 7.8), 400 mM potassium acetate, 12 mM magnesium acetate, 5 mM DTT, 10 mg/ml leupeptin, 10 μg/ml antipain, 5 μg/ml chymostatin, 2 μg/ml pepstatin, 15% glycerol and 4% digitonin. After 10 min on ice the mixture was centrifuged for 1 h at 425,000 × g at 2°C. The supernatant is referred to as the rough microsomal extract.

Antibodies and Western blot analysis

For the detection of protein kinase CK2 we used polyclonal rabbit sera against peptides derived from the α-, α′- and β-subunits [10]. SDS polyacrylamide gel electrophoresis and blotting procedures were described earlier [12]. For the detection of proteins we used the Lumilight system (Roche Diagnostics, Mannheim, Germany).

Immunofluorescence

Cells were grown on cover slips until they were 50–70% confluent. Cells were fixed in 2% formaldehyde in PBS, pH

7.4, for 15 min at 20°C and then washed with PBS, pH 7.4, for 3 × 10 min. Cells were permeabilized with 0.2% Triton X-100 containing 1% normal goat serum (NGS) in PBS for 5 min on ice. Cells were washed again with PBS containing 1% NGS 3 times for 10 min and then incubated with a primary antibody in the appropriate concentration for 1 h at room temperature in a humidified chamber. Cells were washed under the same conditions. Cells were incubated with the secondary antibody (FITC or TRITC conjugated) at room temperature for 1 h in a dark, humidified chamber. Finally, cells were washed again with PBS (4 × 10 min). The cover slips were fixed with a drop of mounting medium and analysed under a fluorescence microscope.

Kinase reaction

Immunoprecipitates from microsomal extracts or rough microsomes with a volume of 20 µl were mixed with 30 µl kinase buffer containing the CK2-specific peptide (0.32 mM peptide RRRDDDSDDD) and $^{32}P\gamma ATP$ and the kinase reaction was performed for 20 min at 37°C. Thirty microliters of the reaction mix were spotted onto Whatman P11 cation exchange paper (Herolab, Wiesloch, Germany), washed 3 times with 10 mM phosphoric acid and once with ethanol. The papers were dried and phosphate incorporation was measured in a scintillation counter.

Immunoprecipitation

For immunoprecipitation we used a polyclonal serum (serum #26) raised against the C-terminus of protein kinase CK2α-subunit. A protein A/G-sepharose mixture was preincubated for 1 h with 50 µl serum #26 and washed 3 times with phosphate buffered saline (PBS), pH 7.4. The microsomal extract was preincubated with a mixture of protein A/G-sepharose (Pharmacia, Freiburg, Germany) to remove non-specifically bound proteins. The supernatant was applied to the preincubated sepharose antibody and incubated for 1 h. The supernatant was removed and the antibody matrix washed 4 times with PBS. The immunoprecipitate was split into two equal aliquots and subjected either to SDS polyacrylamide gel electrophoresis, followed by Western blotting or to a CK2 protein kinase assay.

Results

In a variety of eukaryotic cells protein kinase CK2 was found in almost every compartment. In some cases CK2 was identified by virtue of its kinase activity and the sensitivity towards heparin or polyamines. Most of these locations were verified by immunofluorescence analysis using antibodies against CK2α- or β-subunits (for review see: [13]). Recently, we described anti peptide antibodies which allowed us to discriminate between CK2α- or α′-subunits and a peptide antibody directed against the β-subunit [10]. By using these antibodies we analysed a variety of eukaryotic cells including cos-1 cells for the subcellular distribution of the individual subunits of CK2. Cells were fixed with formaldehyde and then incubated with serum #26, #30 or serum #32 followed by TRITC-labelled goat anti rabbit antibodies. Cells were also incubated with an anti 58 KDa Golgi protein antibody which recognizes a marker protein of the Golgi apparatus (Sigma, Deisenhofen, Germany) followed by FITC labelled goat anti mouse antibodies. Cells were analysed under a fluorescence microscope. As shown in Fig. 1 all three CK2 specific antibodies stained mainly the nucleus. In addition, there is also a strong signal outside of the nucleus. The same structure was also labelled with the antibody specific for the Golgi complex. By merging the green and red immunofluorescence signals we got a yellow signal associated with the Golgi complex indicating that all three proteins are associated with the Golgi complex.

Next, we decided to analyse the association of the CK2 subunits with microsomal subfractions biochemically. For this type of analysis we chose a microsomal sub-fractionation scheme that is well characterized for dog and swine pancreas cells [18]. First, a cytosolic fraction, sER/Golgi complexes as well as a rER fraction were separated on an SDS polyacrylamide gel, proteins were transferred to a PVDF membrane and the membrane was blotted with the antibody directed against the 58 KDa Golgi specific protein. As shown in Fig. 2A there was a weak signal for the 58 KDa protein in the cytosolic fraction and a strong protein band in the sER/Golgi fraction. As expected there was no signal in the rER fraction. Thus, our preparation of the rER was not contaminated by this Golgi protein. Another aliquot of the same preparation was analysed also by SDS polyacrylamide gel electrophoresis followed by Western blot analysis using the antibodies directed against the CK2 subunits. As shown in Fig. 2B the sER/Golgi fraction contained all three subunits of CK2, which is in agreement with the results obtained by immunofluorescence.

Next, we wanted to analyse rough microsomes for the presence of individual subunits of CK2. In order to verify that we are working with CK2 we also prepared a HeLa cell extract. The rough microsomal fraction as well as the HeLa cell extract were analysed on an SDS polyacrylamide gel followed by Western blot analysis starting with antibody #26 directed against a peptide from the unique C-terminal region of the α-subunit followed by antibody #30 directed against a unique peptide derived from the C-terminus of the α′-subunit. Finally, the blot was incubated with serum #32 directed

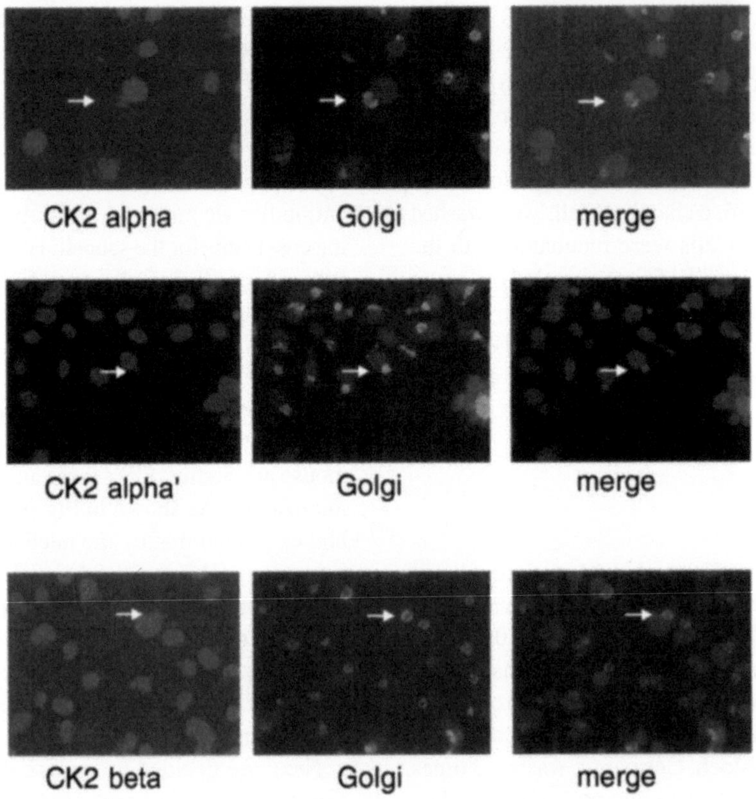

Fig. 1. Indirect immunofluorescence for the detection of individual subunits of CK2. Cos-1 cells were grown on cover slips, fixed with 3.7% formaldehyde and permeabilised with 0.2% Triton X-100. The rabbit sera against the subunits of CK2 were used in a dilution of 1:100, the mouse monoclonal anti 58 KDa Golgi protein antibody was diluted 1:50. As secondary antibodies, either TRITC conjugated goat anti-rabbit or FITC-conjugated goat anti-mouse-antibodies were used. The arrows point to the Golgi apparatus in cos-1 cells. Magnification is 400×.

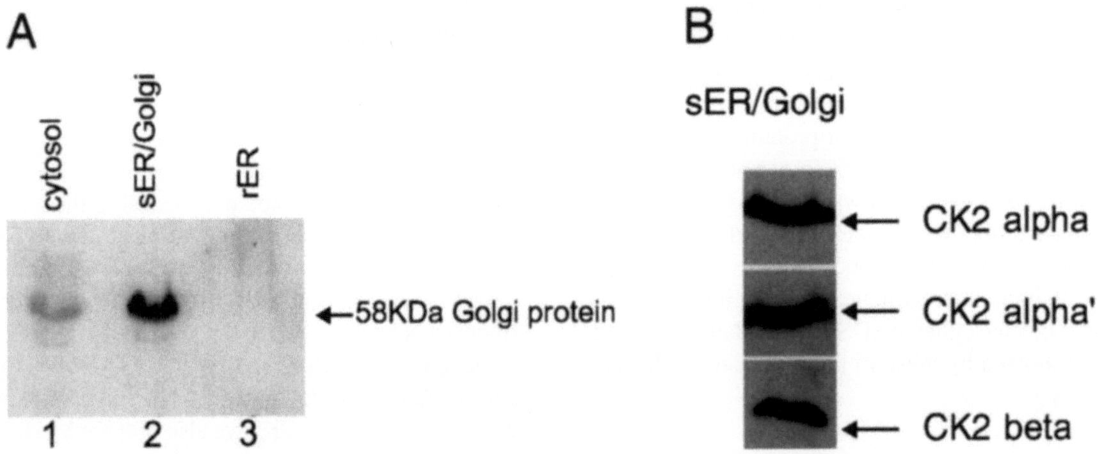

Fig. 2. (A) Detection of the 58 KDa Golgi protein in preparations of dog cytosol (1), sER/Golgi (2) and rER (3). The three different preparations were separated in an SDS polyacrylamide gel, transferred onto a PVDF-membrane and probed with an anti 58 KDa Golgi protein antibody. Finally, proteins were detected by the enhanced chemiluminescence. (B) Detection of the individual subunits of protein kinase CK2 in sER/Golgi from dog pancreas cells. Dog sER/Golgi preparations were separated through an SDS polyacrylamide gel, transferred onto a PVDF-membrane and probed with sera against the α-, α'- and β-subunits (serum #26, #30 and #32) of protein kinase CK2. Proteins were detected by the Lumilight system.

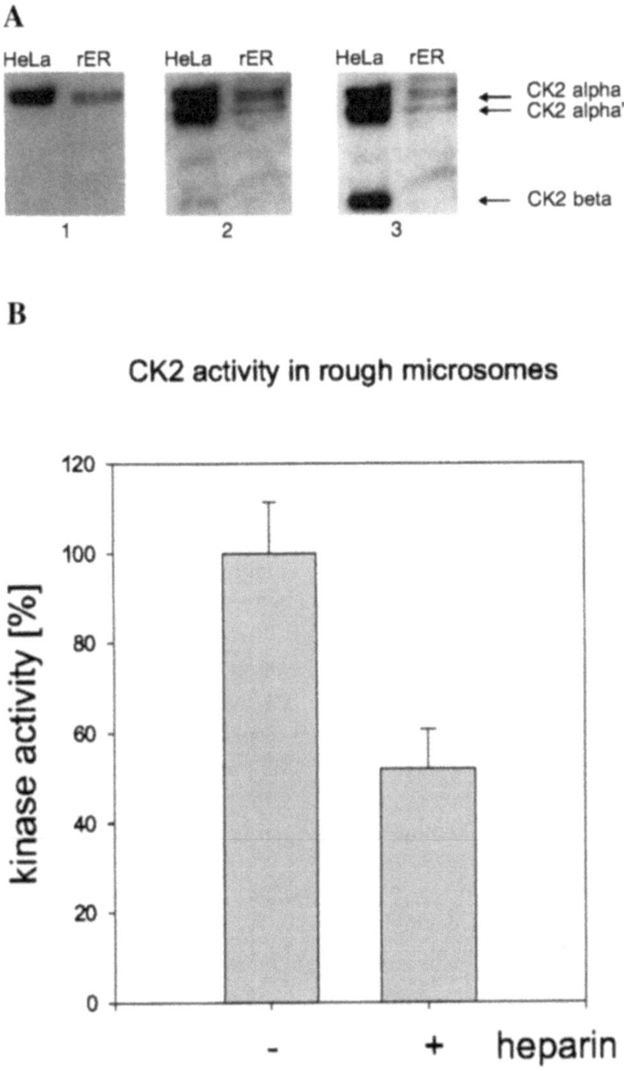

Fig. 3. Detection of the individual subunits of protein kinase CK2 in dog rER and extracts from HeLa cells. (A) HeLa cell lysates as a control and rER from dog pancreas were separated by SDS polyacrylamide gel electrophoresis and blotted onto a PVDF membrane. CK2 was detected with serum #26 against the α- (1), #30 against the α′- (2) and serum #32 against the β-subunit (3). The decoration of the PVDF-membrane with the different sera was successively carried out and proteins were visualized with the Lumilight system. (B) Phosphorylation of the synthetic peptide RRRDDDSDDD by CK2 in the absence or presence of heparin. Extracts of dog rER were mixed with kinase buffer containing the synthetic substrate. ^{32}PγATP was added and a kinase reaction was performed in the absence or presence of the CK2 specific inhibitor heparin. Incorporation of radioactive phosphate was measured in a scintillation counter. The results from three independent experiments are shown (error bars represent S.D.).

against the β-subunit [10]. As shown in Fig. 3A there is a strong signal for CK2α present in HeLa cells and a weaker band in the dog rough microsomal fraction. The same ratio was found when the membrane was decorated with the antibody directed against the α′-subunit. Using the antibody against the β-subunit revealed only a protein band for the β-subunit in the HeLa cell extract but not in the rER fraction. Thus, these results demonstrated that the CK2α- and α′-subunits comigrate with the same subunits from HeLa cells. Furthermore, in contrast to the sER/Golgi fraction the rER fraction only harboured the catalytic and not the regulatory subunits of CK2.

In order to demonstrate not only that at least the catalytic α- and α′-subunits can be detected in rER preparations we also wondered whether we could detect a kinase activity in our microsomal preparations. Therefore, dog rough microsomal extracts were incubated with the commonly used substrate peptide for CK2 [20] and ^{32}PγATP in the absence or in the presence of heparin. As shown in Fig. 3B the microsomal fraction contains a strong CK2 activity, which is inhibited by heparin indicating that the observed kinase activity is most likely CK2. Together with the data shown in Fig. 3A one has to conclude that this kinase activity is due to the catalytic α- and α′-subunits and not to the holoenzyme.

In the next step we asked whether the observed presence of only the α- and α′-subunits of CK2 at the rER might be restricted to dog rER. Therefore, we performed the same type of analysis with swine rER and dog rER in parallel. In this case the rER preparations were divided into three equal aliquots and analysed on an SDS polyacrylamide gel. Proteins were blotted onto PVDF membranes and the membranes were either incubated with serum #26, serum #30 or serum #32. As shown in Fig. 4 we got essentially the same result for dog and swine rER namely an association of the α- and α′-subunits but not the β-subunit with the rER.

Next, we solubilized the rER fraction in order to be able to immunoprecipitate CK2. The immunoprecipitate was divided into three equal aliquots. One aliquot was analysed on an SDS polyacrylamide gel, two other aliquots were used for

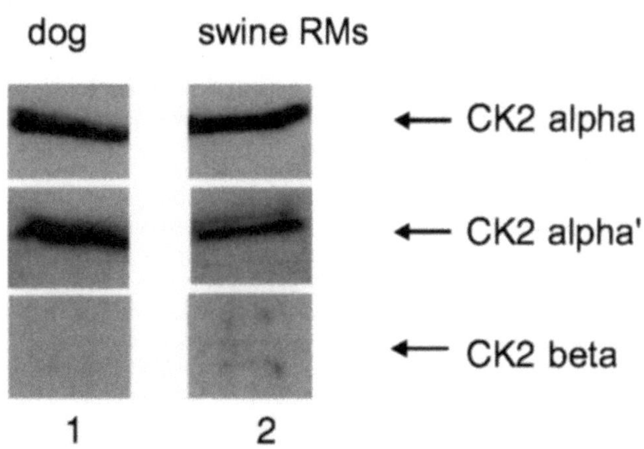

Fig. 4. Detection of the individual subunits of protein kinase CK2 in extracts of dog and swine rER. Extracts from dog (1) and swine (2) rER were separated in an SDS polyacrylamide gel, transferred onto a PVDF-membrane and probed with sera against the α-, α′- and β-subunits (serum #26, #30 and #32) of protein kinase CK2. Proteins were detected by the Lumilight system.

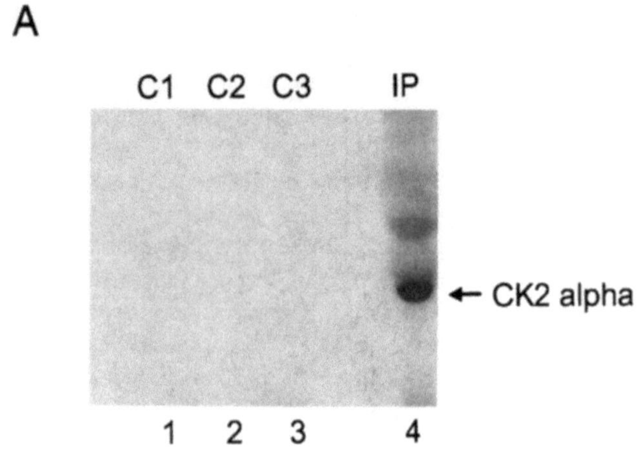

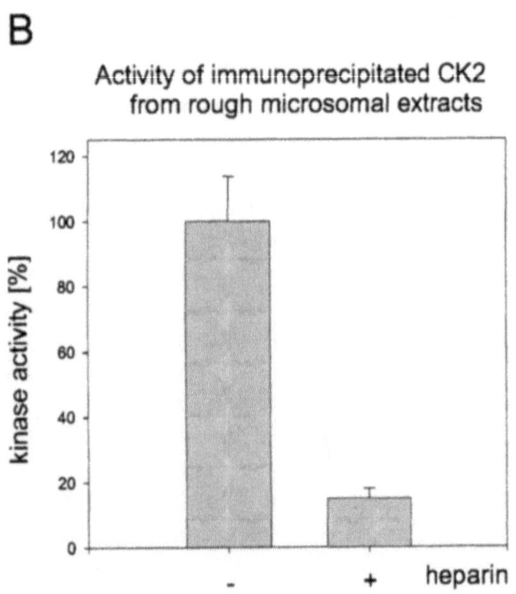

Fig. 5. (A) Immunoprecipitation of CK2 from rough microsomal extracts. For immunoprecipitation we used a polyclonal serum (serum #26) raised against the C-terminus of protein kinase CK2α-subunit. Serum #26 (1), protein A/G sepharose (2), precipitates without antibodies (3) as controls and the immunoprecipitates (4) were separated in an SDS polyacrylamide gel, transferred onto a PVDF-membrane and probed with a monoclonal anti-CK2α antibody (1A5) and proteins were detected by the Lumilight system. (B) Phosphorylation of the synthetic peptide RRRDDDSDDD by immunoprecipitated CK2 in the absence or presence of heparin. The immunoprecipitates from rough microsomal extracts were mixed with kinase buffer containing the synthetic substrate. ^{32}PγATP was added and a kinase reaction in the absence or presence of the CK2 specific inhibitor heparin was performed. Incorporation of radioactive phosphate was measured in a scintillation counter. The results from three independent experiments are shown (error bars represent S.D.).

kinase reactions in the presence or absence of heparin using the synthetic peptide as a substrate. As shown in Fig. 5A antibody #26 precipitated CK2α, which was detected on the PVDF membrane using the CK2α-specific monoclonal antibody 1A5. Lanes C1–C3 were antibody controls to exclude a cross reactivity of the 1A5 antibody with the immunoglobulins of serum #26. Fig. 5B shows the results of the kinase reaction demonstrating a phosphorylation of the synthetic substrate peptide, which is efficiently inhibited by heparin. Thus, we could clearly demonstrate an association of the CK2α-subunit with the rER, which is enzymatically active.

Since the rER contains ribosomes it might be that the α- and α′-subunits of CK2 are associated with the ribosomes and not with the ER. Therefore ribosomes were analysed for the content of CK2. As a control we used the sER/Golgi fractions. Proteins were extracted and analysed on an SDS polyacrylamide gel followed by Western blot analysis either with a mixture of serum #26 and #30 or with serum #32. As shown in Fig. 6 none of the CK2 subunits could be detected associated with ribosomes whereas in the same experiment and on the same gel we clearly detected all three subunits of CK2 associated with sER and Golgi complex.

Next, we asked whether the CK2 subunits are resident at the luminal side or the cytosolic side of the ER. rER and sER/Golgi fractions were prepared exactly as described above. Samples were either taken without any further treatment or incubated with proteinase K that would digest proteins from the cytosolic side of the ER or with proteinase K and Triton X-100 that would totally solubilize the ER fraction. As shown in Fig. 7 the untreated rER only contained the two catalytic subunits as already shown above whereas the final solubilization of the ER fraction did not give a signal for CK2. As expected the sER/Golgi complex contained all three CK2 subunits. After treatment with proteinase K the CK2 subunits were no longer associated with the sER/Golgi complex. The final solubilization of the sER/Golgi fraction did not give a

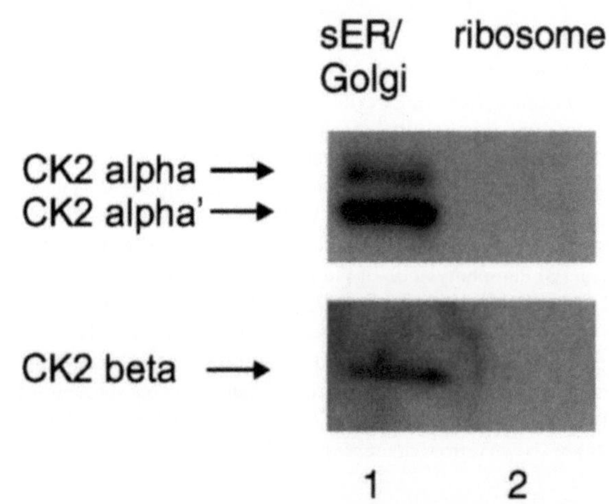

Fig. 6. Detection of the individual subunits of protein kinase CK2 in sER and ribosomes. Dog sER/Golgi apparatus fractions (1) and extracts of dog ribosomes (2) were separated in an SDS polyacrylamide gel, transferred onto a PVDF-membrane and probed with sera against the α-, α′- and β-subunits (serum #26, #30 and #32) of protein kinase CK2 and proteins were detected by the Lumilight system.

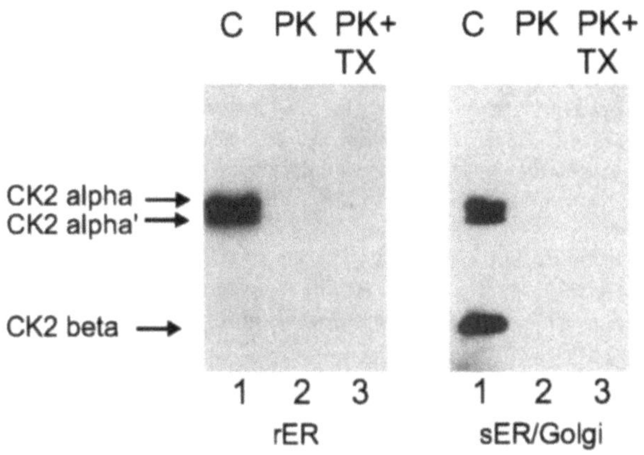

Fig. 7. Protease protection assay with rough microsomes and sER/Golgi preparations. Rough microsomes and sER/Golgi preparations were treated with sucrose (1), proteinase K (final concentration 170 μg/ml) (2) or proteinase K and 0.01% Triton X-100 (3) at 0°C for 1 h. The reaction was stopped and the proteins were separated by SDS polyacrylamide gel electrophoresis and transferred onto a PVDF-membrane. The decoration was performed with sera against the α-, α′- and β-subunits (serum #26, #30 and #32) of protein kinase CK2 and proteins were detected by the Lumilight system.

signal for the CK2 subunits, too. These results indicated that the catalytic α- and α′-subunits are located on the cytosolic side of the rER and also the α-, α′- and β-subunits are located on the cytosolic side of the sER/Golgi complex.

Discussion

Protein kinase CK2 was located by virtue of its enzymatic activity, by immunofluorescence and in some cases biochemically almost everywhere in the eukaryotic cell. Since there is increasing evidence for the presence of individual subunits of CK2 namely the regulatory β-subunit and the two catalytic α- and α′-subunits outside of the holoenzyme we started to analyse the subcellular localization of CK2 subunits with antibodies which allowed us to discriminate between the α- and the α′-subunits as well as antibodies specific for the β-subunits. These antibodies were previously used to detect CK2α- and α′-subunits binding to α-tubulin [10]. Moreover, these antibodies were used to immunoprecipitate an active holoenzyme as well as active α- or α′-subunits from mammalian cells [21].

By immunofluorescence studies with these antibodies we found all three subunits mainly in the nucleus. In addition we found CK2 associated with other cellular compartments or substructures. Here we demonstrate an association of all three subunits of CK2 with the Golgi complex as shown by co-localization with a Golgi specific 58 KDa protein. This staining is different from the one reported recently where CK2α was co-staining with the centrosomes [14]. In order to verify the association of the CK2 subunits with the Golgi complex and to discriminate between Golgi complex and rER we analysed biochemically the association of the individual subunits with these structures. For this purpose we used an established subfractionation scheme for dog and swine pancreas cells [18]. We could clearly demonstrate the association of all three subunits of CK2 with the sER/Golgi complexes confirming our immunofluorescence results. In contrast to the sER/Golgi complex the rER preparations contained only the catalytic α- and α′-subunits but no β-subunits.

The amino acid sequence of CK2 is highly conserved in evolution and our peptide antibodies were generated against peptides which are present in the CK2 subunits in almost all mammalian species [2, 22]. Therefore, it was not surprising that these antibodies recognized the CK2 subunits from dog, swine, monkey and human cells in a Western blot and in immunoprecipitation analysis. Moreover, CK2 from dog microsomal fractions as well as from solubilized ER fractions behave like human CK2, i.e. it phosphorylated a CK2 specific synthetic peptide [20, 23–25] and the kinase was clearly heparin sensitive [7]. Moreover, our results are not limited to dog microsomal fractions because the same results were obtained for swine CK2.

Protein kinase CK2 is a remarkable kinase because so far around 200 substrates are known [26]. Among these substrates are also proteins, which are tightly associated with the endoplasmic reticulum or with the Golgi complex. Digestion experiments demonstrated that the CK2 subunits are located at the cytosolic side of the ER or Golgi complex. This observation is in agreement with earlier results showing that calnexin which was initially identified as an endoplasmic reticulum type I integral membrane protein was phosphorylated on its cytosolic domain by protein kinase CK2 [15]. Another ER associated protein which is phosphorylated by CK2 is the signal sequence receptor α (SSRα/TRAP complex subunit α) protein which acts similar to calnexin as a chaperone for the multimeric complexes consisting of the T-cell receptor, the major histocompatibility complex class I heavy chains and membrane Ig [15]. In the case of calnexin it is known that its phosphorylation by CK2 enhances the association of the protein with ribosomes.

The present study is a big step forward to demonstrate that individual subunits of CK2 exhibit subcellular locations, which are distinct from their localization in the holoenzyme. Evidence was already presented earlier that the β-subunit of CK2 is present in the absence of CK2α in mouse brain and testes [9]. Moreover, asymmetric expression of protein kinase CK2 in favour of the β-subunit was demonstrated in human kidney tumours [27]. On the other hand, fractionated ammonium sulphate precipitation of proteins from a total cell extract we were able to demonstrate CK2α, which is not in complex with CK2β (data not shown). We also demonstrated that the α-subunit from the rER had a kinase activity. Since the α-subunit of CK2 has another substrate specificity than the CK2 holoenzyme [28, 29] it is reasonable to speculate that the asymmetric expression of the α-subunit and the holoenzyme might have a regulatory function.

Acknowledgements

This work was supported by Deutsche Forschungsgemeinschaft grant Mo309/11-3 and SFB 399, B4 and B8. Michael Faust is a supported member of the Graduiertenkolleg 'Zelluläre Regulation und Wachstum' at the Medical Faculty of the University of the Saarland. We are very grateful to Marianne Buchholz for excellent secretarial help. We thank Claudia Götz and Johanna Dudek for many helpful discussions and suggestions.

References

1. Pinna LA: Protein kinase CK2. Int J Biochem Cell Biol 29: 551–554, 1997
2. Allende JE, Allende CC: Protein kinase CK2: An enzyme with multiple substrates and puzzling regulation. FASEB J 9: 313–323, 1995
3. Meggio F, Boldyreff BS, Marin O, Pinna LA, Issinger O-G: CK2: Role of the β-subunit on the stability and specificity of the recombinant reconstituted holoenzyme. Eur J Biochem 204: 293–297, 1992
4. Grankowski N, Boldyreff B, Issinger O-G: Isolation and characterization of recombinant human casein kinase II subunits α and β from bacteria. Eur J Biochem 198: 25–30, 1991
5. Bidwai AP, Hanna DE, Glover CV: Purification and characterization of casein kinase II (CKII) from delta cka1 delta cka2 *Saccharomyces cerevisiae* rescued by Drosophila CKII subunits. The free catalytic subunit of casein kinase II is not toxic *in vivo*. J Biol Chem 267: 18790–18796, 1992
6. Niefind K, Pütter M, Guerra B, Issinger O-G, Schomburg D: CTP plus water mimic ATP in the active site of protein kinase CK2. Nature Struct Biol 6: 1100–1103, 1999
7. Hathaway GM, Lubben TH, Traugh JA: Inhibition of casein kinase II by heparin. J Biol Chem 255: 8038–8041, 1980
8. O'Farrell F, Loog M, Janson IM, Ek P: Kinetic study of the inhibition of CK2 by heparin fragments of different length. Biochim Biophys Acta Protein Struct Mol Enzymol 1433: 68–75, 1999
9. Guerra B, Siemer S, Boldyreff B, Issinger O-G: Protein kinase CK2: Evidence for a protein kinase CK2β subunit fraction, devoid of the catalytic CK2α subunit, in mouse brain and testicles. FEBS Lett 462: 353–357 1999
10. Faust M, Schuster N, Montenarh M: Specific binding of protein kinase CK2 catalytic subunits to tubulin. FEBS Lett 462: 51–56, 1999
11. Prowald A, Schuster N, Montenarh M: Regulation of the DNA binding of p53 by its interaction with protein kinase CK2. FEBS Lett 408: 99–104, 1997
12. Schuster N, Prowald A, Schneider E, Scheidtmann K-H, Montenarh M: Regulation of p53 mediated transactivation by the β-subunit of protein kinase CK2. FEBS Lett 447: 160–166, 1999
13. Faust M Montenarh M: Subcellular localization of protein kinase CK2: A key to its function? Cell Tissue Res 301: 329–340, 2000
14. McKendrick L, Milne D, Meek D: Protein kinase CK2-dependent regulation of p53 function: Evidence that the phosphorylation status of the serine 386 (CK2) site of p53 is constitutive and stable. Mol Cell Biochem 191: 187–199, 1999
15. Ou W-J, Thomas DY, Bell AW, Bergeron JJM: Casein kinase II phosphorylation of signal sequence receptor α and the associated membrane chaperone calnexin. J Biol Chem 267: 23789–23796, 1992
16. Wada I, Rindress D, Cameron PH, Ou W-J, Doherty JJ II, Louvard D, Bell AW, Dignard D, Thomas DY, Bergeron JJM: SSRa and associated calnexin are major calcium binding proteins of the endoplasmic reticulum membrane. J Biol Chem 266: 19599–19610, 1991
17. Chevet E, Wong HN, Gerber D, Cochet C, Fazel A, Cameron PH, Gushue JN, Thomas DY, Bergeron JJM: Phosphorylation by CK2 and MAPK enhances calnexin association with ribosomes. EMBO J 18: 3655–3666, 1999
18. Watts C, Wickner W, Zimmermann R: M13 procoat and a pre-immunoglobulin share processing specificity but use different membrane receptor mechanisms. Proc Natl Acad Sci USA 80: 2809–2813, 1983
19. Walter P, Ibrahimi I, Blobel G: Translocation of proteins across the endoplasmic reticulum. I. Signal recognition protein (SRP) binds to *in vitro*-assembled polysomes synthesizing secretory protein. J Cell Biol 91: 545–550, 1981
20. Sarno S, Vaglio P, Marin O, Issinger O-G, Ruffato K, Pinna LA: Mutational analysis of residues implicated in the interaction between protein kinase CK2 and peptide substrates. Biochemistry 36: 11717–11724, 1997
21. Schuster N, Götz C, Schneider E, Prowald A, Faust M, Montenarh M: Wild-type p53 inhibits protein kinase CK2 activity. J Cell Biochem 81: 172–181, 2001
22. Boldyreff B, Piontek K, Schmidt-Spaniol I, Issinger OG: The beta subunit of casein kinase II: Cloning of cDNAs from murine and porcine origin and expression of the porcine sequence as a fusion protein. Biochim Biophys Acta 1088: 439–441, 1991
23. Kennelly PJ Krebs EG: Consensus sequences as substrate specificity determinants for protein kinases and protein phosphatases. J Biol Chem 266: 15555–15558, 1991
24. Kuenzel EA, Krebs EG: A synthetic peptide substrate specific for casein kinase II. Proc Natl Acad Sci USA 82: 737–741, 1985
25. Kuenzel EA, Mulligan JA, Sommercorn J, Krebs EG: Substrate specificity determinants for casein kinase II as deduced from studies with synthetic peptides. J Biol Chem 262: 9136–9140, 1987
26. Guerra B Issinger OG: Protein kinase CK2 and its role in cellular proliferation, development and pathology. Electrophoresis 20: 391–408, 1999
27. Stalter G, Siemer S, Becht E, Ziegler M, Remberger K, Issinger O-G: Asymmetric expression of protein kinase CK2 in human kidney tumors. Biochem Biophys Res Commun 202: 141–147, 1994
28. Bidwai AP, Reed JC, Glover CV: Phosphorylation of calmodulin by the catalytic subunit of casein kinase II is inhibited by the regulatory subunit. Arch Biochem Biophys 300: 265–270, 1993
29. Guerra B, Götz C, Wagner P, Montenarh M, Issinger O-G: The carboxy terminus of p53 mimicks the polylysine effect of protein kinase CK2-catalyzed MDM2 phosphorylation. Oncogene 14: 2683–2688, 1997

Molecular and Cellular Biochemistry **227**: 81–90, 2001.

Visualization and molecular analysis of nuclear import of protein kinase CK2 subunits in living cells

Véronique Martel,[1] Odile Filhol,[1] Arsenio Nueda,[2] Delphine Gerber,[1] Maria Jose Benitez[3] and Claude Cochet[1]

[1]*INSERM EMI 0104/BRCE, DBMS, CEA, Grenoble, France;* [2]*Almirall Prodesfarma SA, Centro de Investigation, Barcelona, Spain;* [3]*Departamento de Quimica Fisica Aplicada, Universidad Autonoma de Madrid, Madrid, Spain*

Abstract

We have generated fusion proteins between the subunits of CK2 and GFP and characterized their behaviour in living cells. The expressed fusion proteins were functional and interacted with endogenous CK2. Imaging of NIH3T3 cells expressing low level of GFP-CK2α or GFP-CK2β showed that both proteins were mostly nuclear in interphase. Both CK2 subunits contain nuclear localization domains that target them independently to the nucleus. Once in the nucleus, both subunits diffused rapidly in the nucleoplasm. In mitotic cells, CK2 subunits were dispersed throughout the cytoplasm and were not associated to chromatin. Our data are compatible with the idea that each subunit can translocate individually to the nucleus to interact with each other or with important cellular partners. Understanding the molecular mechanisms which regulate the dynamic localization of CK2 subunits will be of central importance. (Mol Cell Biochem **227**: 81–90, 2001)

Key words: protein kinase CK2, green fluorescent protein, nucleo-cytoplasmic transport, protein mobility

Introduction

Protein kinase CK2 is an ubiquitous serine/threonine protein kinase capable of phosphorylating a wide array of substrates *in vitro* [1]. The enzyme is purified from cell extracts as an $\alpha_2\beta_2$ tetramer where α and β are catalytic and regulatory subunits respectively. Essential for cell viability [2], the function and particularly the regulation of CK2 remain unclear: it is required for normal cell cycle progression [3], for the regulation of embryogenesis in Drosophila [4, 5] and it participates in the terminal differentiation of mouse neuroblastoma cells [6] and in the adaptative response to stress in yeast [7]. Antibody microinjection studies demonstrate a differential requirement for each CK2 subunits during the progression through the cell cycle [8]. CK2 is able to phosphorylate multiple substrates found in various subcellular locations; membrane associated such as the CD5 receptor [9], cytoplasmic such as IkBα [10], Drosophila and mouse disheveled protein [11, 12] or nuclear such as RNA polymerase I and III [13, 14], TBP [15], c-Jun [16], SRF [17], nucleolin [18], Antennapedia [4] and ATF-1 [19].

Many putative substrates of CK2 that are critical to the process of cell growth and proliferation have been identified in the nuclear compartment. Therefore, the subcellular targeting of a multifunctional protein kinase such as CK2 may well represents an important aspect of its cellular regulation. Unfortunately, there is only limited information on its regulation and the results dealing with its subcellular distribution are conflicting. The wide use of immunocytochemistry in these studies may explain some of the discrepancies. Green fluorescent protein (GFP) is an intrinsically fluorescent polypeptide that allows a new approach to study proteins in intact cells [20, 21] and recent studies have revealed that GFP is a good candidate as a molecular reporter protein to monitor protein trafficking in living cells.

To better understand the mechanisms behind the dynamic behaviour of CK2 we have generated fusion proteins between the two subunits of CK2 and GFP and characterized their distribution and movement in living cells.

Address for offprints: C. Cochet, INSERM EMI 0104/BRCE, DBMS, CEA, 17 rue des Martyrs, 38054 Grenoble, France

Materials and methods

Construct of plasmids encoding GFP-CK2 subunits

The cDNA encoding CK2α, and CK2β were sub-cloned from pSG5 CK2α or β vectors [22] into pEGFPc1 vectors (Clontech). GFP-CK2α (1–136) and its point-mutated forms were obtained by deleting a XbaI restriction fragment from pEGFPCK2α vector. Details of the mutation constructs will be described (in preparation).

Transient transfection and generation of mammalian cell lines expressing GFP-CK2 subunits

For cell culture and transfection assays, HeLa or NIH3T3 cells were grown at 37°C with 5% CO_2 in DMEM (Life Technologies), with 10% heat-inactivated FBS. For transient expression, NIH3T3 cells were transfected using Fugene 6 reagent (Roche Molecular Biochemicals) according to the manufacturer's instructions.

To generate GFP-CK2 subunits expressing cell lines, NIH3T3 cells were transfected with linearized cDNA using Lipofectin reagent (Life Technologies) as recommended by the manufacturer. Several clones of resistant cells expressing the transfected fusion proteins were selected under the fluorescence microscope, expanded and further sub-cloned by the limiting dilution method to ensure their clonal nature before subsequent analysis.

Microscopic analysis of cells expressing GFP fusion proteins

For standard fluorescence analysis, cells (5×10^3) were plated on glass coverslips and allowed to grow for 24 h. Cells were washed with PBS, fixed with 4% paraformaldehyde for 30 min at 37°C and incubated for 5 min with a solution containing 1 μg/ml of Hoechst 33258 in PBS to label the nuclei. Fixed cells were finally mounted using 5 μl of standard mounting medium (Dako) examined using a fluorescence microscope (Zeiss Axiophot). The nuclear accumulation of GFP fusion proteins was assayed by scoring 300–500 cells on the coverglass. The results were expressed as a nuclear/cytoplasmic ratio (N/C) as previously described [23].

Velocity sedimentation

Linear 5–25% (w/v) sucrose gradients were prepared in TD buffer: 10 mM Tris-HCl, pH 7.5, 1 mM DTT. For each experiment, samples were incubated at 4°C in 0.1 ml of TD buffer for 2 h prior to loading. Gradients were centrifuged at 4°C for 3.5 h at 200,000 × g and fractionated by pipetting 200 μl aliquots.

Immunoblotting and kinase assay

Proteins were transferred to nitrocellulose membrane by electroblotting in a buffer containing 25 mM Tris base, 192 mM glycine pH 8.0, and 20% ethanol. The membrane was blocked in PBS/0.05% Tween 20/5% nonfat dry milk and sequentially probed with the primary antibody for 2 h at room temperature, washed with PBS containing 0.05% Tween 20, probed with horseradish peroxidase-conjugated secondary antibody, and finally incubated for 1 min with ECL substrate (Amersham Pharmacia Biotech). The membrane was then exposed to photographic film (Kodak). GFP antibodies were from Clontech. Antibodies against CK2 subunits (anti-α and anti-β) were used as previously described [24].

CK2 specific kinase activity was determined as previously described [25].

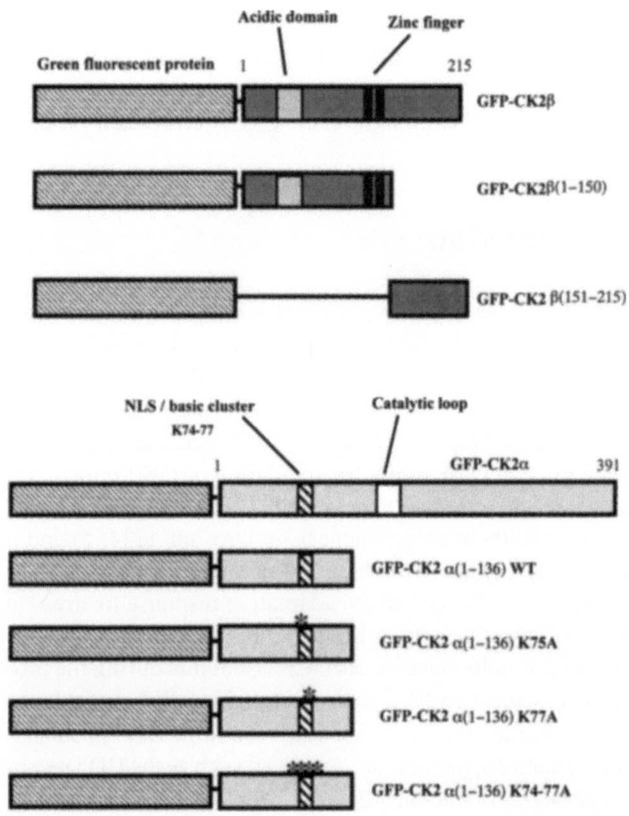

Fig. 1. Constructs used in this work. Both CK2α and CK2β subunits were cloned downstream of green fluorescent protein.

Photobleaching experiments

Fluorescence recovery after photobleaching experiments (FRAP) were performed on a Zeiss confocal microscope (LSM 410), by using the 488 nm laser line and a ax63 oil immersion objective (1.4 numerical aperture). A defined region of the living cells was photobleached at full laser power during 15 sec. Recovery of fluorescence was then monitored by scanning the whole cell. For imaging, the laser power was attenuated to 10% of the bleach intensity. No photobleaching was observed during recovery.

Results

GFP-CK2 fusion proteins are functional

To follow the intracellular localization and dynamics of CK2 we generated transfectants stably expressing chimera between either the catalytic or the regulatory subunits of CK2 and GFP. We fused GFP to the N terminus of CK2α and CK2β (see Material and methods) under the control of the cytomegalovirus (CMV) promoter (Fig. 1). Since the sensitivity of the fluorescence emitted by GFP is very high [26], it is possible to analyze cells expressing relatively low levels of the fusion products. The behaviour of the transfected fusion proteins can then be studied in conditions very close to the physiologic ones, without significantly disrupting the intracellular pool of CK2. The different constructs: GFP, GFP-CK2α and GFP-CK2β were transfected into HeLa or NIH3T3 cells. After transfection, stable pools expressing GFP and GFP-CK2 fusions were established by selection with G418. Flow cytometry-directed sorting was then used to enrich in GFP^+ cells from the transfected pools. Representative clones expressing each of the three proteins were selected.

To characterize the expressed fusion proteins, cellular extracts from NIH3T3 cells were prepared in high salt buffer (1M NaCl) and were analyzed by sucrose gradient ultracentrifugation (Fig. 2). In extract from cells expressing GFP-CK2α, both endogenous CK2 catalytic subunits (α and α′) and GFP-CK2α were peaking in fractions containing the highest CK2 activity (Fig. 2, left panel). The sedimentation of the peak of activity was 7 S, a value expected for the canonical CK2 holoenzyme. The same analysis performed on extract of NIH3T3 cells expressing GFP-CK2β showed a

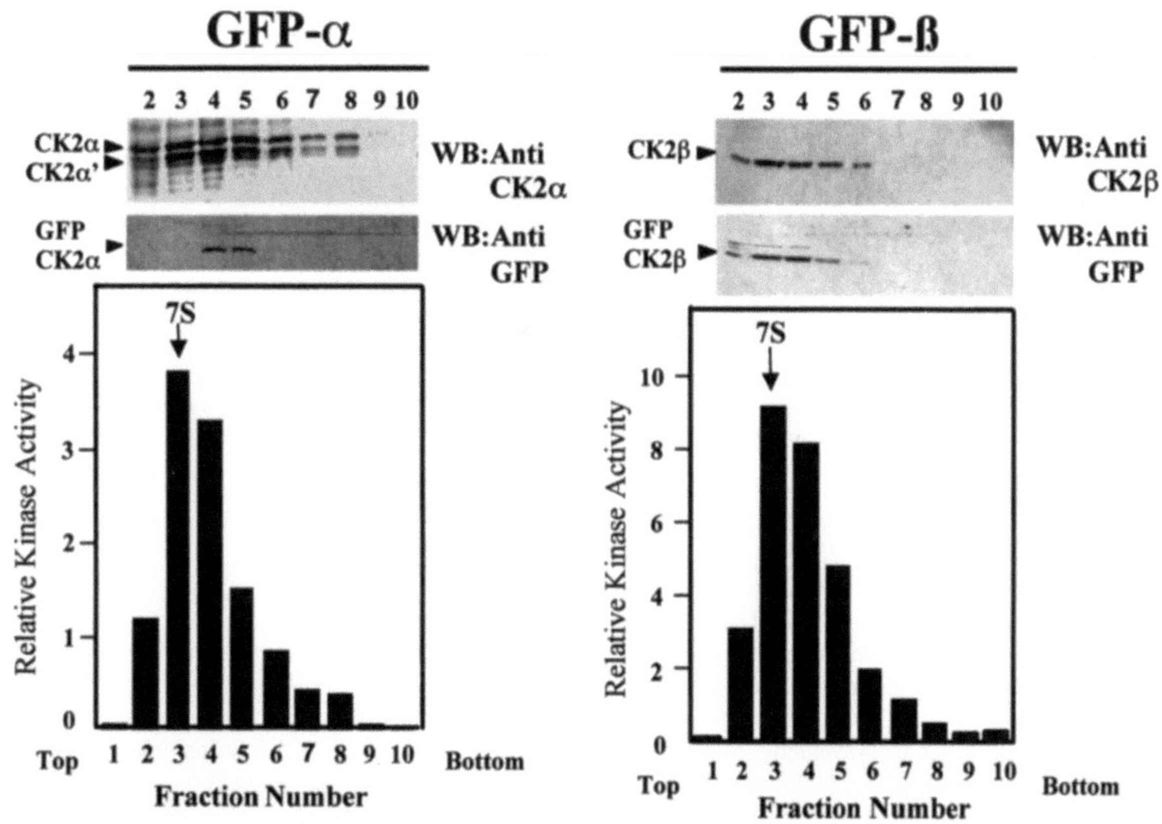

Fig. 2. GFP-CK2 subunits interact with endogenous CK2. Extracts from NIH3T3 cells stably expressing GFP-CK2α or GFP-CK2β subunits were analyzed by ultracentrifugation on sucrose gradients as described in Materials and methods. Collected fractions were tested for kinase activity and analyzed by immunoblot using the indicated antibodies.

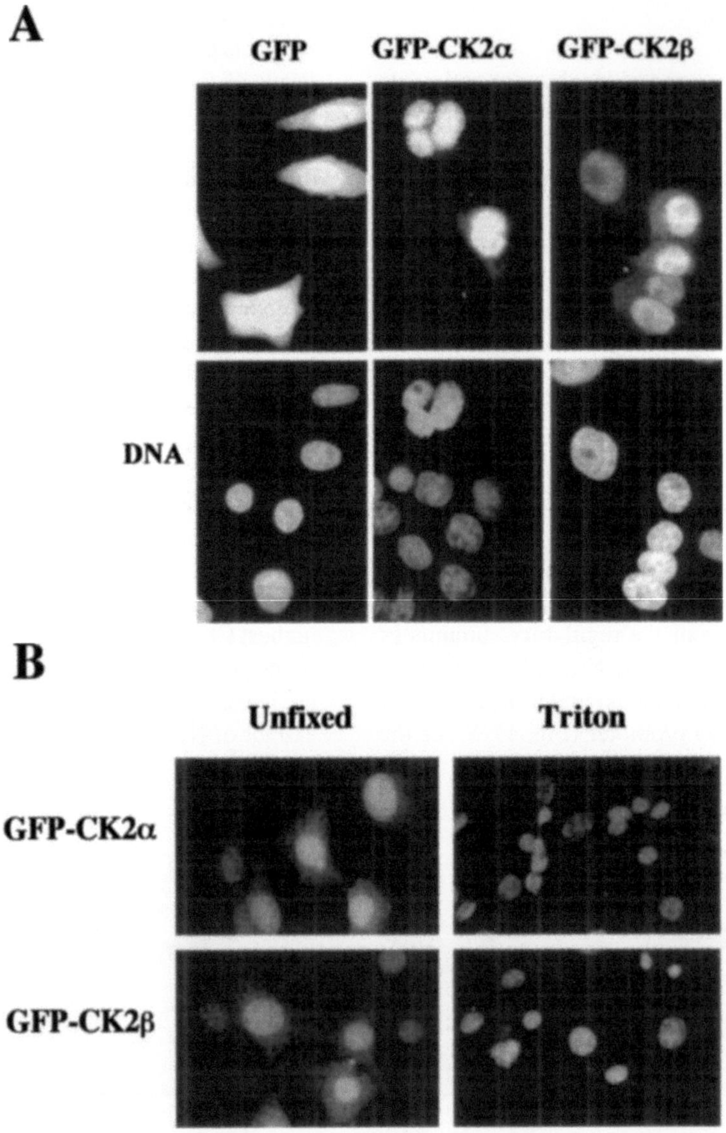

Fig. 3. Fluorescence analysis of cells in interphase. (A) HeLa cells transiently transfected with GFP, GFP-CK2α or GFP-CK2β were analyzed using standard fluorescence microscopy. Cells were stained for DNA with Hoechst. (B) Stably transfected NIH3T3 cells were analyzed without fixation using standard fluorescence microscopy. Cells were also incubated for 10 min at 4°C with 0.1% Triton X-100 in PBS before fixation and analyzed by fluorescence microscopy.

similar co-sedimentation of endogenous CK2β and GFP-CK2β with the peak of activity (Fig. 2, right panel). Thus, despite their difference in apparent molecular mass, the sedimentation coefficient of both GFP-tagged CK2 subunits appeared similar to the value determined for endogenous CK2 subunits integrated into the holoenzyme. Since under our experimental conditions (1 M NaCl), the formation of higher order structures should be prevented [25], these results show that GFP-CK2 subunits may functionally interact with endogenous CK2 subunits most likely generating a mixed pool of CK2 holoenzyme. Thus, the molecular characteristics of the GFP-CK2 subunits, together with their low level of expression, suggest they may act as valid markers to examine the behaviour of endogenous CK2 subunits in living cells.

CK2α and CKβ subunits mostly localize to the nucleus

Analysis by fluorescence microscopy of HeLa cells transiently expressing GFP-CK2 fusion proteins showed that at low to moderate levels of expression, GFP-CK2α and GFP-CK2β fusion proteins localized predominantly to the nucleus with a minor but significant staining in the cytoplasm (Fig. 3A). Thus the normally diffused distribution of GFP was drastically changed by its fusion to either of the CK2 subunits

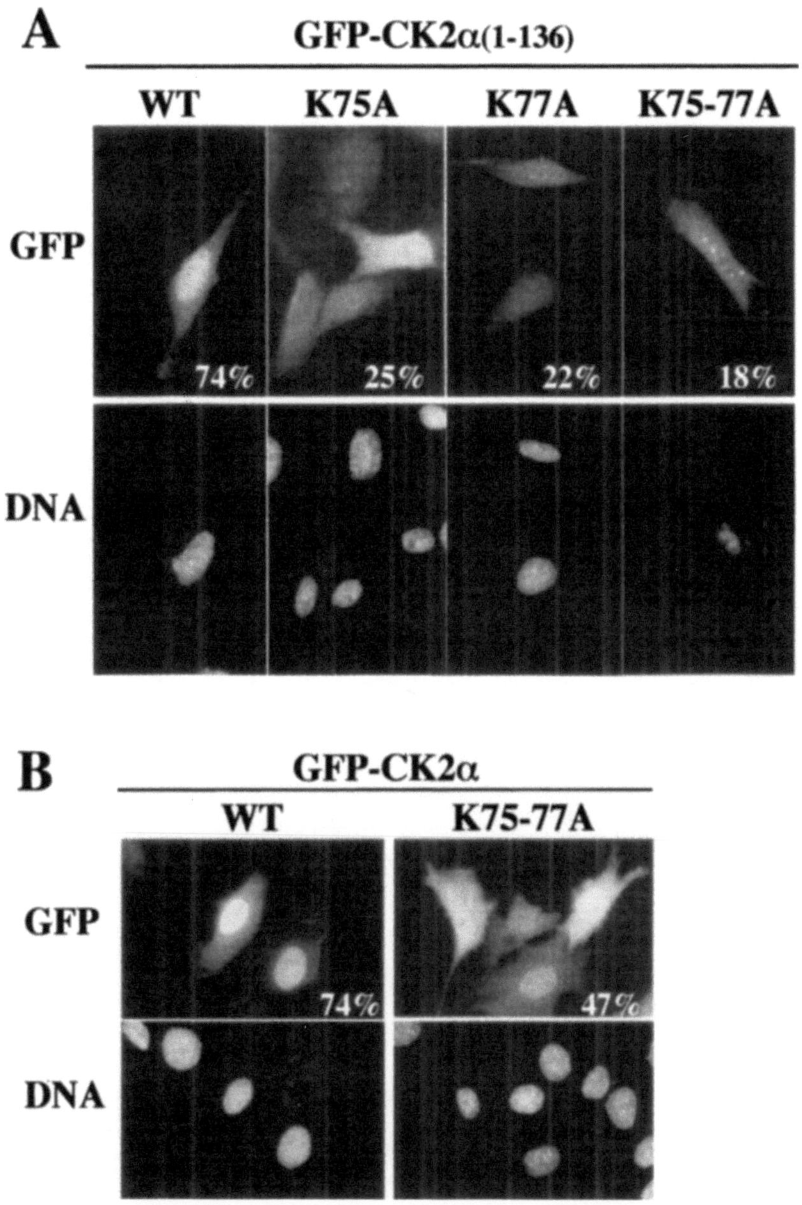

Fig. 4. Identification of a NLS in CK2α. (A) NIH3T3 cells were transiently transfected with GFP-CK2α (1–136) or different GFP-CK2α (1–136) mutants. Cells were stained for DNA with Hoechst and scored for N/C index as indicated. (B) Mutation of K74–77 to A was introduced in full-length GFP CK2α and the construct was transiently transfected. Cells were observed for protein localization.

which were able to trigger nuclear accumulation of the corresponding fusion proteins. Similar observations were obtained using NIH3T3 and K562 cells (not shown). Moreover, in NIH3T3 clones stably expressing the GFP-tagged CK2 subunits, both proteins were also mostly associated to the nuclei as observed in unfixed cells (Fig. 3B). When the same cells were incubated with Triton X-100 before fixation, the majority of the green fluorescence was recovered associated to the nuclei (Fig. 3B). Altogether these observations confirm that both GFP-CK2 subunits behave as endogenous CK2 subunits and are predominantly nuclear proteins in interphase cells. However these results do not distinguish whether the CK2 subunits translocate to the nucleus independently or as part of the same complex.

Minimal domains required for nuclear localization of CK2α subunits

Since the molecular mass of the GFP-CK2 subunits are larger

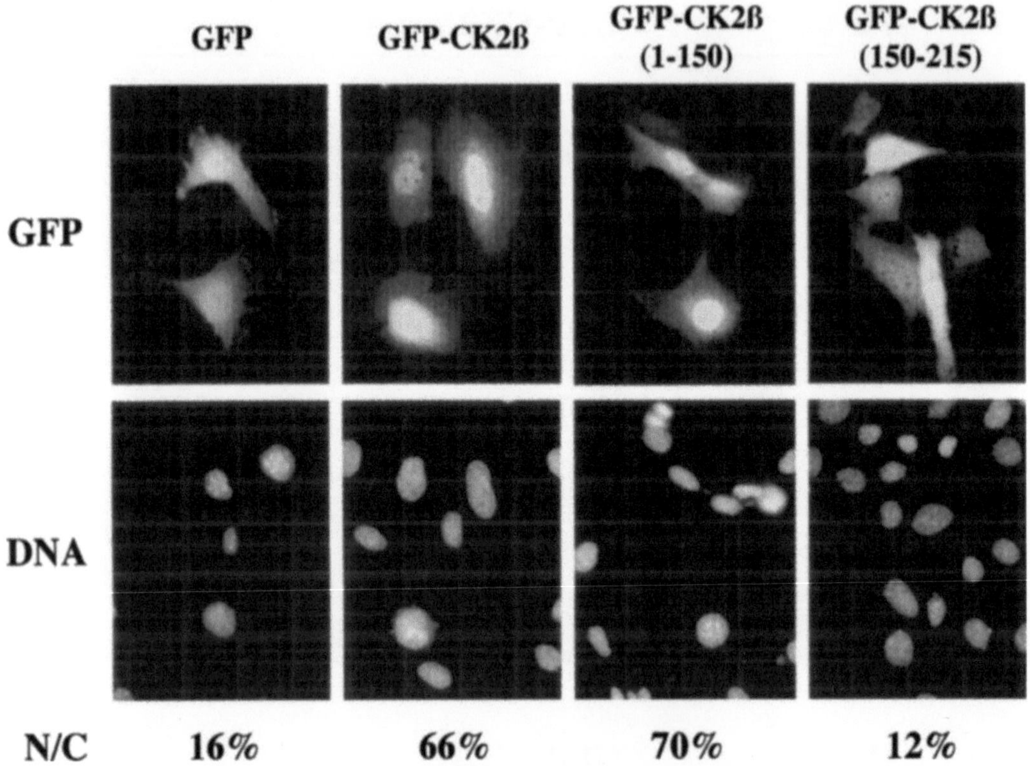

Fig. 5. Subcellular localization of different deletion forms of GFP CK2β. NIH3T3 cells were transiently transfected with the indicated constructs and the expressed proteins were visualized by fluorescence microscopy. Cells were scored for N/C index as indicated.

than the 40–60 kD size limit for passive diffusion of proteins through the nuclear pore complex, the observed nuclear location of these proteins could result from at least three possible mechanisms: (a) each subunit is independently imported via a nuclear targeting signal, (b) one subunit is transported through its association with its counterpart subunit which contains a nuclear targeting signal or (c) each subunit independently or as a complex interact with a protein partner containing a nuclear targeting signal.

A putative nuclear localization signal has been described in the N-terminal region of the CK2α subunit (aminoacids 74–77 in the human sequence), coinciding with a stretch of lysines that have also been involved in substrate recognition, heparin inhibition and down regulation by the CK2β subunit [27]. Four basic aminoacid residues out of a stretch of 6 residues can function as a single basic type NLS signal and the region of CK2α between position 74 and 77 (^{74}**KKKKIK**^{79}) would fit this. Alternatively, two strings of 2 or 3 basic residues (Arg, lys) separated by a stretch of 10–29 aminoacids have been shown to function in some proteins as a bipartite basic type NLS [28]. It is also possible that in CK2α the sequence from position 43–77 could function in this manner. In order to dissect the mechanism responsible for the nuclear localization of CK2α, we fused the first 136 N-terminal aminoacids of this subunit to GFP (Fig. 1). The resulting construct was transiently transfected into NIH3T3 cells and its localization analyzed by fluorescence microscopy (Fig. 4A). The expressed fusion protein, GFP-CK2α (1–136) produced a nuclear staining indistinguishable from that seen for the full length GFP-CK2α protein (compare Figs 3B and 4A) demonstrating that this domain of CK2α is necessary and sufficient to trigger the nuclear accumulation of an heterologous protein (N/C index: 74%). Different point mutations of lysine residues between position 74 and 77 were introduced in the GFP-CK2α (1–136) construct and transiently transfected into NIH3T3 cells. Mutation of the K75 or K77 to A severely impaired the nuclear accumulation of the fusion proteins (N/C index: 25 and 22% respectively). Mutation of four residues at position K74–77 completely abrogated the nuclear import of the corresponding GFP fusion construct (N/C index: 18%). The same mutation in full-length CK2α also impaired the nuclear accumulation of the protein upon transient expression (Fig. 4B). These results show that the nuclear accumulation of CK2α is a signal-mediated process and that the NLS present in this subunit is at least partially responsible for its nuclear translocation.

We then explored the possibility that the CK2α subunit might be also responsible for the nuclear localization of CK2β based on their strong mutual interaction [29]. GFP,

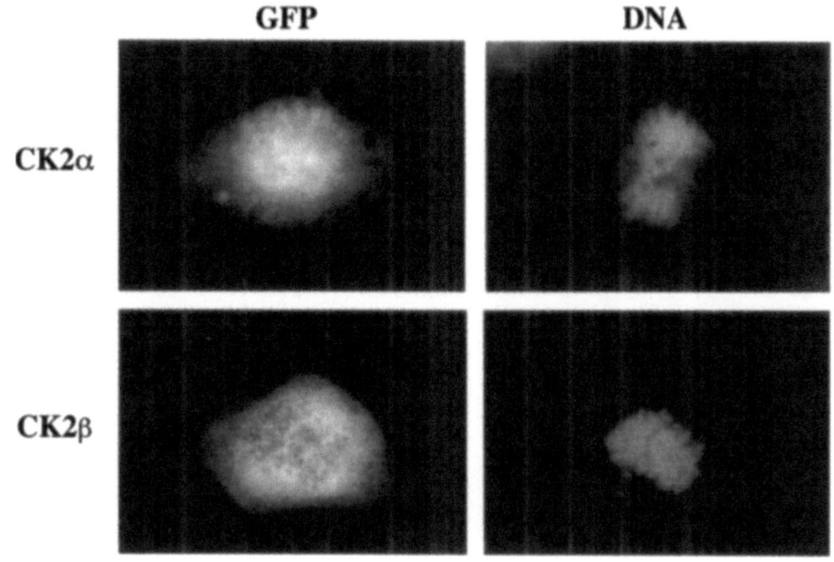

Fig. 6. Changes in the distribution of CK2 subunits during mitosis. Mitotic NIH3T3 cells stably expressing GFP-CK2α or GFP-CK2β were analyzed by standard fluorescence microscopy. Cells were also counterstained for DNA with Hoechst.

GFP-CK2β, the first 150 aminoacids and the last 65 aminoacids of CK2β fused to GFP (GFP-CK2β (1–150) and GFP-CK2β (150–215) were transiently expressed in NIH3T3 cells. Most of the cells transfected with GFP alone as a control exhibited green fluorescence uniformly throughout the cell (Fig. 5). Surprisingly the fusion protein containing the first 150 aminoacids of CK2β also localized to the nucleus in a pattern indistinguishable from that seen for the full length GFP-CK2β (N/C index: 70 and 66% respectively). In contrast, GFP-CK2β (150–215) construct produced a diffuse staining pattern similar to GFP alone (N/C index: 12%). The CK2β structure has revealed that the region between residues 1–150 contains 4 cysteine residues which coordinate a zinc ion and form a zinc finger motif that is responsible for dimerization of the CK2β subunit [30]. Since it is also known that this dimerization domain of CK2β does not associate with CK2α in a stable complex, our results show that this minimal domain contains all the protein determinants necessary for the nuclear localization of CK2β and is not targeted to the nucleus by virtue of its interaction with endogenous CK2α. They also show that GFP-CK2 subunits are independently imported to the nucleus. Altogether, our results show that CK2α and CK2β are predominantly nuclear proteins in interphase cells.

By contrast, in mitotic cells, the fluorescent signal corresponding to CK2α and CK2β were clearly distinct from the DNA signal obtained with Hoechst and both proteins were dispersed throughout the cytoplasm and did not colocalize with DNA, suggesting that they were not associated to condensed chromatin (Fig. 6).

Mobility of CK2 subunit in the nucleus of living cells

Direct visualization of GFP-CK2 subunits in living cells has shown that these proteins localized predominantly in the nucleus during interphase (Fig. 4B). Recently, it has been suggested that the intranuclear transport of some proteins occurs by unconstrained diffusion, leading to high mobility [31]. To address the movement of the CK2 subunits in the nucleus, we performed fluorescence recovery after photobleaching (FRAP) experiments on living stable transfected cell lines expressing GFP-CK2α or GFP-CK2β fusion proteins. In this technique, fluorescence is irreversibly photobleached in a small area of the cell by an intense illumination. Then, the fluorescence recovery through the exchange of bleached for non-bleached proteins is measured using an attenuated laser beam (Fig. 7A). The left images were taken before photobleaching and the right images within 10 sec of the completion of photobleaching. The rapid exchange of bleached for non-bleached proteins during scanning led to the decrease in fluorescence intensity throughout the entire nucleus. Under these conditions, the photobleached area was not detectable in cells expressing either the catalytic or the regulatory subunits of CK2 fused to GFP. However, the size and the location of the bleach area could be verified on fixed cells using identical optical parameters (Fig. 7B). These results indicated that CK2 subunits localized in the nucleus are highly mobile proteins and that the image scan time (10 sec) is too long relative to their mobility.

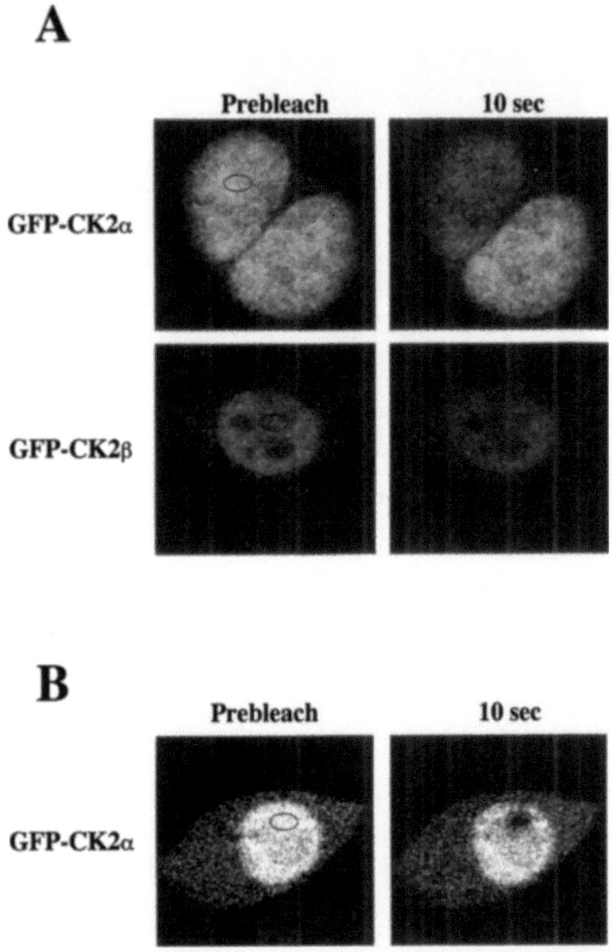

Fig. 7. FRAP analysis of CK2 subunits movement. (A) A surrounded region in the nucleus of NIH3T3 living cells stably expressing GFP-CK2α or GFP-CK2β was bleached for 15 sec. Images correspond to cell before (prebleach) and 10 sec after photobleaching. (B) The same experiment was performed on fixed cells expressing GFP-CK2α fusion protein. Note that bleaching occurred only in the area exposed to intense light illumination and that fluorescence of the whole nucleus remained the same.

Discussion

In this study we describe experiments aimed at elucidating spatial aspects of CK2, focusing on the nuclear import of its subunits. In contrast to earlier analysis of the intracellular distribution of CK2, which relied on classical immunohistochemical and biochemical techniques, the use of GFP permits direct visualization of the fusion protein in intact cells, bypassing the need for cell fractionation or antibodies. Since the sensitivity of the fluorescence emitted by GFP is very high, it is possible to analyze cells expressing relatively low levels of the fusion products. Thus, the behaviour of the transfected fusion proteins can be studied in conditions very close to physiological conditions, without significantly disrupting the intracellular pool of CK2. In this report we present evidence showing that the CK2α and CK2β subunits are mostly localized to the nucleus in interphase cells. However a detailed observation of living cells revealed new insights into the pattern of the CK2 subunits localization. During interphase, CK2 subunits are mainly associated with the chromatin. In contrast, at the onset of mitosis both subunits are not co-localized with condensed DNA and are dispersed into the cytoplasm. Furthermore, our data demonstrate that a unique cluster of predominantly basic aminoacids (aa 74–77) functions as a NLS conferring nuclear localization to CK2α. This cluster localized in subdomain III of the kinase is closely homologous to the NLS of the SV40 T antigen [32]. Our point mutation analysis of this basic cluster shows that specific lysine residues K75 and K77 are instrumental for an efficient nuclear targeting. Interestingly, the basic residues aligned in the 74–83 stretch of subdomain III have also been clearly implicated in substrate recognition [33]. Therefore, one might hypothesize that this basic cluster which folds as a very exposed α helix in the CK2α structure [34] has a dual role and represents the prototype of a bifunctional domain. CK2β does not contain any recognizable NLS; nevertheless it is also imported to the nucleus in interphase cells. The protein is prone to dimerize as revealed by its crystal structure [30] and one might speculate the β–β complexes to be too large for a passive cytoplasmic-nuclear translocation. Our experiments show that the N-terminal half of the protein which is unable to interact with CK2α [35] gains access to the nucleus. Therefore the import reaction of CK2 β is not dependent upon the CK2α subunit. It is possible that the N-terminal domain of CK2β contains a nuclear import signal unrelated to the classical basic NLS domain as previously described [36; 37–39]. Alternatively, CK2β may 'piggy-back' its way into the nucleus [40, 41]. Indeed, *in vivo*, CK2β can interact with various cellular proteins and the binding of NLS-containing partners may not only influence the biochemical properties of CK2β but also its subcellular distribution. Clearly our results indicate that the two CK2 subunits are targeted independently to the nucleus of interphase cells. Such a possibility would fit with the observation that in HeLa cells, the CK2 subunits are separately imported into the nuclei where they can combine to each other generating the canonical holoenzyme [42]. Photobleaching experiments show that like many nuclear proteins, both CK2 subunits are highly mobile nuclear proteins that would screen the nucleus for the presence of potential binding partners. Both CK2 subunits are rapidly displaced from chromatin in mitotic cells. At present, we do not know what molecular mechanisms account for the abrupt nuclear exclusion of these proteins. Since it has been reported that both CK2 subunits are heavily phosphorylated at mitosis [43, 44], a possibility exists that in prophase, both CK2 subunits would lose their binding affinity to nucleoplasmic ligands through phosphorylation.

Given its promiscuous behaviour, CK2 is indeed a good candidate for conveying information from the cytoplasm to the nucleus and our observations raise the intriguing possibility that the nucleocytoplasmic distribution of CK2 subunits may be regulated in a cell cycle-dependent manner. In this context, a dynamic localization of CK2 subunits would underlie the basic processes which may implicate CK2 in the control of cell growth and progression of the cell cycle.

Acknowledgments

This work was supported by grants from the INSERM, the Commisariat à l'Energie Atomique, the Association pour la Recherche contre le Cancer, and the Ligue Nationale contre le cancer.

References

1. Allende JE, Allende CC: Protein kinases. 4. Protein kinase CK2: An enzyme with multiple substrates and a puzzling regulation. FASEB J 9: 313–323, 1995
2. Padmanabha R, Chen-Wu JL, Hanna DE, Glover CV: Isolation, sequencing, and disruption of the yeast CKA2 gene: Casein kinase II is essential for viability in Saccharomyces cerevisiae. Mol Cell Biol 10: 4089–4099, 1990
3. Hanna DE, Rethinaswamy A, Glover CV: Casein kinase II is required for cell cycle progression during G1 and G2/M in Saccharomyces cerevisiae. J Biol Chem 270: 25905–25914, 1995
4. Jaffe L, Ryoo HD, Mann RS: A role for phosphorylation by casein kinase II in modulating Antennapedia activity in Drosophila. Genes Dev 11: 1327–1340, 1997
5. Liu ZP, Galindo RL, Wasserman SA: A role for CKII phosphorylation of the cactus PEST domain in dorsoventral patterning of the Drosophila embryo. Genes Dev 11: 3413–3422, 1997
6. Ulloa L, Diaz-Nido J, Avila J: Depletion of casein kinase II by antisense oligonucleotide prevents neuritogenesis in neuroblastoma cells. EMBO J 12: 1633–1640, 1993
7. Toczyski DP, Galgoczy DJ, Hartwell LH: CDC5 and CKII control adaptation to the yeast DNA damage checkpoint. Cell 90: 1097–1106, 1997
8. Pepperkok R, Lorenz P, Ansorge W, Pyerin W: Casein kinase II is required for transition of G0/G1, early G1, and G1/S phases of the cell cycle. J Biol Chem 269: 6986–6991, 1994
9. Raman C, Kuo A, Deshane J, Litchfield DW, Kimberly RP: Regulation of casein kinase 2 by direct interaction with cell surface receptor CD5. J Biol Chem 273: 19183–19189, 1998
10. McElhinny JA, Trushin SA, Bren GD, Chester N, Paya CV: Casein kinase II phosphorylates I kappa B alpha at S-283, S-289, S-293, and T-291 and is required for its degradation. Mol Cell Biol 16: 899–906, 1996
11. Willert K, Brink M, Wodarz A, Varmus H, Nusse R: Casein kinase 2 associates with and phosphorylates dishevelled. EMBO J 16: 3089–3096, 1997
12. Lee JS, Ishimoto A, Yanagawa S: Characterization of mouse dishevelled (Dvl) proteins in Wnt/Wingless signaling pathway. J Biol Chem 274: 21464–21470, 1999
13. Hockman DJ, Schultz MC: Casein kinase II is required for efficient transcription by RNA polymerase III. Mol Cell Biol 16: 892–898, 1996
14. Hannan RD, Hempel WM, Cavanaugh A, Arino T, Dimitrov SI, Moss T, Rothblum L: Affinity purification of mammalian RNA polymerase I. Identification of an associated kinase. J Biol Chem 273: 1257–1267, 1998
15. Ghavidel A, Schultz MC: Casein kinase II regulation of yeast TFIIIB is mediated by the TATA-binding protein. Genes Dev 11: 2780–2789, 1997
16. Lin A, Frost J, Deng T, Smeal T, al-Alawi N, Kikkawa U, Hunter T, Brenner D, Karin M: Casein kinase II is a negative regulator of c-Jun DNA binding and AP-1 activity. Cell 70: 777–789, 1992
17. Marais RM, Hsuan JJ, McGuigan C, Wynne J, Treisman R: Casein kinase II phosphorylation increases the rate of serum response factor-binding site exchange. EMBO J 11: 97–105, 1992
18. Li D, Dobrowolska G, Krebs EG: The physical association of casein kinase 2 with nucleolin. J Biol Chem 271: 15662–15668, 1996
19. Yamaguchi Y, Wada T, Suzuki F, Takagi T, Hasegawa J, Handa H: Casein kinase II interacts with the bZIP domains of several transcription factors. Nucleic Acids Res 26: 3854–3861, 1998
20. Misteli T, Spector DL: Applications of the green fluorescent protein in cell biology and biotechnology. Nat Biotechnol 15: 961–964, 1997
21. Tsien RY: The green fluorescent protein. Annu Rev Biochem 67: 509–544, 1998
22. Heriche JK, Lebrin F, Rabilloud T, Leroy D, Chambaz EM, Goldberg Y: Regulation of protein phosphatase 2A by direct interaction with casein kinase 2 alpha. Science 276: 952–955, 1997
23. Kakinoki Y, Somers J, Brautigan DL: Multisite phosphorylation and the nuclear localization of phosphatase inhibitor 2-green fluorescent protein fusion protein during S phase of the cell growth cycle. J Biol Chem 272: 32308–32314, 1997
24. Filhol-Cochet O, Loue-Mackenbach P, Cochet C, Chambaz EM: Casein kinase 2 and the cell response to growth factors. Cell Mol Biol Res 40: 529–537, 1994
25. Valero E, De Bonis S, Filhol O, Wade RH, Langowski J, Chambaz EM, Cochet C: Quaternary structure of casein kinase 2. Characterization of multiple oligomeric states and relation with its catalytic activity. J Biol Chem 270: 8345–8352, 1995
26. Patterson GH, Knobel SM, Sharif WD, Kain SR, Piston DW: Use of the green fluorescent protein and its mutants in quantitative fluorescence microscopy. Biophys J 73: 2782–2790, 1997
27. Pinna LA, Meggio F: Protein kinase CK2 ('casein kinase-2') and its implication in cell division and proliferation. Prog Cell Cycle Res 3: 77–97, 1997
28. Dingwall C, Laskey RA: Nuclear targeting sequences – a consensus?. Trends Biochem Sci 16: 478–481, 1991
29. Cochet C, Chambaz EM: Oligomeric structure and catalytic activity of G type casein kinase. Isolation of the two subunits and renaturation experiments. J Biol Chem 258: 1403–1406, 1983
30. Chantalat L, Leroy D, Filhol O, Nueda A, Benitez MJ, Chambaz EM, Cochet C, Dideberg O: Crystal structure of the human protein kinase CK2 regulatory subunit reveals its zinc finger-mediated dimerization. EMBO J 18: 2930–2940, 1999
31. Phair RD, Misteli T: High mobility of proteins in the mammalian cell nucleus. Nature 404: 604–609, 2000
32. Kalderon D, Roberts BL, Richardson WD, Smith AE: A short amino acid sequence able to specify nuclear location. Cell 39: 499–509, 1984
33. Sarno S, Boldyreff B, Marin O, Guerra B, Meggio F, Issinger OG, Pinna LA: Mapping the residues of protein kinase CK2 implicated in substrate recognition: Mutagenesis of conserved basic residues in the al-

pha-subunit. Biochem Biophys Res Commun 206: 171–179, 1995
34. Niefind K, Guerra B, Pinna LA, Issinger OG, Schomburg D: Crystal structure of the catalytic subunit of protein kinase CK2 from Zea mays at 2.1 A resolution. Embo J 17: 2451–2462, 1998
35. Graham KC, Litchfield DW: The regulatory beta subunit of protein kinase CK2 mediates formation of tetrameric CK2 complexes. J Biol Chem 275: 5003–5010, 2000
36. Pollard VW, Michael WM, Nakielny S, Siomi MC, Wang F, Dreyfuss G: A novel receptor-mediated nuclear protein import pathway. Cell 86: 985–994, 1996
37. Pemberton LF, Rosenblum JS, Blobel G: A distinct and parallel pathway for the nuclear import of an mRNA-binding protein. J Cell Biol 139: 1645–1653, 1997
38. Rosenblum JS, Pemberton LF, Blobel G: A nuclear import pathway for a protein involved in tRNA maturation. J Cell Biol 139: 1655–1661, 1997
39. Rout MP, Blobel G, Aitchison JD: A distinct nuclear import pathway used by ribosomal proteins. Cell 89: 715–725, 1997
40. Maridor G, Gallant P, Golsteyn R, Nigg EA: Nuclear localization of vertebrate cyclin A correlates with its ability to form complexes with cdk catalytic subunits. J Cell Sci 106: 535–544, 1993
41. Pines J, Hunter T: The differential localization of human cyclins A and B is due to a cytoplasmic retention signal in cyclin B. Embo J 13: 3772–3781, 1994
42. Chester N, Yu IJ, Marshak DR: Identification and characterization of protein kinase CKII isoforms in HeLa cells. Isoform-specific differences in rates of assembly from catalytic and regulatory subunits. J Biol Chem 270: 7501–7514, 1995
43. Litchfield DW, Luscher B, Lozeman FJ, Eisenman RN, Krebs EG: Phosphorylation of casein kinase II by p34cdc2 *in vitro* and at mitosis. J Biol Chem 267: 13943–13951, 1992
44. Bosc DG, Slominski E, Sichler C, Litchfield DW: Phosphorylation of casein kinase II by p34cdc2. Identification of phosphorylation sites using phosphorylation site mutants *in vitro*. J Biol Chem 270: 25872–25878, 1995

Molecular and Cellular Biochemistry **227**: 91–98, 2001.

Identification and characterization of proteins that interact with *Drosophila melanogaster* protein kinase CK2

Regina L. Trott,[1] Madhavi Kalive,[1] Umesh Karandikar,[1] Rebecca Rummer,[1] Clifton P. Bishop[1] and Ashok P. Bidwai[1,2]

[1]*Department of Biology and* [2]*Mary Babb Randolph Cancer Center, West Virginia University, Morgantown, WV, USA*

Abstract

D. melanogaster CK2 (DmCK2) is a highly conserved protein kinase that is composed of catalytic, α, and regulatory, β, subunits associated as an $\alpha_2\beta_2$ heterotetramer. In order to analyze the functions of CK2 in this metazoan model, we have used the two hybrid approach to identify interacting proteins. One of these cDNAs, DmA24, encodes a novel polypeptide with no homologs in GenBank, and is notable in that it contains a bipartite nuclear localization signal and two sites for phosphorylation by CK2. *In situ* hybridization to polytene chromosomes indicates that the DmA24 gene is located at the 61D interval of chromosome II a region that also harbors 3 additional genes with similar structure. DmA24p interacts with DmCK2α, but not with DmCK2β, demonstrating that this interaction is specific for the catalytic subunit of CK2. In addition, the protein is phosphorylated by the holoenzyme purified from *Drosophila* embryos. These studies identify DmA24p as a potentially new physiological partner of DmCK2. In addition, we also report the results of a large-scale screen that has identified a new set of DmCK2-interacting proteins. Most notable among these are Surf6, a nucleolar protein involved in RNA processing, and Spalt, a homeotic protein. (Mol Cell Biochem **227**: 91–98, 2001)

Key words: protein kinase, CK2, *Drosophila*, two hybrid

Introduction

CK2 is an ubiquitous protein kinase that is highly conserved among eukaryotes [1, 2], and is capable of functioning as an oncogene [3, 4]. In general, CK2 is composed of catalytic (α) and regulatory (β) subunits that combine to form an $\alpha_2\beta_2$ holoenzyme, and distinct isoforms of both subunits have been described in a wide variety of organisms. CK2 is unique in its site selectivity in that it preferentially phosphorylates Ser/Thr residues that are followed by a stretch of acidic residues [2]. However, in at least one case, i.e. yeast Fpr3, the enzyme has been reported to phosphorylate Tyr residues [5]. Analysis of the phosphorylation of peptides suggests that the consensus site for CK2 can best be described as (S/T)xxD/E, and consistent with this, a number of proteins critical for transcription, cell-cycle regulation, and signal transduction contain such site(s) and are known to be phosphorylated *in vitro* and *in vivo* [6]. *In vitro*, CK2 activity is inhibited by polyacidic compounds [7], and stimulated by polybasic compounds [8], but the *in vivo* relevance of these observations remains unclear. Comparisons between recombinant monomeric α subunit and native or reconstituted $\alpha_2\beta_2$ holoenzyme have revealed that the β subunit plays a complex, yet at times paradoxical, role in regulating the basal activity of the α subunit [9–12]. On the one hand, the β subunit stabilizes the α subunit against proteolysis and thermal denaturation and stimulates its activity approximately 5-fold against most substrates [13]; on the other hand, it negatively regulates phosphorylation of selected substrates, notably calmodulin [14, 15]. The β subunit also mediates stimulation of CK2 by poly-

Address for offprints: A.P. Bidwai, Department of Biology, P.O. Box 6057, Brooks Hall, West Virginia University, Morgantown, WV 26506-60571, USA (E-mail: abidwai@wvu.edu)

basic compounds such as polylysine and protamine [15], although studies suggest that these compounds can exert their activating effects via the α subunit as well [14].

Genetic analyses in budding and fission yeast have demonstrated that the enzyme is essential for viability [16, 17] and is required for cell cycle progression in G1 and G2/M [18], for maintenance of cytoskeletal architecture [19], and for cytokinesis [17]. In contrast, analysis of CK2 in *Drosophila* has lagged despite the fact that the cDNAs encoding DmCK2 were the first to be isolated [20]. The absence of mutations in the genes encoding DmCK2 subunits has precipitated the use of alternative approaches. One such approach is the two hybrid strategy [21] to identify interacting proteins (many of which appear to be substrates of the enzyme), followed by genetic analysis using transgenes with alterations in the CK2-interaction domain and/or site(s) of phosphorylation. Such studies have demonstrated that CK2 negatively regulates the homeobox protein ANTP [22], and that this enzyme targets Dishevelled [23], a component of the Wingless signaling pathway, as well as the neurogenic transcriptional repressors such as m5, m7, and m8 [24]. These results suggest that CK2 plays a critical role in embryogenesis and cell-fate determination.

In an attempt to better define the role of DmCK2, we have used the two-hybrid approach to identify and characterize its physiological partners. We describe the isolation and characterization of a cDNA, DmA24, whose product interacts with DmCK2α. The corresponding gene is localized to the 61D interval, a region that harbors three other genes which exhibit evidence of exon shuffling. Although the DmA24 protein is not homologous to other proteins with known functions, it contains a nuclear localization signal, and two sites for phosphorylation by CK2. Consistent with this, we demonstrate that DmA24p is phosphorylated by the embryo holoenzyme. The available data suggest that DmA24 is a newly identified physiological partner and substrate of DmCK2. In addition, we summarize the results of a second two hybrid screen which has identified proteins that are likely to play a role in transcription, RNA processing, and development.

Materials and methods

Yeast two-hybrid screening and β-galactosidase assay

All manipulations involved in construction of the two-hybrid plasmids were carried out as described [25]. The construction of plasmids expressing Dmα/β as fusions with the Gal4 DNA-binding domain (Gal4DB) and the Gal4 Activation-domain (Gal4AD) has been previously described [26]. Yeast strain HF7C [27] expressing GAL4DB-Dmα was used to screen a 3- to 18-h *Drosophila* embryo two hybrid cDNA library [28]. Transformants exhibiting robust His-independent growth were counterscreened for expression of *LacZ* [29]. The library plasmids were recovered, and retested against GAL4DB-Dmα or GAL4DB-Dmβ. Those cDNAs which induced expression of the reporter genes only in response to Dmα (a bait-specific manner) were sequenced on an Applied Biosystems model 373A DNA sequencer using the Prism Dye Terminator Cycle sequencing kit (Applied Biosystems).

Isolation of full-length clones encoding DmA24

One of the cDNAs, DmA24, isolated in this screen was missing its mature 5′ and 3′ ends. We, therefore, screened a commercially available 0–18 h *Drosophila* embryo cDNA library (Stratagene). The DmA24 cDNA was labeled by random hexamer primers (Boehringer Mannheim), and used to screen 200,000 plaques. Four positives were identified, converted into plasmids, and restriction mapped. The largest clone (DmA24-13) was sequenced in order to compile the full length sequence, and for subsequent analysis.

Interaction of full length DmA24 with DmCK2α

Explicit interactions between full length DmA24 and Dmα were tested using the LexA system [30]. DNA corresponding to amino acids 1–387 of DmA24 was amplified from plasmid DmA24-13 by PCR, and subcloned into the EcoRI-BamHI site of plasmid LexA202 [30]. The construction of a plasmid expressing Dmα as an AD-fusion has been previously described [24]. EGY048 containing plasmid pSH18-34 was transformed with a plasmid expressing the AD-alone or AD-Dmα fusion protein using lithium-acetate [25]. A single transformant was subsequently retransformed with a plasmid expressing LexA-DmA24. Three independent transformants were grown in galactose medium, and assayed in triplicate for *LacZ* activity as described [29].

In situ *hybridization to polytene chromosomes*

Salivary glands from third instar larvae were isolated, dissected, and prepared for hybridization essentially as described [31]. The cDNA from clone DmA24 was labeled by nick translation using biotinylated dUTP (Boehringer- Mannheim). Probe binding was visualized using streptavidin coupled to alkaline phosphatase, and slides were photographed at 400× magnification.

Phosphorylation of DmA24 by CK2

Expression of GST-DmA24, purification of the fusion protein and phosphorylations were conducted essentially as described [24]. Phosphorylation reactions were carried out in 50 mM Tris, pH 8.5, 100 mM NaCl, 10 mM $MgCl_2$, 10 μM ATP, 5 μCi [γ-^{32}P]ATP and CK2 (1 μg/ml) in a total volume of 40 μl, at 25°C. The reactions were terminated by boiling for 5 min following the addition of 10 μl of 5 × sample buffer [32]. Samples were separated by electrophoresis in 12% acrylamide gels containing sodium dodecyl sulfate, stained with Coomassie, and the destained gels were exposed to Kodak XAR-5 film at room temperature.

Results and discussion

Two hybrid screen for CK2 interacting proteins

The yeast strain HF7C expressing GAL4DB-Dmα as a bait was used to screen a *D. melanogaster* embryo two hybrid cDNA library and resulted in the isolation of 15 *HIS⁺/LacZ⁺* clones. As expected, these include DmCK2β [33], DmCK2β′ [26], a novel isoform of the β subunit, and DmCK2α [26]. We do not consider our isolation of Dmα (using Dmα as a bait) to imply that this protein can dimerize, but rather that the β/β′ subunits of yeast CK2 mediate this interaction. This interpretation is consistent with the observation that this protein exhibits a strong two-hybrid interaction with the β′ subunit of yeast CK2 (Trott and Bidwai, unpublished). The isolation of DmCK2 subunits thus serves as a valuable internal control. In addition, this screen resulted in the isolation of E(spl)m7, the ribosomal protein DmL22, and two novel proteins, DmA24, and DmA35. E(spl)m7 encodes a neurogenic transcriptional repressor [34], and we have recently demonstrated that m7, and two related proteins m5 and m8, interact with DmCK2, and all three proteins are phosphorylated by the embryo holoenzyme within a highly conserved PEST domain that negatively regulates function of these proteins *in vivo* [24]. Interestingly, DmL22 contains an hyperacidic C-terminal tail that also harbors a site for phosphorylation by CK2, SSNDDEDDDAE (see Table 1). In this regard, yeast and human L22 also exhibit an hyperacidic C-terminal tail, but lack the CK2 site. Given that this enzyme interacts strongly with polyAsp/Glu peptides [7], it is conceivable that this interaction may also be observed in humans and yeast, but the lack of conservation of the CK2 site makes it unclear whether phosphorylation will be meaningful in an evolutionary context. The functions of DmA24 and DmA35 are currently unknown, although both proteins contain consensus sites for phosphorylation by CK2, and are predicted to be nuclear localized with proposed roles in transcription. We describe here the molecular and biochemical characterization of the protein encoded by the two hybrid clone, DmA24.

Nucleotide sequence of the DmA24 cDNA

The cDNA contained in the plasmid DmA24 was completely sequenced as described in Materials and methods. Sequencing revealed that this clone was lacking its 5′ untranslated region (including the initiation codon), as well as its 3′ untranslated region (by virtue of the absence of a polyA-tail). Most relevant to our two hybrid analysis, the encoded protein appeared to be lacking its mature N- and C-terminal ends and, in addition, contained a small peptide introduced at its C-terminus via translational read-through of the polylinker of the AD-vector [28]. In order to establish that the full length protein, rather than the aberrant C-terminal peptide, mediates interaction with Dmα, we have isolated full length clones encoding DmA24, and one of these, DmA24-13, was completely sequenced (Fig. 1). This cDNA (deposited as GenBank accession number AF090440) is ~1354 bp long and contains a major uninterrupted open reading frame of 1164 bp that encodes a 387 amino acid long polypeptide with a calculated molecular mass of 44 kDa, and a predicted pH of 9. In addition, clone DmA24-13 contains 47 nucleotides in the 5′ untranslated region, and 144 nucleotides in the 3′ untranslated region. Although the putative initiation methionine codon is not preceded by a Kozak-type consensus sequence, one in frame stop codon is located 15 nucleotides upstream of the AUG. The presence of an in-frame stop codon 5′ to the initiation codon, perhaps, explains our inability to isolate full-length clones encoding DmA24, as the two hybrid interaction requires a productive fusion between Gal4AD and the cDNA. The 3′ untranslated region terminates in an oligo(A) stretch, suggesting that the cDNA is full length with regard to its 3′ untranslated region. However, because the potential polyA-addition signal, AATAAA, is not spaced at the expected distance with respect to the poly A-tail (see Fig. 1), it is unclear whether this is the preferred signal *in vivo*.

Two hybrid interaction of DmA24 with DmCK2α

The two hybrid cDNA contained in clone DmA24 was tested for bait specificity via transformation of HF7C expressing Gal4DB-DmCK2α or Gal4BD-DmCK2β, and testing for induction of *HIS3* expression. As shown in Fig. 2A, DmA24 interacts with Dmα, but not with Dmβ. The inability of the β subunit to interact with DmA24 is not attributable to a lack of expression, instability, or toxicity of Gal4DB-Dmβ fusion protein in yeast as this very construct exhibits a robust interaction with Dmα [26]. Although not explicitly tested for, the inability of DmA24 to induce *HIS3* expression in combina-

```
TCGTGCCGAATTCGGCACGAGTGGAAGTCTAAGTTGTTGTTAAATTTATGGAAAAGGAATCCAATTTATCGGACAACTTGCAGTCCCGGT
                                               M  E  K  E  S  N  L  S  D  N  L  Q  S  R
TATTTGCACTGGGTCCTGCCCCGCCAGGATTTCGTCCACCGGATGCTACCGCCTTTGCCACAGCGGAGTCAAATCTTCGTAATGCCATCC
L  F  A  L  G  P  A  P  P  G  F  R  P  P  D  A  T  A  F  A  T  A  E  S  N  L  R  N  A  I
GCTACCTTAATGATACCTATGGCGAGCTCGATCAGCTGCGCACTGTCATCCAGAAGGAGAAGAATAACTCGAAGCGTACGTCCTACAAGG
R  Y  L  N  D  T  Y  G  E  L  D  Q  L  R  T  V  I  Q  K  E  K  N  N  S  K  R  T  S  Y  K
CGCAGTCCATGATGCTTGCCATGCAAGTGATACGTTTCAAAGGACTTATTGGGAAGTACAGGGACATTCTGATTGGCCTGCAGGATGCCG
A  Q  S  M  M  L  A  M  Q  V  I  R  F  K  G  L  I  G  K  Y  R  D  I  L  I  G  L  Q  D  A
AAATCATCAAGGTCAAACAATACATAGAGCAAATGGTGGAGATGAGGAACAAAATCCAGGCAAAGACCGATCCCACACAGCGCGAGTTCC
E  I  I  K  V  K  Q  Y  I  E  Q  M  V  E  M  R  N  K  I  Q  A  K  T  D  P  T  Q  R  E  F
AAGTGGGCAGCTCAATCTCACTGGTGCAGGAACAGCGCAAGACCAAGAGAATCTGCCGCAGGTGGCGGGATCGCCTGGTCCAGCTGACGG
Q  V  G  S  S  I  S  L  V  Q  E  Q  R  K  T  K  R  I  C  R  R  W  R  D  R  L  V  Q  L  T
AGGCCAAGATGGACAAAACTGAGCAACTTACTGAGAAACCAAAGCTGATCAATAAACCTAGAAATATGTTAAATGCAAAAACCAACAAGT
E  A  K  M  D  K  T  E  Q  L  T  E  K  P  K  L  I  N  K  P  R  N  M  L  N  A  K  T  N  K
GGAAGGAAGCTATAAACTTGAAGCAAAGCGGTCTAATTAAAAGTGGCTGTAAGAGGTGTGAATATCTAAGAAAGGTGACTAAGAATCGTC
W  K  E  A  I  N  L  K  Q  S  G  L  I  K  S  G  C  K  R  C  E  Y  L  R  K  V  T  K  N  R
CATCCAAACTTTGTAAGGTGTGTAAAAAATGGTCTGATGATGAAAAGGCAAAAGCGGCAGGAAATAAGAAAGCCAACAAAACACCAATAA
P  S  K  L  C  K  V  C  K  K  W  S  D  D  E  K  A  K  A  A  G  N  K  K  A  N  K  T  P  I
TGGCGAAGTCGAAAAAATTAGAAGTACGCGTGCGTACGGATTCTGAAGAAGCTATAATTAAAATAGAGATAGAGCATGGTGAATTTACCG
M  A  K  S  K  K  L  E  V  R  V  R  T  D  S  E  E  A  I  I  K  I  E  I  E  H  G  E  F  T
ATGAAGAGGAAATTGGGTCATTGGAAAAGGAAACGACCTTAAAAGATGAGATCAAGGAGACTGTTGAATCTCACGACACGGCAGCAGTTG
D  E  E  E  I  G  S  L  E  K  E  T  T  L  K  D  E  I  K  E  T  V  E  S  H  D  T  A  A  V
AGGAGGAAAATGAGACCAGTCAAGACGATGTGGACCCCAAAGTAGTCGCTTCCTTACAAAATTCACTGAAATCCATCGACTGTAAAAATA
E  E  E  N  E  T  S  Q  D  D  V  D  P  K  V  V  A  S  L  Q  N  S  L  K  S  I  D  C  K  N
TGGCCTTGGCCTACGTGAAGGTCTTACAGGAGTACGCCATGTTAAATGATAACTATCGGGCCATTGGACTACTTACTGAAGTTGAAAAAT
M  A  L  A  Y  V  K  V  L  Q  E  Y  A  M  L  N  D  N  Y  R  A  I  G  L  L  T  E  V  E  K
CAGTTTTAAATCCGCCTCAGAACGAGGATTTCGATCTATGATCTTATTGACGCCACAAAAATAAAGTTAAGTTTTTATATGTAATCTACA
S  V  L  N  P  P  Q  N  E  D  F  D  L  * 387

ATTAAAATGTTAGTGTGTCCTGCCTTTGAGAGTCAAGACCCAAAATGTGTAAATAATTGTAAAAAAAAAAAAAAAAAAA...
```

Fig. 1. Nucleotide sequence and deduced amino acid sequence of DmA24. The nucleotide sequence of the cDNA from clone DmA24-13 is shown above the deduced amino acid sequence (shown in one-letter code), and the asterisk denotes the end of the open reading frame. An in-frame stop codon upstream of the initiating methionine is underlined, the potential polyA-addition signal, AATAAA, is denoted by a double line, and the dotted line indicates the polyA-tail. Arrows indicate the limits of the two hybrid clone, DmA24.

tion with Gal4BD-Dmβ, would also suggest that this protein does not interact with Gal4DB by itself.

Since the protein contained within the two hybrid clone was missing its mature N- and C-terminal ends, we generated a construct expressing the full length DmA24 protein, and tested for interaction with Dmα in the LexA-based two hybrid system [30]. We transformed yeast strain EGY048 with plasmids expressing either the AD-alone, or the AD-Dmα fusion protein, and the resulting strains were subsequently retransformed with a plasmid expressing LexA-DmA24. As expected, the parental strain, EGY048 by itself, or one expressing the AD-alone, AD-Dmα, or LexA-DmA24 did not exhibit any observable induction of *LacZ* (Fig. 2B). On the other hand, a strain coexpressing LexA-DmA24 in combination with AD-Dmα exhibited a robust expression of *LacZ*, whereas one coexpressing LexA-DmA24 in combination with AD-alone exhibited marginal induction of the reporter gene. It should be noted that the original two-hybrid interaction between Dmα and DmA24 was detected using the former protein as a DB-fusion and the latter as an AD-fusion (Fig. 2A), whereas the explicit testing in the LexA system involved the inverse orientation, i.e. DmA24 as a LexA-fusion and Dmα as an AD-fusion (Fig. 2B). These results demonstrate that the interaction is neither orientation-specific, nor is it dependent on a particular version (Gal4 or LexA) of the two-hybrid system. Collectively, these results demonstrate that the DmA24 protein exhibits a robust and specific interaction with Dmα.

Cytological location of the DmA24 gene

The DmA24 cDNA was used to localize the corresponding gene via *in situ* hybridization to polytene chromosomes. The analysis suggests that the gene is located on the left arm of chromosome H at the 61D polytene interval (data not shown). This result is in agreement with the recent completion of the *Drosophila* genome sequence demonstrating identical cytology, and indicating that the DmA24 gene is single copy. A comparison of our cDNA to the genomic sequence suggests that the DmA24 protein is encoded on 7 exons (data not shown). Most interestingly, the last exon, encoding amino acids 316–387 of the DmA24 polypeptide, exhibits conservation with regards to the corresponding exon in three other genes, CG13896, CG13895, and CG13894 (see below). That these four genes colocalize to the 61D region, and appear to contain a similar exon as well, suggest that they may represent a gene family.

Motifs in DmA24 and phosphorylation by Drosophila *embryo CK2*

Extensive Blast searches of the amino acid sequence of DmA24 (Fig. 1) against all available sequences in GenBank revealed no significant homologies, suggesting that orthologs or homologs of this protein have yet to be described. It should,

A

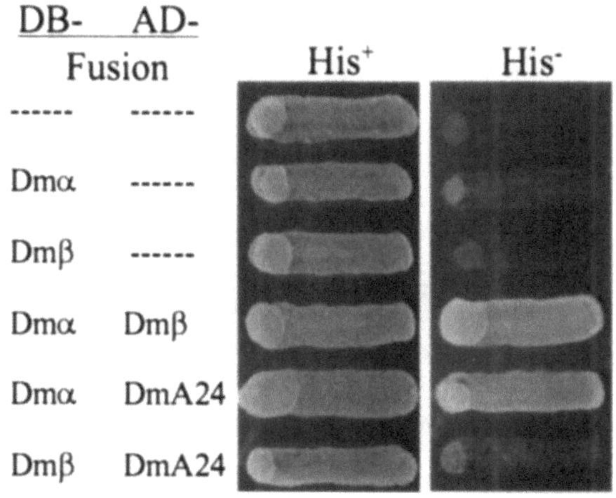

B

LexA-	AD-	LacZ
-----	-----	1.4 + 0
-----	AD	4.5 + 0.7
-----	Dmα	3.0 + 0.4
DmA24	-----	2.5 + 0.4
DmA24	AD	13 + 3
DmA24	Dmα	332 + 13

Fig. 2. Two hybrid interactions of DmA24 with DmCK2 subunits. (A) *S. cerevisiae* strain HF7C was transformed with plasmids expressing the indicated fusions with GAL4DB or GAL4AD. A dashed line indicates the absence of a plasmid. Following growth in dropout medium, cells were plated onto complete (His+) or minimal media lacking histidine (His-) and incubated at 29°C for 3–4 days. (B) Yeast EGY048 was transformed with plasmids expressing the LexA-DmA24 in combination with either the AD alone or the AD-DmCK2α fusion protein. Transformants were grown in galactose media, and *LacZ* activity was determined in triplicate as described [29], and is expressed in Miller Units.

however, be noted that a low level of sequence homology was found with the non-histone chromosomal protein Prod, which is associated with centromeric regions of *Drosophila* chromosomes, although the level of sequence conservation is, at best, modest (data not shown). The predicted DmA24p sequence contains no signal sequences suggesting that it is neither secreted nor sequestered into cellular organelles. However, a search for motifs indicate the presence of a bipartite nuclear localization motif, and 2 consensus sites for phosphorylation by CK2, WS^{235}DDE and FT^{283}DEEE (Fig. 3A), both of which satisfy the requirement for an Asp/Glu at the n+1 and n+3 positions [35, 36].

We were also interested in determining whether the interaction of DmA24 with DmCK2 elicits phosphorylation of this protein. Phosphorylations were conducted using two isoforms of CK2, i.e. the monomeric α subunit purified from a yeast expression system [9], and the holoenzyme purified from embryos [37]. The former isoform is relevant to our two hybrid analysis (see Fig. 2), whereas the latter isoform mimics the environment likely to be encountered *in vivo*. The results demonstrate that GST-DmA24 is phosphorylated by the embryo-holoenzyme, but very weakly, or not at all, by the monomeric α subunit (Fig. 3B, compare lanes 1 and 2). We have also tested for the effects of poly(DL)lysine, and find that this polybasic activator stimulates α subunit mediate phosphorylation of DmA24 modestly, whereas it dramatically stimulates phosphorylation by the embryo holoenzyme (data not shown). We have deferred from conducting parallel analysis on the phosphorylation of GST, because we have previously demonstrated that this affinity tag is not phosphorylated by either isoform of CK2 in the absence or the presence of polybasic activators [24]. These results suggest that insofar as phosphorylation of DmA24 is concerned, the β subunit plays a positive role. Comparative kinetic analysis with the two isoforms will be needed to address whether the β subunit enhances phosphorylation by modulating the k_m and/or the V_{max}. However, our observation that β alone does not exhibit a two hybrid interaction with DmA24, taken together with the phosphorylation analysis, suggest that the physical interaction of CK2α with DmA24 may be greatly enhanced in the presence of CK2β, i.e. in the context of the holoenzyme. However, such analysis is currently precluded by the unavailability of a yeast strain that produces DmCK2 holoenzyme suitable for two hybrid analysis. That DmA24p interacts with CK2 and is phosphorylated as well is in line with similar analysis from a number of laboratories demonstrating that this enzyme exists in a complex *in vivo* with many of its substrates such as, topoisomerase II [38–40], HSP90 [41], ANTP [22], m5/7/8 [24], etc.

As mentioned above, the full length DmA24 sequence did not exhibit significant homology to other proteins. During this analysis, we discovered that the C-terminal 50 amino acids of the DmA24 polypeptide exhibit an inordinate level of conservation with the C-terminal regions of proteins encoded by the CG13896, CG13895, and CG13894 genes. We have, therefore, aligned this region in order to emphasize the overall level of sequence similarity (Fig. 3C). This region does not exhibit similarity to any other proteins, thus precluding any predictions of its potential functions. That DmA24 forms a direct complex with DmCK2α, and is phosphorylated by the embryo holoenzyme strongly suggest that this protein represents a new physiological partner and substrate of CK2. Although the functions of the corresponding gene are unclear at present, the recently initiated effort to isolate systematic mutations in all of the genes will clarify the phenotypes associated with loss of DmA24, and should facilitate analysis of the role of its phosphorylation by DmCK2.

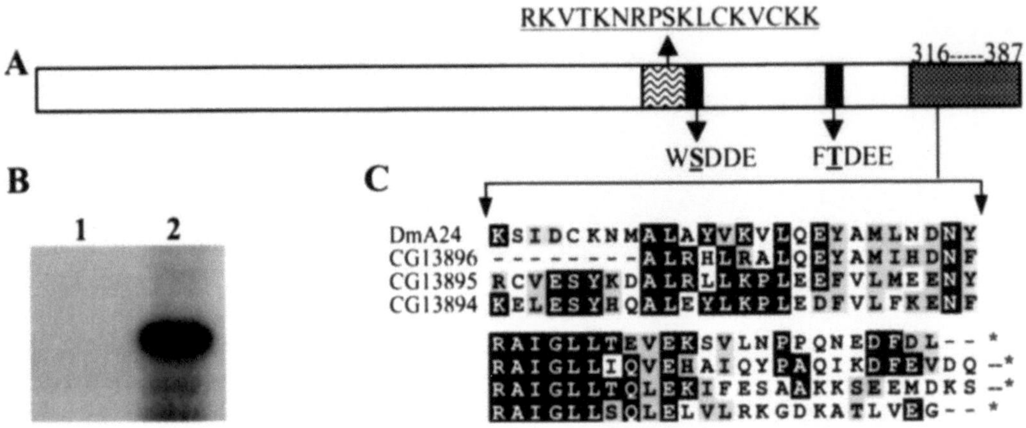

Fig. 3. Functional motifs in DmA24, and phosphorylation by CK2. (A) The motifs contained in DmA24 are a nuclear localization signal (cross-hatched box), two sites for phosphorylation by CK2 (black box), and the shaded box corresponds to sequences exhibiting a high level of conservation to the C-termini of proteins encoded by three other genes, CG13896, CG13895, and CG13894. (B) GST-DmA24 was subjected to phosphorylation using the monomeric α subunit (lane 1) or the embryo holoenzyme (lane 2). (C) Sequences encompassing the penultimate 50 amino acids of DmA24 were aligned to the corresponding regions of the indicated proteins. Identical residues = white letters on a black background; shaded letters = conservative substitutions; *C-terminal end.

Summary of second round of two hybrid screens

As mentioned above, our preliminary two hybrid screens identified a limited repertoire of proteins. We have, therefore, rescreened the embryonic two hybrid cDNA library at greater saturation. We have screened 40 × 10^6 independent transformants, and have isolated approximately 250 *His*$^+$/*LacZ*$^+$ clones. From this pool, we have so far analyzed 75 clones that activate transcription of *HIS3* and *LacZ* at varying levels in a bait-specific manner. Because β/β′ cDNAs are isolated at a relatively high frequency when Dmα is used as a bait [26, 33], we culled the available pool for clones encoding β/β′ via PCR using cDNA-specific primers. The remaining clones were sequenced in order to identify the encoded protein (see Table 1).

As expected, we have identified the α and β/β′ subunits of DmCK2, as well as L22 and E(spl)m7. In addition, the second screen yielded the ribosomal protein L7, which does not contain consensus sites for phosphorylation by CK2. Three proteins with predicted signal transduction functions are CGI5784, CG8159, and CG1603, and all three proteins contain CK2 sites. The most noteworthy of these are the

Table 1. Proteins that interact with the catalytic subunit of DmCK2

Protein[a]	Function	CK2 site[b]
DmCK2β	CK2 subunit	M^{1}SSSEE
DmCK2β′	CK2 subunit	M^{1}TDSDE
DmCK2α	CK2 subunit	None
L22	Ribosomal protein	SSNDDEDDDAE*
L7	Ribosomal protein	None
E(spl)m7	Repressor of neurogenesis	S^{154}DCD
CG15784	Unknown	T133SSSSSSSSSSSSSSSSSSSSSSSSSEDEE S376SSSSSSSSSSSSSSSSSSSSSDGE
CG8159	Putative transcription factor	S^{154}SSTEDD
CG1603	Putative transcription factor	T^{150}DDDD
Surf6	RNA processing in nucleolus	EDEET^{59}YEE DS^{87}DDVEDE
Spalt	Homeotic	E^{274}EDTEEDADQEQDQEQFIDTYEEEE

[a]Protein names beginning with the prefix CG refer to expressed genes whose functions are presently unknown. The names used have been assigned by Berkeley Drosophila Genome Project. [b]The CK2 consensus motif is shown with a superscript indicating the position of the site relative to the full length protein. *C-terminus.

$T^{133}S_{22}EDEE$ and $S^{376}S_{21}DGE$ sites in CG15784, a protein predicted to contain coiled-coils and three nuclear localization signals. Experiments to address phosphorylation of this protein are currently underway . It should be noted that if this is in fact the case, the former site could theoretically account for phosphorylation stoichiometry of ≥25 on a mole/mole basis. Two proteins identified in this screen that were not previously known to interact with CK2 are Surf6 and Spalt. Although the function of the former protein is not known in *Drosophila*, its human homolog is localized to the nucleolus, and is involved in RNA processing [42]. Spalt is an evolutionarily conserved protein which provides homeotic gene function in the head and tail region of the *Drosophila* embryo [43]. Like the case with ANTP, Spalt contains 3 consensus sites for CK2 in the vicinity of its homeobox domain, suggesting that the activity of this protein may also be restricted to regions of the embryo via phosphorylation . Future studies aimed at biochemical and molecular genetic analysis of phosphorylation of Surf6 may provide clues to the role of this enzyme in regulating nucleolar functions, whereas analysis with E(spl)m7 and Spalt should clarify the role of this enzyme in a developmental context.

Interestingly, most of the proteins identified contain CK2 sites and are predicted to be localized to the nucleus or the nucleolus. This not surprising given that CK2 functions extensively in both of these cellular locales [44]. Given our results with the E(spl) family [24], wherein ability to be phosphorylated is the basis for interaction with CK2, one would predict that the proteins we have isolated may in part be due to the presence of sites for modification by CK2. However, none of the potentially best substrates for DmCK2 that were predicted from a genome-wide analysis in *Drosophila* [45], have so far been identified in our two hybrid screens. A notable example is *Drosophila* D1 [46], which is an excellent substrate for DmCK2 *in vitro* and *in vivo*, but does not exhibit a two hybrid interaction with either subunit (Bidwai, unpublished). In addition, results with yeast Fpr3 [5, 47] suggest that secondary structure constraints do influence interaction and/or phosphorylation, and suggest that the mechanism of interaction of this enzyme with these potential substrates may be distinct in each case. Surprisingly, our two hybrid screens appear to be lacking for many highly conserved proteins that have been identified in similar screens with the human enzyme [48, 49]. One possible reason underlying the isolation of distinct proteins in our screens could be the utilization of embryonic libraries, which might be underrepresented for transcripts encoding proteins that are maternally deposited, or represent genes that do not become transcriptionally active until, perhaps, the larval, pupal, or adult stages. Future screens utilizing libraries spanning a wider developmental time-scale will, therefore, be needed to identify those genes that are stage specific, and whose products interact with CK2.

Acknowledgements

We thank Claiborne Glover for providing valuable resources. This work was supported by American Cancer Society Grant (RPG 9918901) to A.P.B.

References

1. Allende JE, Allende CC: Protein kinase CK2: An enzyme with multiple substrates and a puzzling regulation. FASEB J 9: 313–322, 1995
2. Pinna LA: A historical view of protein kinase CK2. Cell Mol Biol Res 40: 383–390, 1994
3. Kelliher MA, Seldin DC, Leder P: Tal-1 induces T cell acute lymphoblastic leukemia accelerated by casein kinase II alpha. EMBO J 15: 5160–5166, 1996
4. Seldin DC, Leder P: Casein kinase II alpha transgene-induced murine lymphoma: Relation to Theileriosis in cattle. Science 267: 894–897, 1995
5. Wilson LK, Dhillon N, Thomer J, Martin GS: Casein kinase II catalyzes tyrosine phosphorylation of yeast nucleolar immunophilin Fpr3. J Biol Chem 272: 12961–12967, 1997
6. Glover CVC: On the physiological role of casein kinase II in *Saccharomyces cerevisiae*. Prog Nucleic Acid Res Mol Biol 59: 95–133, 1998
7. Meggio F, Pinna LA, Marchiori F, Borin G: Polyglutamyl peptides: A new class of inhibitors of type-2 casein kinases. FEBS Lett 162: 235–238, 1983
8. Hathaway GM, Traugh JA: Interaction of polyamines and magnesium with casein kinase II. Arch Biochem Biophys 233: 133–138, 1984
9. Bidwai AP, Hanna DE, Glover CVC: Purification and characterization of casein kinase II (CKII) from ΔCKA1 ΔCKA2 *S. cerevisiae* rescued by *Drosophila* CKII subunits: The free catalytic subunit of casein kinase II is not toxic *in vivo*. J Biol Chem 267: 18790–18796, 1992
10. Hu E, Rubin CS: Expression of wild-type and mutated forms of the catalytic (alpha) subunit of *Caenorhabditis elegans* casein kinase II in *Escherichia coli*. J Biol Chem 265: 20609–20615, 1990
11. Lin WJ, Tuazon PT, Traugh JA: Characterization of the catalytic subunit of casein kinase II expressed in *Escherichia coli* and regulation of activity. J Biol Chem 266: 5664–5669, 1991
12. Meggio F, Boldyreff B, Marin O, Pinna LA, Issinger OG: Role of the beta subunit of casein kinase-2 on the stability and specificity of recombinant reconstituted holoenzyme. Eur J Biochem 204: 293–297, 1992
13. Guerra B, Boldyreff B, Sarno S, Cesaro L, Issinger OG, Pinna LA: CK2: A protein kinase in need of control. Pharmacol Ther 82: 303–313, 1999
14. Bidwai AP, Reed JC, Glover CVC: The phosphorylation of Calmodulin by the catalytic subunit of casein kinase II is inhibited by the regulatory subunit. Arch Biochem Biophys 300: 265–270, 1993
15. Meggio F, Boldyreff B, Marin O, Marchiori F, Perich JW, Issinger OG, Pinna LA: The effect of polylysine on casein-kinase-2 activity is influenced by both the structure of the protein/peptide substrates and the subunit composition of the enzyme. Eur J Biochem 205: 939–945, 1992
16. Padmanabha R, Chen-Wu JLP, Hanna DE, Glover CVC: Isolation, sequencing, and disruption of the yeast *CKA2* gene: Casein kinase II is essential for viability in *Saccharomyces cerevisiae*. Mol Cell Biol 10: 4089–4099, 1990
17. Snell V, Nurse P: Genetic analysis of cell morphogenesis in fission yeast – a role for casein kinase II in the establishment of polarized growth. EMBO J 13: 2066–2074, 1994

18. Hanna DE, Rethinaswamy A, Glover CVC: Casein kinase II is required for cell cycle progression during G1 and G2/M in *Saccharomyces cerevisiae*. J Biol Chem 270: 25905–25914, 1995
19. Rethinaswamy A, Birnbaum MJ, Glover CVC: Temperature-sensitive mutations of the CKAI gene reveal a role for casein kinase II in maintenance of cell polarity in *Saccharomyces cerevisiae*. J Biol Chem 273: 5869–5877, 1998
20. Saxena A, Padmanabha R, Glover CVC: Isolation and sequencing of cDNA clones encoding alpha and beta subunits of *Drosophila melanogaster* casein kinase II. Mol Cell Biol 7: 3409–3417, 1987
21. Fields S, Sternglanz R: The two-hybrid system: An assay for protein–protein interactions. Trends Genet 10: 286–292, 1994
22. Jaffe L, Ryoo H-D, Mann RS: A role for phosphorylation by casein kinase II in modulating Antennapedia activity in *Drosophila*. Genes Dev 11: 1327–1340, 1997
23. Willert K, Brink M, Wodarz A, Varmus H, Nusse R: Casein kinase 2 associates with and phosphorylates Dishevelled. EMBO J 16: 3089–3096, 1997
24. Trott RL, Kahve M, Paroush Z, Bidwai AP: *Drosophila melanogaster* casein kinase II interacts with and phosphorylates the basic-helix-loop-helix (BHLH) proteins m5, m7, and m8 derived from the Enhancer of split complex. J Biol Chem 276: 2159–2167, 2001
25. Ausubel FM, Brent R, Kingston RE, Moore DD, Seidman JG, Smith JA, Struhl K: Current Protocols in Molecular Biology. Wiley Interscience, New York, 1989, pp 2003–2039
26. Bidwai AP, Zhao WF, Glover CVC: A gene located at 56F1-2 in *Drosophila melanogaster* encodes a novel metazoan beta-like subunit of casein kinase II. Mol Cell Biol Res Commun 1: 21–28, 1999
27. Feilotter HE, Hannon GJ, Ruddell CJ, Beach D: Construction of an improved host strain for two hybrid screening. Nucleic Acids Res 22: 1502–1503, 1994
28. Durfee T, Becherer K, Chen PL, Yeh SH, Yang Y, Kilburn AE, Lee WH, Elledge SJ: The retinoblastoma protein associates with the protein phosphatase type 1 catalytic subunit. Genes Dev 7: 555–569, 1993
29. Miller JH: Experiments in Molecular Genetics. Cold Spring Harbor Laboratory Press, New York, 1972, pp 352–355
30. Gyuris J, Golemis E, Chertkov H, Brent R: Cdi1, a human G1 and S phase protein phosphatase that associates with cdk2. Cell 75: 791–803, 1993
31. Langer-Safer PR, Levine M, Ward DC: Immunological method for mapping genes on *Drosophila* polytene chromosomes. Proc Natl Acad Sci (USA) 79: 4381–4385, 1982
32. Laemmli UK: Cleavage of structural proteins during the assembly of the head of bacteriophage T4. Nature (London) 227: 680–685, 1970
33. Bidwai AP, Saxena A, Zhao W, McCann RO, Glover CVC: Multiple, closely spaced alternative 5′ exons in the DmCKIIbeta gene of *Drosophila melanogaster*. Mol Cell Biol Res Commun 3: 283–291, 2000
34. Campos-Ortega JA: Neurogenesis in *Drosophila*: An historical perspective and some prospects. Perspect Dev Neurobiol 4: 267–271, 1997
35. Kuenzel EA, Mulligan JA, Sommercom J, Krebs EG: Substrate specificity determinants for casein kinase H as deduced from studies with synthetic peptides. J Biol Chem 262: 9136–9140, 1987
36. Litchfield DW, Arendt A, Lozeman FJ, Krebs EG, Hargrave PA, Palczewski K: Synthetic phosphopeptides are substrates for casein kinase II. FEBS Lett 261: 117–120, 1990
37. Glover CVC, Shelton ER, Brutlag DL: Purification and characterization of a type II casein kinase from *Drosophila melanogaster*. J Biol Chem 258: 3258–3256, 1983
38. Ackerman P, Glover CVC, Osheroff N: Phosphorylation of DNA topoisomerase II *in vivo* and in total homogenates of *Drosophila* Kc cells: The role of casein kinase II. J Biol Chem 263: 12653–12660, 1988
39. Bojanowski K, Filhol O, Cochet C, Chambaz EM, Larsen AK: DNA topoisomerase H and casein kinase II associate in a molecular complex that is catalytically active. J Biol Chem 268: 22920–22926, 1993
40. Cardenas ME, Walter R, Hanna D, Gasser SM: Casein kinase II copurifies with yeast DNA topoisomerase II and re-activates the dephosphorylated enzyme. J Cell Biol 104: 533–543, 1993
41. Miyata Y, Yahara I: The 90-kDa heat shock protein, HSP90, binds and protects casein kinase II from self aggregation and enhances its kinase activity. J Biol Chem 267: 7042–7047, 1992
42. Magoulas C, Fried M: The Surf-6 gene of the mouse surfeit locus encodes a novel nucleolar protein. DNA Cell Biol 15: 305–316, 1999
43. Kuhnlein R, Frommer G, Friedrich M, Gonzalez-Gaitan M, Weber A, Wagner- Bernholz J, Gehring W, Jackle H, Schuh R: Spalt encodes an evolutionarily conserved zinc finger protein of novel structure which provides homeotic gene function in the head and tail region of the *Drosophila* embryo. EMBO J 13: 168–179, 1994
44. Ahmed K: Nuclear matrix and protein kinase CK2 signaling. Crit Rev Eukaryot Gene Expr 9: 329–336, 1999
45. Bidwai AP: Structure and function of casein kinase II. Rec Res Devel Mol Cell Biol 1: 51–82, 2000
46. Ashley CT, Pendleton CG, Jennings WW, Saxena A, Glover CVC: Isolation and sequencing of cDNA clones encoding *Drosophila* chromosomal protein D1: A repeating motif in proteins which recognize AT DNA. J Biol Chem 264: 8394–8401, 1989
47. Marin O, Meggio F, Sarno S, Cesaro L, Pagano MA, Pinna LA: Tyrosine versus serine/threonine phosphorylation by protein kinase casein kinase-2. J Biol Chem 274: 29260–29265, 1999
48. Kusk M, Ahmed R, Thomsen B, Bendixen C, Issinger OG, Boldyreff B: Interactions of protein kinase CK2beta subunit within the holoenzyme and with other proteins. Mol Cell Biochem 191: 51–58, 1999
49. Grein S, Raymond K, Cochet C, Pyerin W, Chambaz EM, Filhol O: Searching interaction partners of protein kinase CK2beta by two hybrid screening. Mol Cell Biochem 191: 105–109, 1999

Molecular and Cellular Biochemistry **227**: 99–105, 2001.

A gene located at 72A in *Drosophila melanogaster* encodes a novel zinc-finger protein that interacts with protein kinase CK2

Madhavi Kalive,[1] Regina L. Trott[1] and Ashok P. Bidwai[1,2]
[1]*Department of Biology and* [2]*Mary Babb Randolph Cancer Center, West Virginia University, Morgantown, WV, USA*

Abstract

Drosophila melanogaster protein kinase CK2 (DmCK2) is a Ser/Thr protein kinase composed of catalytic α and regulatory β subunits associated as an $\alpha_2\beta_2$ heterotetramer. Using the two hybrid system, we have screened a *Drosophila* embryo cDNA library in order to identify proteins that interact with DmCK2α. One of these cDNAs encodes a novel previously undescribed zinc-finger protein, which we call ZFP47. ZFP47 interacts with DmCK2α but not with DmCK2β, indicating that this interaction is specific for the catalytic subunit of CK2. *In situ* hybridization to polytene chromosomes indicates that the corresponding gene is located at the 72A interval of chromosome III. Sequence analysis indicates that ZFP47 contains a consensus site for phosphorylation by CK2, 4 C_2H_2-type zinc-fingers, and a bipartite nuclear localization signal. Consistent with the prediction of a site for phosphorylation by CK2, we demonstrate that ZFP47 is phosphorylated by CK2 purified from *Drosophila* embryos. These studies demonstrate that ZFP47 is a new physiological partner and substrate of CK2. (Mol Cell Biochem **227**: 99–105, 2001)

Key words: protein kinase, CK2, *Drosophila*, two hybrid

Introduction

Protein kinase CK2 is a highly conserved enzyme that is ubiquitous among eukaryotes. CK2 is composed of catalytic (α) and regulatory (β) subunits that combine to form an $\alpha_2\beta_2$ holoenzyme [1]. CK2 from most organisms is composed of two distinct α subunit isoforms. The only exceptions identified to date are *D. melanogaster* [2], *Caenorhabditis elegans* [3], and *Schizosaccharomyces pombe* [4], which contain a single α subunit. On the other hand, β subunit heterogeneity has so far been documented in *Saccharomyces cerevisiae* [5], *D. melanogaster* [6], and *Arabidopsis thaliana* [7].

Studies utilizing synthetic peptides suggests that the sequence S/T-x-x-D/E defines a minimal consensus for phosphorylation by CK2, and that acidic residues at the n+1 and n+2 positions relative to the phosphoacceptor dramatically stimulate phosphorylation [8, 9]. In addition, CK2 has also been reported to phosphorylate yeast Fpr3 at a Tyr residue [10], although it has been suggested that structural features, rather than the sequence *per se*, favor this non-canonical phosphorylation [11]. Consistent with the preference of CK2 for hyperacidic domains, a number of proteins critical for transcription, cell-cycle progression, and signal transduction contain such sites, and are known to be phosphorylated by the enzyme *in vitro* and *in vivo* (reviewed in [1]. Comparisons between the monomeric α subunit and the $\alpha_2\beta_2$ holoenzyme demonstrate that the β subunit increases the V_{max} approximately 5–50-fold with marginal effects on the k_m for the protein substrate [12–14]. These results suggest that substrate specificity is likely to be integral to the α subunit. CK2 is stimulated by polybasic compounds such as polylysine and protamine [15, 16], and inhibited by polyacidic compounds such as polyAsp/Glu [17], but the *in vivo* relevance of these observations remains unknown.

Genetic analyses in yeast have demonstrated that CK2 activity is essential for cell cycle progression [18], cytokinesis

Address for offprints: A.P. Bidwai, Department of Biology, P.O. Box 6057, Brooks Hall, West Virginia University, Morgantown, WV 26506-6057, USA (E-mail: abidwai@wvu.edu)

[19], and cell polarity [20]. In addition, overexpression of CK2 elicits leukemias and lymphomas in transgenic mice [21, 22] suggesting that the enzyme is a potential oncoprotein. In contrast to these model systems, analysis of *D. melanogaster* CK2 (DmCK2) has lagged despite the fact that the cDNAs encoding DmCK2 were the first to be isolated [2], and the enzyme from this organism has been subjected to extensive biochemical [23–26] and molecular [6, 27] analysis. This deficiency is due to the lack of mutations in the genes encoding DmCK2 subunits. However, an alternative strategy that has been successful applied, is the identification of CK2-interacting proteins via the two hybrid approach. Because many of these proteins appear to be substrates, misexpression of transgenes encoding non-phosphorylated and/or constitutively phosphorylated variants via the Gal4-system [28], followed by phenotypic analysis have begun to clarify functions of DmCK2. Studies along these routes have demonstrated that the Homeobox protein ANTP is negatively regulated by CK2 during embryogenesis [29], and that transcriptional repressors derived from the *Enhancer of Split Complex* such as m5, m7, and m8, which regulate neurogenesis, and the segment polarity protein, Dishevelled, are targets of this enzyme [30, 31]. These results suggest that CK2 is likely to be a critical regulator of embryogenesis as well as cell-fate determination.

We describe here the isolation of a cDNA that encodes a novel zinc-finger protein, which we call ZFP47. Sequence analysis suggests that ZFP47 has a predicted mass of 47 kDa, and contains 4 C_2H_2 zinc-fingers, a bipartite nuclear localization signal, and a consensus site for phosphorylation by CK2. Consistent with the latter prediction, we demonstrate that ZFP47 is phosphorylated by the holoenzyme purified from *Drosophila* embryos. These studies suggest that ZFP47 is a new physiological partner and substrate of DmCK2.

Materials and methods

Construction of two hybrid plasmids and Gal4-based yeast two hybrid screening

The construction of plasmids expressing DmCK2α and DmCK2β as C-terminal fusions with the DNA-binding (DB), and activation-domain (AD) of *S. cerevisiae* Gal4 has been described previously [6]. Two hybrid plasmids suitable for expressing the α, α′, β, and β′ subunits of yeast CK2 were made by PCR amplification of the respective open reading frames, and subcloning in frame with Gal4AD contained in the vector, pGAD424 [32]. The PCR amplified open reading frames encoding all 4 subunits of yeast CK2 were completely sequenced on an Applied Biosystems, Inc. model 373A DNA sequencer using the Prism Dye Terminator Cycle sequencing kit (Applied Biosystems) using custom primers to ensure that no errors were introduced.

Yeast strain HF7C [33] expressing Gal4DB-DmCK2α (the bait) was used to screen a 3–18-h *D. melanogaster* embryo two hybrid cDNA library. A total of 2×10^6 transformants were plated on glucose dropout medium lacking tryptophan, leucine, and histidine [34], and colonies exhibiting rapid growth were counterscreened for expression of *LacZ* [35]. The library plasmids were recovered from His+/LacZ+ clones, and used to retransform HF7C expressing Gal4DB-alone, GAL4DB-Dmα and GAL4DB-Dmβ. Those cDNAs which induced expression of *HIS3* and *LacZ* only in response to GAL4DB-Dmα were identified by sequencing. One of these cDNAs, DmA35, encodes a novel zinc-finger protein, which we call ZFP47. This cDNA and its encoded protein are the subject of this study.

In situ *hybridization to polytene chromosomes*

In situ hybridization to polytene chromosomes was carried out as described [36]. Briefly, the DmA35 cDNA was labeled by nick translation with biotin-dUTP (BioRad Laboratories) and hybridized at high stringency to salivary gland squashes prepared from third instar larvae of *D. melanogaster*. Bound probe was visualized with a streptavidin-alkaline phosphatase conjugate (BioRad Laboratories). Slides were examined by phase-contrast microscopy and photographed at 400 ×.

Phosphorylation of ZFP47 by CK2

ZFP47 was expressed as a fusion protein with *Schistosoma japonicum* glutathione-S-transferase (GST), and purified as previously described [30]. Following purification, the fusion protein was exchanged into storage buffer (50 mM Tris, pH 8.0, 0.5 mM EDTA, 10% glycerol, 200 mM NaCl, 1 mM PMSF) using a Biomax-10K centrifugal filter device (Millipore). Phosphorylations of GST-ZFP47 fusion proteins were carried out in 50 mM Tris, pH 8.5, 100 mM NaCl, 10 mM $MgCl_2$, 10 μM ATP, 5 μCi [γ-^{32}P]ATP in a total vol. of 40 μl, at 25°C. The reactions were initiated by the addition of 5 μl of the enzyme at a concentration of 4 μg/ml. For studying the effects of poly(DL)lysine on phosphorylation, reactions were supplemented at a final concentration of 100 μg/ml. The reactions were terminated by adding 10 μl of 5 × sample buffer [37], boiled for 5 min, and subjected to electrophoresis on 12% acrylamide gels containing sodium dodecylsulfate. Gels were stained with Coomassie Blue, and exposed to Kodak XAR-5 film at room temperature.

Results and discussion

Isolation of cDNAs encoding ZFP35

The yeast strain HF7C expressing Gal4DB-DmCK2α as a bait was used to screen a *D. melanogaster* embryonic two hybrid cDNA library. From approximately 2 × 10^6 transformants, 15 clones that activated transcription of *HIS3* and *LacZ* were recovered. All 15 clones induced the two reporter genes only when co-transformed with DmCK2α (data not shown). Sequencing of the cDNAs revealed that seven of the clones encode DmCK2β [27], one encodes DmCK2α [6], two encode DmCK2β′, a novel isoform of the β subunit [6], three encode the neurogenic transcriptional repressor, E(spl)m7 [30], one (DmA24) encodes a potential transcription factor, and one (DmA35) encodes a novel zinc-finger protein (this report).

The library plasmid was recovered from the yeast clone DmA35 (henceforth referred to as pA35), and retested for its ability to induce transcription of the reporter genes, *HIS3* and *LacZ*. As expected, no observable induction of the *HIS3* or *LacZ* genes was observed when HF7C expressed pA35 by itself (Figs 1A and 1B). On the other hand, a robust transcription of *HIS3* and *LacZ* was observed when cells coexpressed pA35 and DmCK2α, suggesting that reporter gene induction was dependent on the presence of a Gal4DB-fusion protein. Surprisingly, coexpression of pA35 and DmCK2β weakly elicited transcription of the *HIS3* gene (based on colony size on minimal medium lacking histidine, see Fig. 1A), whereas that of *LacZ* was undetectable using either ONPG or Xgal as a substrate (Fig. 1B). At face value, the former result would imply that the protein encoded by the pA35 cDNA directly interacts, albeit weakly, with DmCK2β. We, however, consider this possibility unlikely for a number of reasons. Explicit testing of the interactions between the subunits of DmCK2 via the two hybrid system indicates that the heterotypic interaction, i.e. α vs. β, is very robust [6]. On the other hand, the α-α homotypic interaction appears to elicit *HIS3* gene expression (see Fig. 1A), but a counterscreen for *LacZ* activity is negative with ONPG, and exhibits a weak blue color with Xgal as a substrate (Se Fig. 1B). This apparent discrepancy in the transcriptional readout of the two reporter genes is attributable to a lower threshold requirement of the *HIS3* gene product for growth of yeast in the absence of histidine, combined with our observation that Xgal is at least 10-fold more sensitive than ONPG for detection of *LacZ* activity (Trott and Bidwai, unpublished). Furthermore, the histidine-independent growth of an α-α strain is eliminated by 1 mM 3-aminotriazole, an inhibitor of the *HIS3* gene product, whereas growth of the α-β combinations is unaffected at concentrations exceeding 20 mM (data not shown). Consistent with this, the His-independent growth of a strain expressing ZFP47 +

A

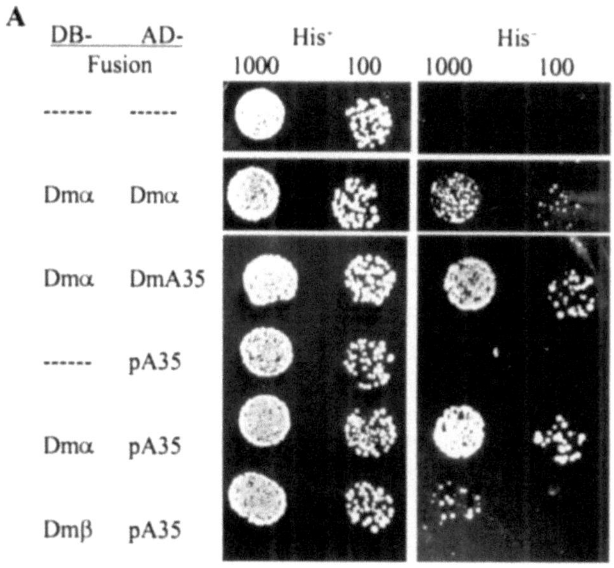

B

DB-	AD-	*LacZ* Activity	
Fusion		ONPG	Xgal
------	------	0	White
Dmα	Dmα	0	Light Blue
Dmα	DmA35	2.3±0.1	Blue
------	pA35	0	White
Dmα	pA35	2.2±0.3	Blue
Dmβ	pA35	0	White

C

Interaction of *Drosophila* and yeast CK2 subunits

	Scα	Scα'	Scβ	Scβ'
Dmα	N.D.	N.D.	+ +	+++++
Dmβ	++	+++++	N.D.	N.D.

Fig. 1. Isolation of ZFP47 and interaction with DmCK2 subunits. (A) *S. cerevisiae* strain HF7C was transformed with plasmids expressing the indicated fusions with GAL4DB or GAL4AD. Untransformed HF7C was used as a control; a dashed line indicates the absence of a plasmid. Following growth in dropout medium, aliquots corresponding to 1000 and 100 cells were plated onto complete (His⁺) or minimal media lacking histidine (His⁻) and incubated at 29°C for 3–4 days. (B) Transformants expressing the indicated fusion proteins were grown in glucose medium, and *LacZ* activity was determined in triplicate as described [35], and is expressed in Miller units. In addition, a qualitative assessment of *LacZ* activity was determined via a filter-based assay using Xgal as a substrate. (C) Yeast strain HF7C was transformed with plasmids expressing the *Drosophila* CK2 subunits as fusions with GAL4DB, and yeast CK2 subunits as fusions with GAL4AD. Transformants were selected, and tested for induction of *HIS3* expression by plating on minimal medium lacking histidine, whereas induction of *LacZ* was tested qualitatively using Xgal, and quantitatively using ONPG as substrates. The interaction data is represented qualitatively, and N.D. indicates not determined.

Dmβ is more sensitive to this inhibitor than one expressing ZFP47 + Dmα (data not shown). In addition, we have observed that Dmβ interacts modestly with the yeast α subunit, but very strongly with the α′ subunit in the two hybrid system, whereas Dmα exhibits a modest interaction with the yeast β subunit, and a strong interaction with yeast β′ (Fig. 1C). The former results are in line with genetic analysis demonstrating that overexpression of Dmβ suppresses the phenotype associated with loss of yeast β/β′ [38]. Taking into account these observations, it is, therefore, conceivable that the apparent weak interaction of pA35 and Dmβ may be bridged by the yeast α subunits. The only rigorous way to exclude this possibility would be to carry out the analysis in a strain lacking α and α′, but such a strain cannot be constructed because CK2 activity is essential for viability [39]. In any event, the analysis demonstrates that pA35 exhibits a robust interaction with DmCK2α and, perhaps, a weak interaction with DmCK2β, although the precise mechanism may be confounded by the interplay of the α/β subunits of yeast CK2.

Nucleotide sequence of the pA35 cDNA, and cytological location and genomic organization of the corresponding gene

The cDNA contained in the plasmid pA35 was completely sequenced as described in Materials and methods. This cDNA (deposited as GenBank Accession No. U44206) is 1251 bp long, contains a single open reading frame of 1230 bp, and based upon the absence of a polyA-tail is lacking its 3' untranslated region (Fig. 2). The putative initiation codon is located within the context CAAA**ATG**, which is an excellent match to the start site consensus for *D. melanogaster*, and satisfies the requirement for an adenosine residue at the n-3 position relative to the initiation codon [40]. Conceptual translation suggests that this cDNA encodes a 410 amino acid polypeptide (which we call ZFP47), with a calculated molecular mass of 47 kDa and a theoretical pI of 8.2.

The ZFP47 cDNA was subsequently used to localize the corresponding gene via *in situ* hybridization to polytene chro-

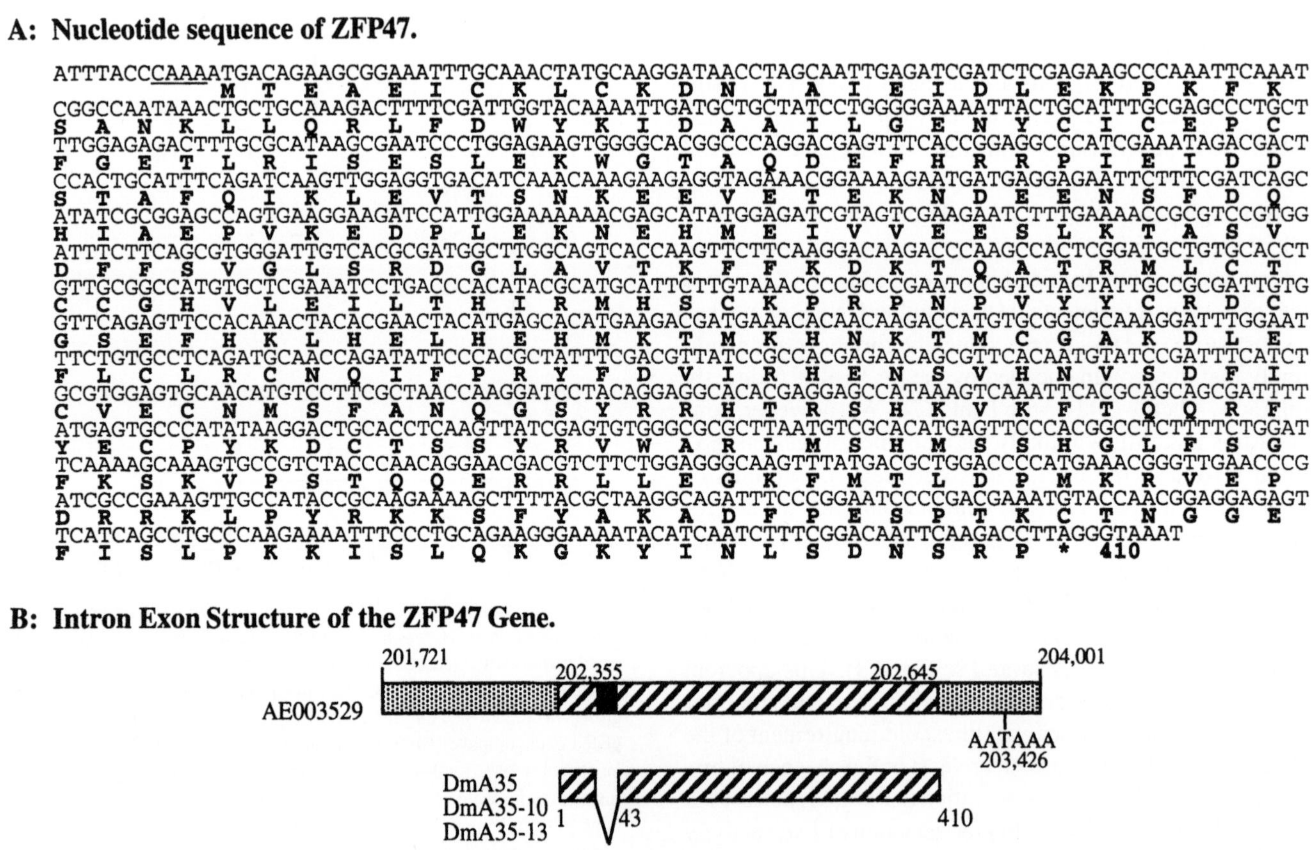

Fig. 2. Nucleotide sequence of DmA35 cDNA, and intron/exon structure of the corresponding gene. (A) The nucleotide sequence of the 1251 bp cDNA (deposited as GenBank Accession # U44206) is shown above the deduced amino acid sequence of ZFP47 (shown in one letter code). The potential Kozak consensus sequence 5' to the initiation codon, ATG, is underlined, and the dotted region at the 3' end of the cDNA indicates the absence of the polyA-addition signal and the polyA-tail. (B) The start and end points of the cDNA sequence are shown as a crosshached box within the context of bp 202,021–203,520 of a genomic clone (GenBank Accession # AE003529, shaded box). The open reading frame encoding ZFP47 is encoded on two exons (crosshatched boxes), separated by an intron (black box), and numbers indicate the amino acids of the protein encoded on the respective exons. The location of the the most likely polyA addition signal AATAAA in the genomic sequence is indicated.

mosomes of third instar larvae. The analysis suggests that the ZFP47 gene is located on the left arm of chromosome III at the 72A polytene interval (data not shown). This result is in full agreement with data from the *Drosophila* genome sequencing project demonstrating identical cytology, and indicating that the ZFP47 gene is single copy. With the recent completion of the *Drosophila* genome sequence, we have used the ZFP47 cDNA sequence to identify the corresponding genomic sequence and its organization. To this end, we have identified a 250 kb genomic clone (Gen Bank Accession # AE003529) that harbors sequences corresponding to the ZFP47 gene from nucleotide positions 202,355–202,645 (see Fig. 2B). No additional genomic clones corresponding to ZFP47 were identified in this analysis, suggesting that this gene is, in fact, single copy in the *Drosophila* genome, in agreement with our cytological analysis (see above). A comparison of this region of the genomic clone to the isolated cDNAs suggests that the ZFP47 protein is encoded on two exons (Fig. 2B). The first exon encompassing amino acids 1–42, and the second exon encompassing amino acids 43–410 are separated by a single 57 bp intron which satisfies the requirement for GT at the donor splice site, and an AG at the acceptor site (data not shown). In addition, the genomic sequence contains a single polyA-addition signal, AATAAA, that is 781bp 3' to the end of the coding region. In this regard, neither the two hybrid clones (pA35), nor two additional clones isolated from a commercially available *Drosophila* embryonic library (DmA35-10 and DmA35-13) contain a polyA-tail. It is, therefore, difficult to predict whether the polyA-addition signal in the corresponding genomic sequence is the preferred site *in vivo*. In addition, the 3' region of the genomic sequence does not exhibit the presence of polyA-like sequences, thus precluding the possibility that the ZFP47 polypeptide is encoded by a processed pseudogene that arose via retrotransposition during *Drosophila* evolution.

Motifs in ZFP47 and phosphorylation by CK2

The deduced amino acid sequence of ZFP47 was used to conduct a BlastP analysis against all available sequences (GenBank release of March 2001). The closest similarity of ZFP47 observed by this analysis was to the 4 zinc-finger domains of a human [41] and murine [42] kruppel-like protein, PLZF, that has been proposed to encode the promyelocytic leukemia associated zinc-finger protein (Gen Bank Accession # AAD03619 and AAB33814, respectively). This conservation is, however, limited to the putative zinc-finger-domains, both with regards to the zinc-ligands and the highly conserved aromatic residue, Phe/Tyr, between ligand 1 and

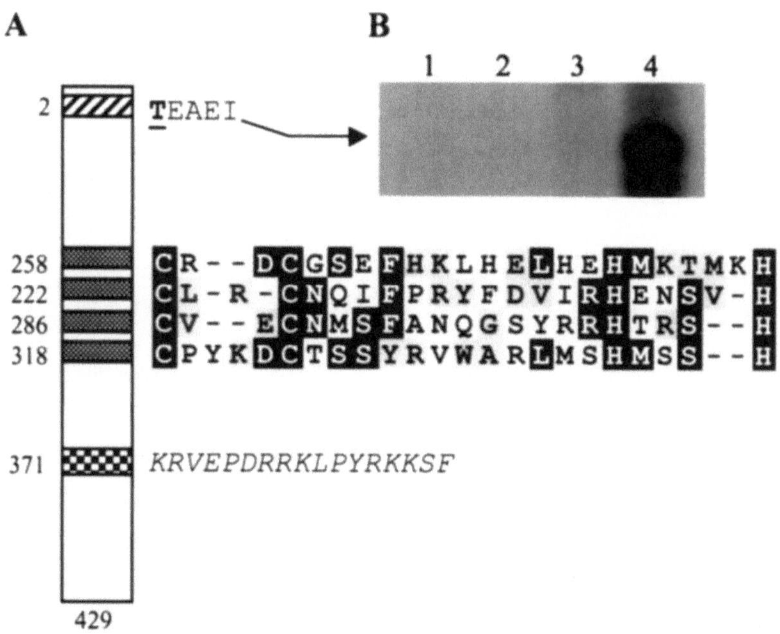

Fig. 3. Functional motifs in ZFP47 and phosphorylation by CK2. (A) The amino acid sequence of ZFP47 was analyzed for the presence of functional motifs using algorithms available via the world wide web. The motifs contained in ZFP47 are an N-terminal consensus site for phosphorylation by CK2 (cross-hatched box), 4 C_2H_2-type zinc-finger domains (shaded boxes), and a bipartite nuclear localization signal (checkerboard box). The numbers next to the boxes correspond to the position of the motifs relative to the predicted N-terminus. (B) GST-ZFP47 fusion protein was purified, and subjected to phosphorylation using monomeric α subunit (lanes 1 and 2), and embryo CK2 (lanes 3 and 4). Samples were phosphorylated in either the absence (lanes 1 and 3), or presence of 100 μg/ml poly(DL)lysine (lanes 2 and 4). Samples were electrophoresed in 12% SDS-polyacrylamide gels, stained with Coomassie blue, and autoradiographed. The arrow indicates the mobility of full-length GST-ZFP47.

2 (data not shown). Because zinc-finger motifs are, in general, highly conserved with regard to the spacing of the putative zinc-ligands, Cys and/or His [43], and that the conservation of the intervening residues is weak, suggests that the ZFP47 gene encodes a novel zinc-finger protein. In the absence of a clear homolog in other organisms that could be used to predict a possible function for ZFP47, we have used the deduced amino acid sequence of this protein to predict the presence of functional and/or structural motifs. This analysis suggests that the encoded polypeptide contains 3 such motifs (Fig. 3A). The first of these is N-terminal sequence, M^{1}TEAE, which fits the consensus for phosphorylation by CK2 and satisfies the requirement of the enzyme for a phosphoacceptor (Ser/Thr) with acidic residues at the n+1, and n+3 positions [11]. The second motif, present at 4 copies, is predicted to form a C_2H_2-type zinc-finger, and satisfies the requirement for an aromatic residue, in this case Phe/Tyr, between the second (Cys) and third (His) ligands that are predicted to coordinate the zinc atom [43]. The third motif, K^{318}RVEPDRRKLPYRKKSF, is highly homologous to bipartite nuclear localization sequences [44].

Analysis of the primary sequence of ZFP47 suggests the presence of a consensus site for phosphorylation by CK2. In order to ascertain whether ZFP47 is, in fact, phosphorylated by CK2, we have bacterially expressed this protein as a GST-fusion, purified it to homogeneity, and subjected it to phosphorylation using two isoforms of DmCK2, i.e. the monomeric α subunit purified from a yeast expression system [24], and the $\alpha_2\beta_2$ holoenzyme purified from *Drosophila* embryos [26]. Although the former isoform is closer to the scenario of the two hybrid interaction, the latter isoform may reflect more accurately the situation *in vivo*. We have not explicitly tested for phosphorylation of unfused GST, as we have previously demonstrated that GST is not a substrate for either isoform of CK2 [30]. The results demonstrate that ZFP47 does not appear to be phosphorylated by Dmα (Fig. 3B, lane 1), or by the holoenzyme purified from embryos (Fig. 3B, lane 3). We, therefore, tested whether phosphorylation of ZFP47 by either isoform was responsive to poly(DL)lysine. The results demonstrate that while phosphorylation by the monomeric α subunit remains unresponsive to this polybasic activator, phosphorylation by the holoenzyme is markedly stimulated in its presence (Fig. 3B, compare lanes 2 and 4). The phosphorylation of GST-ZFP47 in a poly(DL)lysine-dependent manner is not due to a non specific phosphorylation, perhaps, of the GST affinity tag itself, as we have previously demonstrated that neither BSA [12] nor GST [30], are targeted by CK2 upon polybasic activation. This result appears consistent with the proposal that the β subunit of CK2 largely mediates interaction with polybasic activators [12, 45]. Therefore, analogous to our previous result with Calmodulin [12], and the actin bundling protein, Sac6 (Bidwai, unpublished), phosphorylation of ZFP47 appears to be dependent on the presence of the regulatory (β) subunit of DmCK2, and maximal phosphorylation appears to also be dependent on prior activation of the enzyme by poly(DL)lysine. ZFP47 thus represents a substrate whose phosphorylation is regulated by the β subunit in a polylysine-dependent manner. However, its role in modulating physical association of CK2 with ZFP47 is somewhat unclear for reasons outlined above. Given the very weak interaction of ZFP47 with Dmβ, and a relatively robust interaction with Dmα, it is conceivable that the interaction of this protein with CK2 may be greatly enhanced when tested against the holoenzyme, i.e. the protein may contain regions of interaction that involve both subunits. However, such analysis is currently precluded by the unavailability of a yeast strain that produces DmCK2 holoenzyme suitable for two hybrid analysis. That CK2 interacts with, and phosphorylates this protein is also consistent with the observation that this kinase exists in a complex with many of its *in vivo* substrates such as Topoisomerase II [46], HSP90 [47], ANTP [29], m5/7/8 [30], and Dishevelled [31] to name a few. However, unlike the case with ANTP, Topoisomerase II, etc., the consequences of CK2-mediated phosphorylation of ZFP47 on its *in vivo* functions are difficult to predict *a priori*, because no mutations that colocalize at the 72A interval have so far been reported in this organism. However, the recently initiated effort to generate systematic mutations in all of the genes in this model organism, should elaborate on the biological functions of this gene, and will permit a functional analysis of the phosphorylation of this protein by CK2 *in vivo*.

Acknowledgements

We thank Claiborne Glover for providing valuable resources. This work was supported by American Cancer Society Grant (RPG-9918901-DDC) to A.P.B.

References

1. Glover CVC: On the physiological role of casein kinase II in *Saccharomyces cerevisiae*. Prog Nucleic Acid Res Mol Biol 59: 95–133, 1998
2. Saxena A, Padmanabha R, Glover CVC: Isolation and sequencing of cDNA clones encoding alpha and beta subunits of *Drosophila melanogaster* casein kinase II. Mol Cell Biol 7: 3409–3417, 1987
3. Hu E, Rubin CS: Casein kinase II from *Caenorhabditis elegans*: Properties and developmental regulation of the enzyme; cloning and sequence analysis of cDNA and the gene for the catalytic subunit. J Biol Chem 265: 5072–5080, 1990
4. Roussou I, Dretta G: The *Schizosaccharomyces pombe* casein kinase II alpha and beta subunits: Evolutionary conservation and positive role for the beta subunits. Mol Cell Biol 14: 576–586, 1994
5. Bidwai AP, Reed JC, Glover CVC: Casein kinase II of *Saccharomyces cerevisiae* contains two distinct regulatory subunits, beta and beta¢. Arch Biochem Biophys 309: 348–355, 1994
6. Bidwai AP, Zhao WF, Glover CVC: A gene located at 56F1-2 in *Drosophila melanogaster* encodes a novel metazoan beta-like subunit of casein kinase II. Mol Cell Biol Res Comm 1: 21–28, 1999

7. Collinge MA, Walker JC: Isolation of an *Arabidopsis thaliana* casein kinase II beta subunit by complementation in *Saccharomyces cerevisiae*. Plant Mol Biol 25: 649–658, 1994
8. Kuenzel EA, Krebs EG: A synthetic substrate specific for casein kinase II. Proc Natl Acad Sci 82: 737–741, 1985
9. Kuenzel EA, Mulligan JA, Sommercorn J, Krebs EG: Substrate specificity determinants for casein kinase II as deduced from studies with synthetic peptides. J Biol Chem 262: 9136–9140, 1987
10. Wilson LK, Dhillon N, Thorner J, Martin GS: Casein kinase II catalyzes tyrosine phosphorylation of yeast nucleolar immunophilin Fpr3. J Biol Chem 272: 12961–12967, 1997
11. Marin O, Meggio F, Sarno S, Cesaro L, Pagano MA, Pinna LA: Tyrosine versus serine/threonine phosphorylation by protein kinase casein kinase-2. J Biol Chem 274: 29260–29265, 1999
12. Bidwai AP, Reed JC, Glover CVC: The phosphorylation of calmodulin by the catalytic subunit of casein kinase II is inhibited by the regulatory subunit. Arch Biochem Biophys 300: 265–270, 1993
13. Hu E, Rubin C: Casein kinase II from *Caenorhabditis elegans*. Cloning, characterization, and developmental regulation of the gene encoding the beta subunit. J Biol Chem 266: 19796–19802, 1991
14. Lin WJ, Tuazon PT, Traugh JA: Characterization of the catalytic subunit of casein kinase II expressed in *Escherichia coli* and regulation of activity. J Biol Chem 266: 5664–5669, 1991
15. Hathaway GM, Traugh JA: Kinetics of activation of casein kinase II by polyamines and reversal of 2,3-bisphosphoglycerate inhibition. J Biol Chem 259: 2850–2855, 1984
16. Hathaway GM, Traugh JA: Interaction of polyamines and magnesium with casein kinase II. Arch Biochem Biophys 233: 133–138, 1984
17. Meggio F, Pinna LA, Marchiori F, Borin G: Polyglutamyl peptides: A new class of inhibitors of type-2 casein kinases. FEBS Lett 162: 235–238, 1983
18. Hanna DE, Rethinaswamy A, Glover CVC: Casein kinase II is required for cell cycle progression during G1 and G2/M in *Saccharomyces cerevisiae*. J Biol Chem 270: 25905–25914, 1995
19. Snell V, Nurse P: Genetic analysis of cell morphogenesis in fission yeast – a role for casein kinase II in the establishment of polarized growth. EMBO J 13: 2066–2074, 1994
20. Rethinaswamy A, Birnbaum MJ, Glover CVC: Temperature-sensitive mutations of the CKA1 gene reveal a role for casein kinase II in maintenance of cell polarity in *Saccharomyces cerevisiae*. J Biol Chem 273: 5869–5877, 1998
21. Kelliher MA, Seldin DC, Leder P: Tal-1 induces T cell acute lymphoblastic leukemia accelerated by casein kinase II alpha. EMBO J 15: 5160–5166, 1996
22. Seldin DC, Leder P: Casein kinase II alpha transgene-induced murine lymphoma: Relation to Theileriosis in cattle. Science 267: 894–897, 1995
23. Ackerman P, Glover CVC, Osheroff N: Phosphorylation of DNA topoisomerase II *in vivo* and in total homogenates of *Drosophila* Kc cells: The role of casein kinase II. J Biol Chem 263: 12653–12660, 1988
24. Bidwai AP, Hanna DE, Glover CVC: Purification and characterization of Casein Kinase II (CKII) from ΔCKA1 ΔCKA2 *S. cerevisiae* rescued by Drosophila CKII subunits: The free catalytic subunit of casein kinase II is not toxic *in vivo*. J Biol Chem 267: 18790–18796, 1992
25. Birnbaum MJ, Wu J, O'Reilley DR, Rivera-Marrero CA, Hanna DE, Miller LK, Glover CVC: Expression, purification and characterization of Drosophila casein kinase II using the baculovirus system. Prot Express Pur 3: 142–150, 1992
26. Glover CVC, Shelton ER, Brutlag DL: Purification and characterization of a type II casein kinase from *Drosophila melanogaster*. J Biol Chem 258: 3258–3256, 1983
27. Bidwai AP, Saxena A, Zhao W, McCann RO, Glover CVC: Multiple, closely spaced alternative 5' exons in the DmCKIIbeta gene of *Drosophila melanogaster*. Mol Cell Biol Res Comm 3: 283–291, 2000
28. Brand AH, Perrimon N: Targeted gene expression as a means of altering cell fates and generating dominant phenotypes. Development 118: 401–415, 1993
29. Jaffe L, Ryoo H-D, Mann RS: A role for phosphorylation by casein kinase II in modulating Antennapedia activity in Drosophila. Genes Dev 11: 1327–1340, 1997
30. Trott RL, Kalive M, Paroush Z, Bidwai AP: *Drosophila melanogaster* casein kinase II interacts with and phosphorylates the basic-helix-loop-helix (bHLH) proteins m5, m7, and m8 derived from the Enhancer of split complex. J Biol Chem 276: 2159–2167, 2001
31. Willert K, Brink M, Wodarz A, Varmus H, Nusse R: Casein kinase 2 associates with and phosphorylates Dishevelled. EMBO J 16: 3089–3096, 1997
32. Fields S, Sternglanz R: The two hybrid system: An assay for protein–protein interactions. Trends Genet 10: 286–292, 1994
33. Feilotter HE, Hannon GJ, Ruddell CJ, Beach D: Construction of an improved host strain for two hybrid screening. Nucleic Acids Res 22: 1502–1503, 1994
34. Ausubel FM, Brent R, Kingston RE, Moore DD, Seidman JG, Smith JA, Struhl K: Current Protocols in Molecular Biology. Wiley Interscience, New York, 1989, pp 2003–2039
35. Miller JH: Experiments in Molecular Genetics. Cold Spring Harbor Laboratory Press, New York, 1972, pp 352–355
36. Langer-Safer PR, Levine M, Ward DC: Immunological method for mapping genes on *Drosophila* polytene chromosomes. Proc Natl Acad Sci 79: 4381–4385, 1982
37. Laemmli UK: Cleavage of structural proteins during the assembly of the head of bacteriophage T4. Nature (London) 227: 680–685, 1970
38. Bidwai AP, Reed JC, Glover CVC: Cloning and disruption of CKB1, the gene encoding the 38-kDa beta subunit of *Saccharomyces cerevisiae* casein kinase II. J Biol Chem 270: 10395–10404, 1995
39. Padmanabha R, Chen-Wu JLP, Hanna DE, Glover CVC: Isolation, sequencing, and disruption of the yeast *CKA2* gene: Casein kinase II is essential for viability in *Saccharomyces cerevisiae*. Mol Cell Biol 10: 4089–4099, 1990
40. Cavener DR, Cavener BA: Translation start sites and mRNA leaders. In: G. Maroni (ed). An Atlas of Drosophila Genes: Sequences and Molecular Features. Oxford University Press, New York, 1993, pp 157–183
41. Zhang T, Xiong H, Kan LX, Zhang CK, Jiao XF, Fu G, Zhang QH, Lu L, Tong JH, Gu BW, Yu M, Liu JX, Licht J, Waxman S, Zelent A, Chen E, Chen SJ: Genomic sequence, structural organization, molecular evolution, and aberrant rearrangement of promyelocytic leukemia zinc finger gene. Proc Natl Acad Sci 96: 11422–11427, 1999
42. Cook M, Gould A, Brand N, Davies J, Strutt P, Shaknovich R, Licht J, Waxman S, Chen Z, Gluecksohn-Waelsch S, Krumlauf R, Zelent A: Expression of the Zinc-Finger Gene PLZF at Rhombomere Boundaries in the Vertebrate Hindbrain. Proc Natl Acad Sci 92: 2249–2253, 1995
43. Aitken A: Protein consensus sequence motifs. Mol Biotechnol 12: 241–253, 1999
44. Kaffman A, O'Shea EK: Regulation of nuclear localization: A key to a door. Annu Rev Cell Dev Biol 15: 291–339, 1999
45. Meggio F, Boldyreff B, Issinger OG, Pinna LA: Casein kinase 2 down regulation and activation by polybasic peptides are mediated by acidic residues in the 55–64 region of the β-subunit: A study with calmodulin as phosphorylatable substrate. Biochem 33: 4336–4342, 1994
46. Cardenas M, Dang Q, Glover CVC, Gasser SM: Casein kinase II phosphorylates the eukaryote-specific c-terminal domain of topoisomerase II *in vivo*. EMBO J 11: 1785–1796, 1992
47. Miyata Y, Yahara I: The 90-kDa heat shock protein, HSP90, binds and protects casein kinase II from self aggregation and enhances its kinase activity. J Biol Chem 267: 7042–7047, 1992

Molecular and Cellular Biochemistry **227**: 107–112, 2001.

Environmental reprogramming of the expression of protein kinase CK2β subunit in fish

Marco Alvarez,[1] Gudrun Kausel,[2] Jaime Figueroa[2] and María Inés Vera[1]

[1]*Laboratorio de Biología Celular y Molecular, Facultad de Ciencias de la Salud, Universidad Andrés Bello and MIFAB, Santiago;* [2]*Instituto de Bioquímica, Facultad de Ciencias, Universidad Austral de Chile, Valdivia, Chile*

Abstract

The dramatic segregation of the nucleolar components in winter-acclimatized carp is the most striking cellular-phenotypical feature observed during the seasonal adaptation of this fish toward the circannual changes in its habitat. Our studies also show that the carp habitat temperature and photoperiod winter conditions provoke a remarkable reduction of both rRNA transcription and the processing of their precursors. To gain knowledge on the mechanisms involved in the regulation of nucleolar activity during the seasonal adaptation process, we studied the behavior of some genes, specifically snoRNA U3 and protein kinase CK2.

Consistent with the reduction in the synthesis and processing of pre-rRNA observed during the cold season, the level of CK2β expression decreases in winter when compared to that attained in summer. Similarly, in winter, liver and kidney cells contain lower levels of CK2β subunit protein compared to summer. CK2 is associated with or modifies different factors and enzymes involved in the nucleolar activity; therefore, its higher or lower content could be part of the molecular mechanisms underlying the nucleolar seasonal changes that occur during the compensatory acclimatization process. (Mol Cell Biochem **227**: 107–112, 2001)

Key words: protein kinase CK2β subunit, nucleolus, fish acclimatization, carp

Introduction

Acclimatization of a eurythermal fish, i.e. the cyclical adaptation to the seasonal changes of its habitat, involves among other cellular and molecular changes, a clear modulation of gene expression [1–6]. We have studied the mechanisms behind this kind of functional adjustment in *Cyprinus carpio* (carp) and observed that phenotypically, one of the most dramatic cellular modifications occurs during winter when the cytoplasm of the hepatocyte is full of glycogen granules and the nucleolar components appear segregated [7]. Concurrently, the transcription of ribosomal RNA (rRNA) and the processing of their precursor (pre-rRNA) decrease notoriously compared to the summer-adapted fish [8, 9]. During the warm season, however, the hepatocyte cytoarchitecture exhibits all the features typical of a metabolically-active cell [7]. The cytoplasm is therefore empty of glycogen, the nucleolar components are intermingled, and the transcription and processing of rRNA occur at a high rate [7–9].

When carp are treated with insulin, either by oral administration or intra-peritoneal injections, a clear reversion of the winter cell phenotype takes place, rendering all of the ultra-structural features that characterize the summer-adapted hepatocyte [10, 11]. Phosphorylation is needed for rRNA synthesis in the nucleolus of eukaryotic cells [12]. Furthermore, insulin enhancement of rRNA synthesis is correlated with protein kinase CK2 phosphorylation of nucleolin as well as of other protein nucleolar factors [13, 14]. At the cellular level, the nucleolar activity in the carp seems to play a central role in the functional adjustment that the seasonal acclimatization imposes upon the fish, and we therefore deemed it important to examine the possible association of protein kinase CK2 expression events with the remarkable nucleolus reorganization that occurs between summer and winter.

Protein kinase CK2, a tetramer composed of the α catalytic and the β regulatory subunits, is known to be a highly conserved serine/threonine phosphorylating enzyme present in the nucleus and cytoplasm of all eukaryotic cells [15]. To

Address for offprints: M.I. Vera, Laboratorio de Biología Celular y Molecular, Universidad Andrés Bello, República 217, Santiago, Chile (E-mail: mvera@abello.unab.cl)

study the expression of this protein kinase in a eurythermal fish undergoing seasonal acclimatization, we isolated the cDNA of the carp CK2β subunit and found a significant differential gene expression between winter- and summer-acclimatized carp [16, 17]. We report herein that the CK2β subunit level clearly changes between seasons in carp liver and kidney cells which, in addition to our previous observations [16], provides evidence that the expression of the CK2β subunit is regulated in this fish.

Materials and methods

Animals and tissues

Seasonal-acclimatized male carp were captured and maintained as described previously [2]. The water temperature was 20–22°C in summer and 8–10°C in winter. Tissue sections for immunocytochemistry experiments were prepared and stored as reported [2]. Briefly, the tissue blocks were fixed in Bouin for 24 h at room temperature, then dehydrated in a graded series of ethanol, ending with 1-butanol and then embedded in Paraplast. Sections 5–7 μm thick were placed on glass slides.

Western blot and immunocytochemical detection

Carp liver and kidney protein extracts for the western blot were prepared as described by Figueroa *et al.* [2]. The tissue was homogenized in 50 mM phosphate buffer, pH 7.4, containing 1 mM EDTA and 1 mM EGTA, precipitated at –20°C after the addition of 5 vol. acetone, and the precipitate was recovered by centrifugation. This step was repeated twice and the precipitate was dissolved in 25 mM Tris-HCl, pH 6.8 containing 5% glycerol, 1% SDS, 0.01% bromophenol blue and 1.5% β-mercaptoetanol and it was then fractionated by PAGE, and electrotransferred to nitrocellulose membranes. The filters were then immunostained according to Stenberger *et al.* [18] and Inostroza *et al.* [19]. The antibody was a kind gift from the laboratory of Dr. David Litchfield, and it corresponds to a rabbit polyclonal antiserum against the C-terminal end (198–215) of a consensus sequence of CK2β subunit [20, 21]. The label was quantified using an automated image digitizing system as described earlier [22], in tissue sections from four different carp for each season. The Student's *t*-test was used to assess differences, $p < 0.05$.

Modelization of carp CK2β subunit

A 3-dimensional homology modeling of the carp CK2β subunit was performed with the derived amino acid sequence from carp cDNA, GeneBank accession number AF133088 [16], using the Swiss Model program, version 3.5 [23]. The comparative protein modeling was performed with the ProMod/ProMod II program and the energy minimization with Gromos96. The 3D-structures used as templates were human CK2β (Protein Data Bank accession numbers 1QF8A and 1QF8B, respectively). Figure 1 was prepared with the MOLMOL program [24].

Results and discussion

During our recent studies on the involvement of CK2β in the complex mechanisms that comprise the cellular and molecular compensatory responses that a eurythermal fish requires to survive the cyclical seasonal changes of its habitat, we found that the carp CK2 regulatory subunit cDNA exhibits a high degree of evolutionary conservation [16]. The homology observed in the amino acid derived sequence with other vertebrates [16] suggested a 3-dimensional structure similar to that described by Chantalat *et al.* [25] by crystallographic analyses of a human CK2β dimer lacking 33 residues at the C-terminus. Upon structural protein modeling of the carp CK2β, as shown in Fig. 1, the destruction box, the acidic cluster, and the zinc finger motif can be easily recognized. The carp CK2β cDNA sequence displays a 94 and 87% identity when compared with the zebrafish and human sequences, respectively. The differences primarily map the third codon's nucleotide. In brief, carp CK2β cDNA codes for a 215-amino acid protein completely homologous to the zebrafish and distinctive with respect to the human CK2β primary structure in only three carboxy terminal amino acids.

When we assessed the level of expression of the carp CK2β subunit gene [16] during the seasonal acclimatization of *C. carpio*, hybridizing liver and kidney tissue sections with a specific 20-mer antisense oligonucleotide probe, most of the CK2β mRNA was found in the nuclei. Lower signals were also detected in the cytoplasm [16, 17]. The identification of CK2β transcripts by *in situ* hybridization was visibly more intensive in the liver and kidney sections from summer-acclimatized fish when compared to those from the cold-season adapted carp. Competitive RT-PCR using total RNA from winter- and summer-acclimatized carp liver as a template and the competitor, an RNA template, confirmed the striking CK2β subunit gene transcriptional differences that the *in situ* assays revealed. Quantitatively, the CK2 regulatory subunit gene transcription is 7.2 times higher in the liver of summer-acclimatized fish than that observed in the cold season-acclimatized carp.

To address the question of whether the differential CK2β transcription followed a similar pattern at the translational level, the carp CK2β subunit content was measured by immunocytochemistry analyses. A polyclonal anti-CK2β antise-

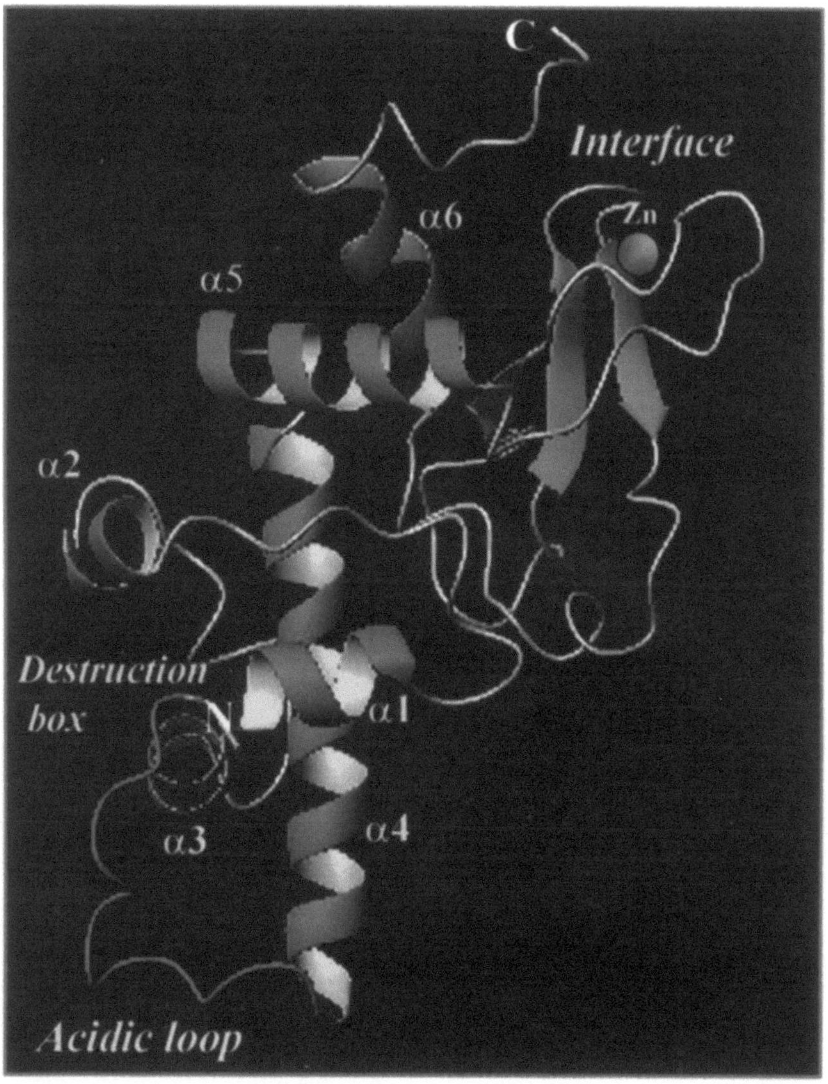

Fig. 1. Ribbon representation of the carp CK2β subunit structure, modeled according to the amino acid sequence derived from carp CK2β cDNA (Gene-Bank, Accession Number AF133088). A homology model was constructed using the Swiss Model program version 3.5 [23] and the comparative protein modeling was performed with the ProMod/ProMod II program and the energy minimization with Gromos96. The 3D-structures used as templates were human CK2β 1QF8A and 1QF8B (Protein Data Bank). The figure was prepared with the MOLMOL program [24].

rum directed toward the C-terminal synthetic peptide $\beta^{198-215}$ [20, 21], kindly provided by David Litchfield, was probed against the total protein extracted from carp liver and kidney tissue. As depicted in Fig. 2A, the antiserum recognizes only one band in both the carp kidney and liver cells, with a molecular mass between 26 and 27 kDa. Thus the antibody exhibited a clear specificity, identifying a protein of the size expected for the animal CK2 regulatory subunit [15]. In addition to the immunoblotting analyses, we immunostained carp liver sections with the aim of gaining knowledge about the cellular localization of the CK2β subunit protein. Consistent with the reports that map CK2 as a predominantly nuclear enzyme [26–28, Ahmed *et al.*, this issue], we also found that cells were stained mainly at the nuclear and perinuclear levels (Fig. 2B). The results attained with the polyclonal anti-CK2β antiserum allowed the quantification of the carp CK2β subunit *vis à vis* the seasonal adaptive state of the fish. Figure 3 depicts the immunostaining attained when liver (A) and kidney tissue (B) from summer- and winter-acclimatized carp were probed. The stained sections clearly show differences in carp CK2β content between summer and winter. The significant differences in CK2 regulatory subunit content were validated when the digitized signals were quantified. Figure 3 shows that liver cells from summer-acclimatized fish exhibited twice the amount of CK2β subunit when compared to the cold season counterparts. Concurrently, Fig.

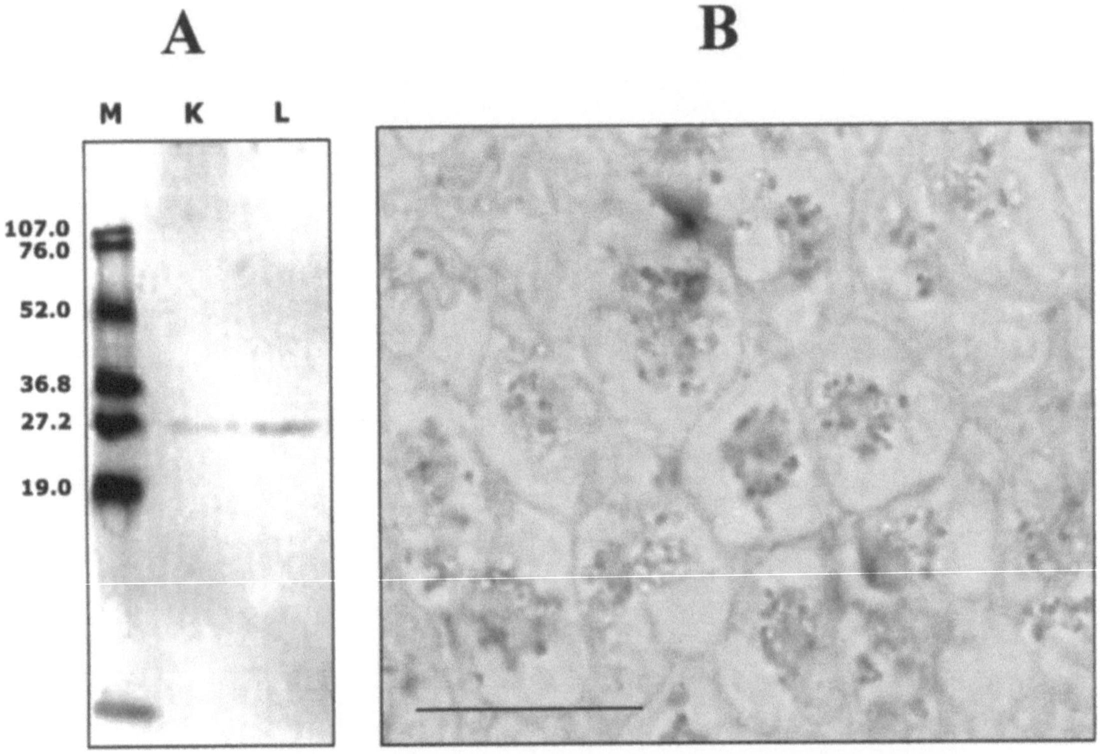

Fig. 2. Carp CK2β immunodetection with a rabbit polyclonal anti-CK2β antiserum directed against the C-terminal synthetic peptide $\beta^{198-215}$ kindly provided by David Litchfield. (A) Immunoblotting of fractionated protein extracts of liver and kidney (1:100). Lane M: pre-stained molecular weight standard, lane K: 60 μg carp kidney total protein, lane L: 60 μg carp liver total protein. (B) Immunocytochemical staining of liver section with anti-CK2β antiserum, dilution 1:100. (Bar: 20 μm).

3 depicts the significant variation attained upon quantification of the CK2β level in kidney cells where those from the warm-season-adapted carp display 3 times more than the winter-acclimatized carp immunostained cells. Thus the seasonal modulation of the CK2β subunit gene transcription exhibits a consistent correlation with the pattern of appearance of the CK2 regulatory subunit within the framework of the complex adaptive process that implies a physiologically coordinated response of the fish toward the stress that the changes in the physical parameters of their habitat imposes.

Several reports have shown shifts of CK2 expression primarily affecting a variety of tumor and leukemic cells [14, 29, Seldin *et al.*, this issue]. Thus protein kinase CK2 does not behave as a classical constitutive enzyme [Pyerin *et al.*, this issue]. More recently, Gerber *et al.* [27] have approached the study of the cellular dynamics of protein kinase CK2 subunits using specific stressors. Interestingly, they found that CK2α and CK2β are differentially targeted inside the nucleolus, whereas the CK2α subunit is preferentially accumulated.

Although our studies were performed assessing the transcription and content of the protein kinase CK2β subunit, the results obtained determined that the enzyme is regulated under physiological conditions, particularly in our model system that demands a completely different cellular strategy than homeotherms to attain the homeostasis needed for survival. We are studying the carp pituitary gland to assess its possible role in the complex process that transduces the periodic external physical milieu changes into the needed signaling molecules that build the compensatory response of the fish [30]. Nucleolar activity appears to be of the utmost importance for the proper cell-functional adjustment in the seasonal adaptive process of the fish. Assessing small nucleolar RNA (snoRNA) expression, we determined a remarkable decrease in winter compared to summer, a drop off that could be linked to the reduction of the processing of pre-rRNA observed in winter [8]. We have also documented a consistent behavior of genes related to the nucleolar activity such as 5.8S rRNA and U3 snoRNA [8, 9].

The localization of CK2 in the nucleolus, along with its kinase activity, appears to play a key role in the regulation of rDNA transcription [28]. We are currently involved in the study of carp nucleolin, a major substrate of CK2. Nucleolin is implicated in the control of ribosome biogenesis through the regulation of rRNA transcription [31]. Additionally, this nucleolar protein plays a role in the processing of pre-rRNA and the packaging of ribosome particles.

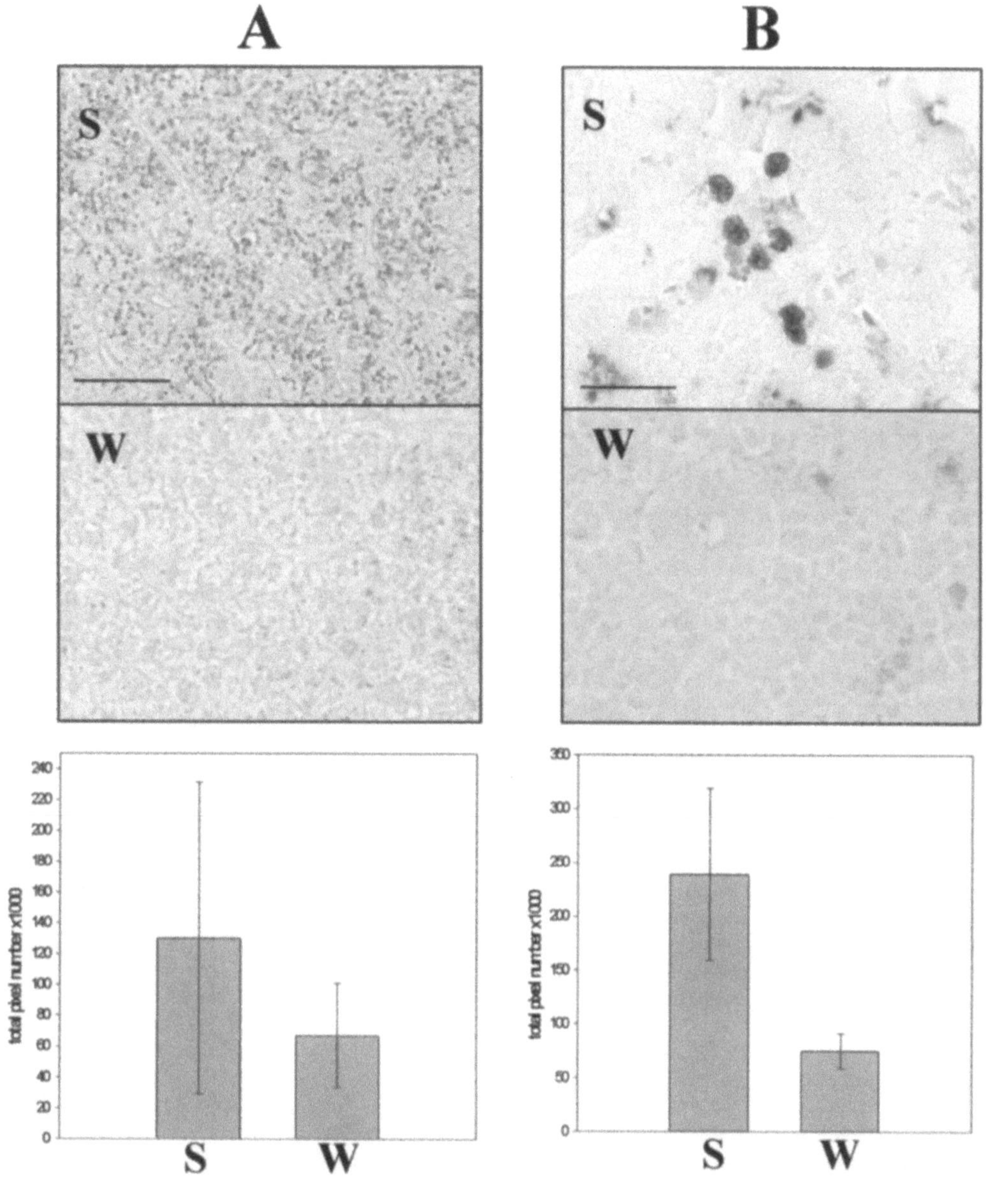

Fig. 3. Immunocytochemical staining and semi-quantitative analyses of CK2β subunit content in liver (A) and kidney (B) sections from summer- (S) and winter-acclimatized (W) carp. The detection was performed with a rabbit polyclonal anti-CK2β antiserum [19, 20], dilution 1:100. (Bar: 100 μm). Semi-quantitative analyses of the immunolabeling are represented graphically. The histograms depict the mean total pixel values (± S.D.) in four different individuals corresponding to each season. In both tissues, the analyses rendered significant differences at Student *t*-test $p < 0.05$.

Acknowledgements

Our thanks to Manuel Krauskopf and Margaret Snook for their critical reading of the manuscript. This work has been supported by the grants: FONDECYT 1000061 and DI-A 61/99 UNAB. The Millenium Institute of Fundamental and Applied Biology (MIFAB) is financed in part by the Ministerio de Planificación y Cooperación (Chile).

References

1. Krauskopf M, Amthauer R, Araya A, Concha M, León G, Ríos L, Vera MI, Villanueva J: Temperature acclimatization of the carp. Cellular and molecular aspects of the compensatory response. Arch Biol Med Exp 21: 151–157, 1988
2. Figueroa J, Molina A, Alvarez M, Villanueva J, Reyes A, León G, Krauskopf M: Prolactin gene expression and changes of prolactin pituitary level during the seasonal acclimatization of the carp. Comp Biochem Physiol 108B: 551–560, 1994

3. Goldspink G: Adaptation of fish to different environmental temperature by qualitative and quantitative changes in gene expression. J Therm Biol 20: 167–174, 1995
4. Tiku PE, Gracey AY, Macartney AI, Beynon RJ, Cossins AR: Cold-induced expression of Δ^9-desaturase in carp by transcriptional and posttranscriptional mechanisms. Science 271: 815–818, 1996
5. Kausel G, Vera MI, San Martín R, Figueroa J, Molina A, Muller M, Martial J, Krauskopf M: Transcription factor Pit-1 expression is modulated upon seasonal acclimatization of eurythermal ectotherms: Identification of two Pit-1 genes in the carp. J Cell Biochem 75: 598–609, 1999
6. Sarmiento J, Leal S, Quezada C, Kausel G, Figueroa J, Vera MI, Krauskopf M: Environmental acclimatization of the carp modulates the transcription of β-actin. J Cell Biochem 80: 223–228, 2000
7. Sáez L, Zuvic T, Amthauer R, Rodríguez E, Krauskopf M: Fish liver protein synthesis during cold acclimatization. Seasonal changes of the ultrastructure of the carp hepatocyte. J Exp Zool 230: 175–186, 1984
8. Vera MI, Norambuena L, Alvarez M, Figueroa J, Molina A, León G, Krauskopf M: Reprogramming of nucleolar gene expression during the acclimatization of the carp. Cell Mol Biol Res 39: 665–674, 1993
9. Vera MI, Ríos HM, de la Fuente E, Figueroa J, Krauskopf M: Seasonal acclimatization of the carp involves differential expression of 5.8S ribosomal RNA in pituitary cells. Comp Biochem Physiol 118B: 777–781, 1997
10. Sáez L, Amthauer R, Rodríguez E, Krauskopf M: Effects of insulin on the fine structure of hepatocytes from winter-acclimatized carp: Studies on protein synthesis. J Exp Zool 230: 187–197, 1984
11. Vera MI, Romero F, Figueroa J, Amthauer R, León G, Villanueva J, Krauskopf M: Oral administration of insulin in winter-acclimatized carp (*Cyprinus carpio*) induces hepatic ultrastructural changes. Comp Biochem Physiol 106A: 677–682, 1993
12. Hannan R, Taylor L, Cavanaugh A, Hannan K, Rothblum LI: UBF and the regulation of ribosomal DNA transcription. In: M.R. Paule (ed). Transcription of Ribosomal RNA Genes by Eukaryotic RNA Polymerase I. Springer-Verlag, Berlin, Heidelberg, 1998, pp 221–228
13. Csermely P, Schnaider T, Chatman B, Olson MOJ, Kahn CR: Insulin induces the phosphorylation of nucleolin. J Biol Chem 268: 9747–9752, 1993
14. Guerra B, Issinger O-G: Protein kinase CK2 and its role in cellular proliferation, development and pathology. Electrophoresis 20: 391–408, 1999
15. Allende JE, Allende CC: Protein kinase CK2: An enzyme with multiple substrates and puzzling regulation. FASEB J 9: 313–323, 1995
16. Vera MI, Kausel G, Barrera R, Leal S, Figueroa J, Quezada C: Seasonal adaptation modulates the expression of the protein kinase CK2β subunit gene in the carp. Biochem Biophys Res Com 271: 735–740, 2000
17. Leal S, Kausel G, Figueroa J, Escobar J, Pinto R, San Martín R, Vera MI: Quantification of the expression of CK2β subunit during the acclimatization of the fish *C. carpio*. Biol Res: R-90, 2000
18. Stenberger LA, Hardy PH, Cuculis JJ, Meyer HG: The unlabeled antibody enzyme method of immunochemistry. J Histochem Cytochem 18: 315–333, 1970
19. Inostroza J, Vera MI, Goicoechea O, Amthauer R, Krauskopf M: Apolipoprotein A-I synthesis during the acclimatization of the carp (*Cyprinus carpio*). J Exp Zool 256: 8–15, 1990
20. Litchfield DW, Lozeman FJ, Cicirelli MF, Harrylock M, Ericsson LH, Piening CJ, Krebs EG: Phosphorylation of the beta subunit of casein kinase II in human A431 cells. Identification of the autophosphorylation site and a site phosphorylated by $p34^{cdc2}$. J Biol Chem 266: 20380–20389, 1991
21. Litchfield DW, Luscher B, Lozeman FJ, Eisenman RN, Krebs EG: Phosphorylation of casein kinase II by p34cdc2 *in vitro* and at mitosis. J Biol Chem 267: 13943–13951, 1992
22. Kausel G, Vera M-I, Figueroa J, Hernández J, San Martin R, Molina A, Chavez VM, Muller M, Martial J, Krauskopf M: Effect of seasonal acclimatization on the expression of the carp transcription factor Pit-1. Biochem Mol Biol Int 45: 813–821, 1998
23. Peitsch MC: ProMod and Swiss-Model: Internet-based tools for automated comparative protein modeling. Biochem Soc Trans 24: 274–279, 1996
24. Koradi R, Billeter M, Wüthrich K: MOLMOL: A program for display and analysis of macromolecular structures. J Mol Graphics 14: 51–55, 1996
25. Chantalat L, Leroy D, Filhol O, Nueda A, Benitez MJ, Chambaz EM, Cochet C, Dideberg O: Crystal structure of the human protein kinase CK2 regulatory subunit reveals its zinc finger-mediated dimerization. EMBO J 18: 2930–2940, 1999
26. Krek W, Maridor G, Nigg EA: Casein kinase II is a predominantly nuclear enzyme. J Cell Biol 116: 43–55, 1992
27. Gerber D, Souquere-Besse S, Puvio F, Dubois M-F, Bensaude O, Cochet C: Heat-induced relocalization of protein kinase CK2. Implication of CK2 in the context of cellular stress. J Biol Chem 275: 23919–23926, 2000
28. Faust M, Montenah M: Subcellular localization of protein kinase CK2. A key to its function? Cell Tissue Res 301: 329–340, 2000
29. Faust RA, Gapany M, Tristani P, Davis A, Adams GL, Ahmed K: Elevated protein kinase CK2 activity in chromatin of head and neck tumors: Association with malignant transformation. Cancer Lett 101: 31–35, 1996
30. Figueroa J, Reyes A, Ríos M, Vera, MI, Krauskopf M: Effect of temperature and photoperiod on prolactin transcription in *Cyprinus carpio*. Zool Sci 14: 353–357, 1997
31. Ginisty H, Sicard H, Roger B, Bouvet P: Structure and functions of nucleolin. J Cell Sci 112: 761–772, 1999

Molecular and Cellular Biochemistry **227**: 113–117, 2001.

Mutation at the CK2 phosphorylation site on Cdc28 affects kinase activity and cell size in *Saccharomyces cerevisiae*

Gian Luigi Russo,[1] Christian van den Bos[2] and Daniel R. Marshak[3]

[1]*Istituto di Scienze dell'Alimentazione, Consiglio Nazionale delle Ricerche, Avellino, Italy;* [2]*Martek Biosciences Inc., Columbia, MD, USA;* [3]*Cambrex Biosciences Group, Walkersville, MD, USA*

Abstract

We have recently reported that protein kinase CK2 phosphorylates both *in vivo* and *in vitro* residue serine-46 of the cell cycle regulating protein Cdc28 of budding yeast *Saccharomyces cerevisiae*, confirming a previous observation that the same site is phosphorylated in Cdc2/Cdk1, the human homolog of Cdc28. In addition, *S. cerevisiae* in which serine-46 of Cdc28 has been mutated to alanine show a decrease of 33% in both cell volume and protein content, providing the genetic evidence that CK2 is involved in the regulation of budding yeast cell division cycle, and suggesting that this regulation may be brought about in G1 phase of the mammalian cell cycle. Here, we extended this observation reporting that the mutation of serine-46 of Cdc28 to glutamic acid doubles, at least *in vitro,* the H1-kinase activity of the Cdc28/cyclin A complex. Since this mutation has only little effects on the cell size of the cells, we hypothesize multiple roles of yeast CK2 in regulating the G1 transition in budding yeast. (Mol Cell Biochem **227**: 113–117, 2001)

Key words: protein kinase CK2, *CDC28, Saccharomyces cerevisiae,* cell division cycle

Introduction

Protein kinase CK2 (formally known as casein kinase 2) is a serine/threonine kinase ubiquitously present in all tissues and cell lines so far investigated [1–5]. The enzyme is composed of two catalytic subunits named α and α′, and a regulatory subunit β differently assembled among them [6–8]. CK2 is involved in transcription, signal transduction, proliferation and in development following stimulation by a variety of extracellular signals [1, 9, 5]. In the budding yeast *Saccharomyces cerevisiae*, CK2 is composed of at least four different subunits α, α′, β and β′ which are encoded by four distinct genes, *CKA1, CKA2, CKB1* and *CKB2*, respectively [10]. Using temperature-sensitive alleles of the *CKA2* gene, it has been demonstrated that *S. cerevisiae* CK2 is required for cell cycle progression during G1, prior to DNA synthesis, and at G2/M phases [11]. In addition, deletion of the gene for either catalytic subunit alone has little effect, but deletion of both is inviable with cells substantially increased in size prior to growth arrest [12].

Among the large number of substrates phosphorylated by CK2 [3], we have, previously, shown that human serine/threonine kinase Cdc2/Cdk1 is phosphorylated *in vitro* by CK2 on residue Ser39, at the same site also phosphorylated *in vivo* in G1 phase of HeLa cell division cycle, suggesting a potential role of CK2 in cell cycle regulation [13, 14]. Cdc2/Cdk1 is one of the critical protein kinase regulate cell division in eukaryotic cells. In budding yeast *Saccharomyces cerevisiae*, Cdc2/Cdk1 homolog, encoded by the *CDC28* gene [15], is required at two crucial points: at START, when the cells entry into S phase, and at the G2/M transition [16–18]. The catalytic activity of Cdc28 depends on the physical binding to its regulatory subunits called cyclins [18–20]. Expression of G1-phase cyclins (*CLN1, CLN2, CLN3*) are essential for START, while B-type cyclins (encoded by *CLB1* to *CLB6* genes) are necessary for S phase and mitosis [19, 20]. The transition between low and high Clb/Cdc28 kinase regulates exit from mitosis in budding yeast cell cycle inducing activation of APC (anaphase-promoting factors complex) components [18–22].

Address for offprints: G.L. Russo, Istituto di Scienze dell'Alimentazione, via Roma 52 A/C, 83100 Avellino, Italy (E-mail: russogl@inwind.it)

The catalytic activity of Cdc28, as well as that of other Cdk family members, is regulated not only by interaction with ligands, but also by phosphorylation and dephosphorylation events [23–28]. We recently reported that residue serine-46 of *S. cerevisiae* Cdc28, embedded in a consensus sequence for CK2, is phosphorylated at this site both *in vitro* and *in vivo* [29]. The mutant *cdc28-S46A* exhibited a H1 kinase activity, and a growth rate similar to that of wild-type cells; however, the Ser46Ala mutation did result in a decrease of 33% in cell size, confirming a functional role for the phosphorylation of Cdc28 by CK2 in *S. cerevisiae* and suggesting a functional role for this phosphorylation in human cells [13, 29].

The present work describes the effects of a different Cdc28 mutant, namely Ser46Glu on cyclin binding and kinase activity.

Materials and methods

Strains and site-specific mutagenesis

The relevant genotypes of strains used in this study are described in Table 1. YPD medium consisted of 1% yeast extract, 2% peptone, 2% D-glucose. Standard genetic methods for strain construction in *S. cerevisiae* were as reported [30]. The construction of *cdc28-S46E* allele was performed by mutagenesis of Cdc28 residue Ser46 to glutamic acid using exactly the same procedure recently described to produce the *cdc28-S46A* mutant [29]. The oligonucleotide used for mutagenesis of Ser46 to Glu had the following sequence: 5'-ATA AGA CTA GAG GAG GAA GAC GAG GGT G-3'. Both strands of DNA surrounding Ser46 to Glu mutation were sequenced by the dideoxy method [31]. The pRS-*cdc28-S46E* plasmid was transformed into strain CY3086. Loss of the YCp50 plasmid containing wild type *CDC28* was selected for by plating transformants on 5-fluoro-orotic acid [30]. The resulting strains SY129, SY130 and SY131 are as reported in Table 1.

Cyclin A binding and H1-kinase assay

pGEX plasmid expressing GST-cyclin A was kindly provided by Dr. David Beach (Cold Spring Harbor Laboratory). GST-cyclin A expressed and purified on GST affinity column (Amersham-Pharmacia) following the manufacturer's instructions, was cross-linked to CNBr-activated sepharose (Amersham-Pharmacia) following the manufacturer's instructions and using 8.5 μg of recombinant protein per 1 ml of gel. Different amount of yeast cell extracts prepared exactly as described [29] from strains SY129, SY130 and SY131, were added to 40 μl of 1:1 suspension of GST-cyclin A beads in tris buffered saline solution and incubated for 1 h at 4°C with gentle rocking. Subsequently, the complexes were washed 3 times in buffer A (100 mM Tris-HCl, pH 7.5; 200 mM NaCl; 1 mM EDTA; 5% glycerol; 0.5 mM dithiothreitol; 1 mM phenylmethylsulfonyl fluoride; 1 μg each of chymostatin, leupeptin, antipain, pepstatin per ml; 1 mM sodium pyrophosphate, 1 mM sodium vanadate; 10 mM sodium fluoride; 20 mM *p*-nitrophenylphosphate), three in Cdc28 kinase buffer and subsequently used as enzymatic source to phosphorylate histone H1 (Boeringher).

Cdc28 kinase assay was performed essentially as described [32, 29] except for the use of histone H1 instead of the peptide substrate. Briefly, assays contained in a final vol. of 50 μl, 10 μM [γ-^{32}P]ATP (20,000 cpm/pmol), 1 μg/ml histone H1. The reaction was incubated for 30 min at 30°C, and subsequently blocked by adding 50 μl of 2 × Laemmli sample buffer, heated 5 min at 95°C, and the supernatants loaded on 15 % SDS-PAGE. The amount of radioactivity incorporated on H1 was quantified by phosphoimager (Fuji) and expressed as arbitrary units. In some cases, the gel was blotted to nitrocellulose in order to determine the amount of Cdc28 bound to GST-cyclin A using an anti-Cdc28 antibody (antibody Ab182, a generous gift of Drs M. Tyers and B. Futcher, Cold Spring Harbor Laboratory) diluted 1:1,000. Specific bands were detected using sheep anti-rabbit immunoglobulin conjugated to horseradish peroxidase (Amersham-Pharmacia) as secondary antibody at the dilution recommended by manufacturer and the ECL-plus detection system (Amersham-Pharmacia).

Measurement of cell volumes

Cells were grown in 5 ml of YPD medium, in a roller at 30°C to mid-log phase. Cultures were then placed on ice and briefly sonicated. Cells were diluted into Isoton buffer, and their volume measured using a Coulter Channelyzer Model 256 (manufactured by Coulter Electronics). Mean cell vol. was

Table 1. Yeast strains and their relevant genotype used in the present study

Strain	Relevant chromosomal genotye	Plasmid genotype
CY3086	*MATα CDC28::HIS3*[WT *CDC28* on YCp50] *ura3-1 leu2-3,112 his3-11,15 trp1-1 ade2-1 can1-100*	YCp50-*CDC28*
SY129	*MATα CDC28::HIS3*[WT *CDC28* in pRS415] *ura3-1 leu2-3,112 his3-11,15 trp1-1 ade2-1 can1-100*	pRS-*CDC28* [29]
SY130	*MATα CDC28::HIS3*[*CDC28-ala* in pRS415] *ura3-1 leu2-3,112 his3-11,15 trp1-1 ade2-1 can1-100*	pRS-*cdc28-S46A* [29]
SY131	*MATα CDC28::HIS3*[*CDC28-glu* in pRS415] *ura3-1 leu2-3,112 his3-11,15 trp1-1 ade2-1 can1-100*	pRS-*cdc28-S46E*

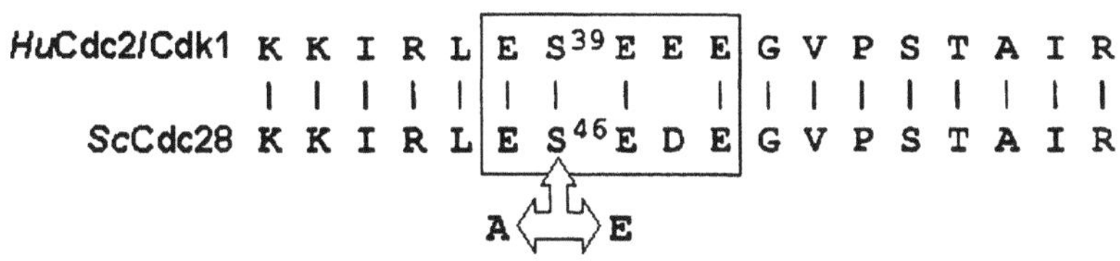

Fig. 1. CK2 phosphorylation site on Cdc28. Homology between human (*Hu*) Cdc2/Cdk1 and *S. cerevisiae* (*Sc*) Cdc28 in the region containing the CK2 phosphorylation site. Amino acids are indicated with single letter code. Arrows indicate the mutation of Ser46 to alanine (A), or glutamic acid (E).

estimated from the cell volume distribution obtained by a Coulter Channalyzer C-1000, as described [33].

Results and discussion

Effect of cdc28-S46 and cdc28-E46 mutations on cyclin binding

S. cerevisiae Cdc28 contains a canonical CK2 phosphorylation site at Ser46 that is phosphorylated both *in vitro* and *in vivo* [29]. The amino acid sequence surrounding Ser46 is highly homologous to that surrounding the CK2 phosphorylation site on human Cdc2/Cdk1 (Fig. 1). We also reported that mutation of Ser46 to alanine (*cdc28-S46A*) does not affect Cdc28 kinase activity, nor its expression, neither the level of Cdc28 tyrosine phosphorylation in strains SY129 compared to SY130 [data not reported; 29]. However, *S. cerevisiae* cells in which Ser46 was mutated to alanina, showed a decrease in both cell volume and protein content of 33% [29].

In order to verify if the CK2 phosphorylation site in *S. cerevisiae* was involved in cyclin binding activity, we measured the Cdc28 kinase activity of strains reported in Table 1 after precipitation with cyclin A-sepharose beads. In fact, the amino acid cluster surrounding Ser39 in human Cdc2/Cdk1 is essential for cyclin A binding and kinase activity [34–36]. Therefore, we used GST-cyclin A cross-linked to sepharose to precipitate Cdc28 and compare the histone H1 kinase activity of wild-type Cdc28 (SY129) vs. Ser46Ala and Ser46Glu mutants. As shown in Fig. 2A, the kinase activity of Cdc28-S46E mutant was significantly higher compared to those of strains SY129 and SY130. Anti-Cdc28 immunoblotting performed on the same samples presented in Fig. 2A did not detect any significant difference in the total amount of Cdc28 precipitated (Fig. 2B).

Mutations in the acidic cluster (E38, E40, E41, E42) surrounding Ser39 decreases enormously human Cdc2/Cdk1 kinase activity, while they interfered much less with cyclin A binding [35]. Accordingly, the increase in H1 kinase activity showed by *cdc28-S46E* mutant might be related to the presence of an extra negative charge (Glu) in the acidic cluster. Similarly, both alanine and glutamic acid mutations did

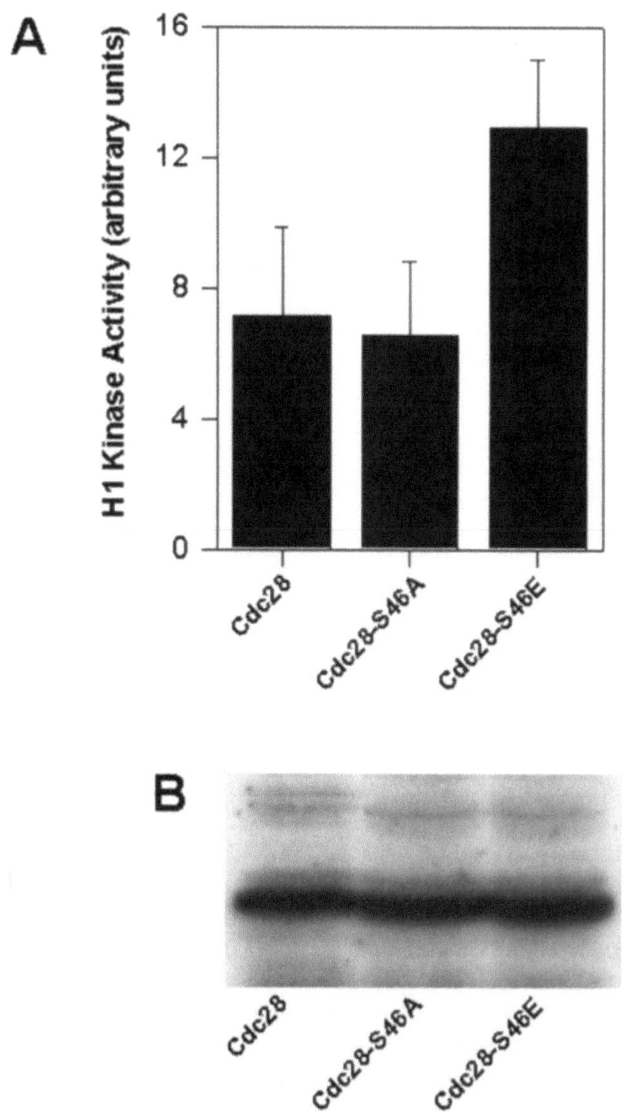

Fig. 2. Histone H1 kinase activity of stains SY129, SY130 and SY131 after cyclin A precipitation. (A) Exponentially growing cells were broken and wild-type Cdc28 and both Cdc28-S46A and Cdc28-S46E mutants were precipitated using GST-cyclin A as described in Materials and methods. The complexes obtained were used as enzymatic source to phosphorylate histone H1. Bar indicates standard error; $p < 0.005$. (B) Immunoblotting of the same precipitates used in (A) to determine Cdc28 kinase activity to demonstrate that the amount of protein present in the cyclin A complexes was indeed the same. The immunoblotting was performed using an anti-Cdc28 polyclonal antibody.

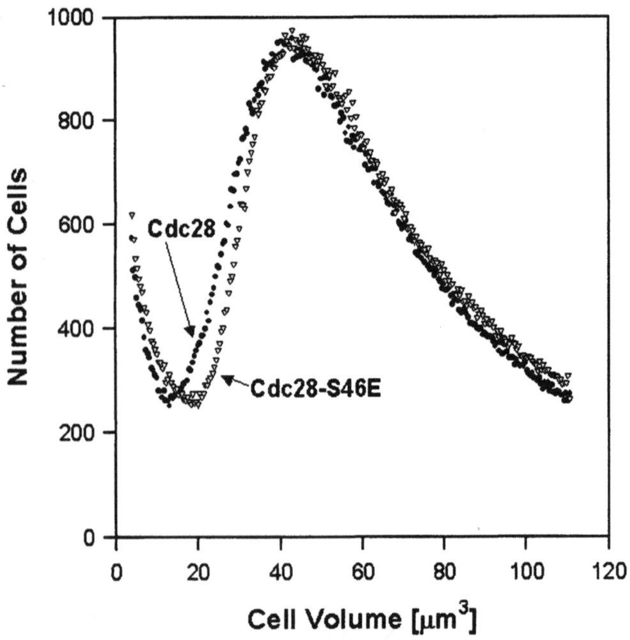

Fig. 3. Cell volume measurements of strain SY131. Strain 131 (*cdc28-S46E*) exhibits an increase of cell volume in the smaller population of cells compared to wild-type (strain 129), while no difference was observed in the larger population of cells. Both measurements were performed in YPD medium using a coulter counter as described in Materials and methods; each curve presents the average of 5 different clones with three repeats.

not show any significant change in cyclin A binding indicating that, at least *in vitro*, CK2 phosphorylation of Ser46 plays a role in regulating the kinase activity of Cdc28 complex(es) rather than its binding activity to cyclin A.

Effect of cdc28-S46E mutation on cell size

Recently we reported that, at stationary phase, *cdc28-S46A* mutant was greatly reduced in cell size compare to the control strain (SY129). The decrease of cell size in *cdc28-S46A* was of 33%. The difference was related to both small (left part of the graph), and big cells (right part of the graph), and it does resemble the *cdc25-1* phenotype observed in cells with a defect in Ras/cAMP pathway [29]. Based on the result reported in Fig. 2, we measured the cell size of strain SY131. Our prediction was that if the H1 kinase activity of Cdc28-S46E was higher compared to the control strain and to Cdc28-S46A, the cells would enter S phase at a larger size. Fig. 3 shows that among the population of large cells, no difference was present between wild-type and mutant (right part of the graph). On the opposite, the population of small cell (left part of the graph) is larger in strain SY131 compared to the control. This result partially confirms our hypothesis, and suggests that other factor(s) are involved in the effect of Ser46 mutation in the regulation of G1/S transition in *S. cerevisiae.*

In summary, we demonstrated that mutation of the CK2 phosphorylation site, Ser46, to glutamic acid in *S. cerevisiae* Cdc28 increases its *in vitro* kinase activity in a complex with cyclin A. In addition, this mutation has a small, but significant effect on *S. cerevisiae* cell volume, causing an enlargement of the cell size accordingly to our prediction. In fact, we have recently reported that the mutation of Ser46 to a non-phosphorylable residue (Ser46Ala) causes a strong reduction of the cell volume [29], while the addition of an extra negative charge (Ser46Glu) produces an opposite effect. Since in an asynchronous culture of wild-type *S. cerevisiae,* only a fraction of Cdc28 molecules is phosphorylated on Ser46, this can explain why we did not observe a significant difference between the H1 kinase activities of the GST-cyclin A-Cdc28 complex compared to that of Cdc28-S46A mutant.

Our data let to speculate on possible multiple roles of CK2 in *S. cerevisiae* cell cycle. Perhaps, CK2 phosphorylation of Cdc28 prevents binding to G1 cyclins in response to adverse external conditions or until cells have reached a critical size. In the absence of this inhibitory phosphorylation (Ser46Ala mutant), the association of Cdc28 with these cyclins may occur earlier in G1 than in wild-type cells, hence accelerating transition into S phase and rendering cells smaller; the opposite is observed when we strengthen the CK2 effect with the Ser46Glu mutation. Mutating Ser46 to Ala caused a decrease in cell size of approximately 30% of that caused by overexpression of a hyper stable variant of Cln3 [37]. This difference may be explained as follows: accumulation of the G1 cyclins might be rate limiting for controlling progression from G1 into S. In such a scenario, binding of cyclins to Cdc28 may not only be governed by the presence or absence of Ser46 phosphorylation, but also by the concentration of cyclins, therefore Cdc28 unphosphorylated on residue Ser46 may nonetheless remain unbound to cyclins until the concentration of cyclins is sufficiently elevated. Hence, transition into S phase would occur only after sufficient levels of cyclins have accumulated so the phenotype caused by abolishing phosphorylation of Cdc28 on Ser46 should be less pronounced than that caused by over expression of Cln3. After START, it might be that the Ser46 phosphorylation by CK2 has a positive effect on the catalytic activity of Cdc28-B type cyclin complexes towards specific substrates at G2/M transition, explaining the data presented in Fig. 2 of this work. Accordingly, the temperature-sensitive mutant, cka2ts, exhibits defects in both G1/S and G2/M transitions [11].

Future works will devote to shed light o the multifunctional role of CK2 in budding yeast cell cycle.

Acknowledgements

G.L.R. has been partially supported by a grant from Italian Consiglio Nazionale delle Ricerche (# 106040).

References

1. Issinger O-G: Casein kinases: Pleiotropic mediators of cellular regulation. Pharmac Ther 59: 1–30, 1993
2. Allende JE, Allende CC: Protein kinase CK2: An enzyme with multiple substrates and a puzzling regulation. FASEB J 9: 313–323, 1995
3. Pinna LA: Protein kinase CK2. Int J Biochem Cell Biol 29: 551–554, 1997
4. Dobrowolska G, Lozeman FJ, Li D, Krebs EG: CK2, a protein kinase of the next millennium. Mol Cell Biochem 191: 3–12, 1999
5. Guerra B, Boldyreff B, Sarno S, Cesaro L, Issinger OG, Pinna LA: CK2: A protein kinase in need of control. Pharmacol Ther 82: 303–313, 1999
6. Chester N, Yu IJ, Marshak DR: Identification and characterization of protein kinase CKII isoforms in HeLa Cells. J Biol Chem 270: 7501–7514, 1995
7. Boldyreff B, Mietens U, Issinger O-G: Structure of protein kinase CK2: Dimerization of the human β-subunit. FEBS Lett 379: 153–156, 1996
8. Korn I, Gutkind S, Srinivasan N, Blundell TL, Allende CC, Allende JE: Interactions of protein kinase CK2 subunits. Mol Cell Biochem 191: 75–83, 1999
9. Litchfield DW, Dobrowolska G, Krebs EG: Regulation of casein kinase II by growth factors – a reevaluation. Cell Mol Biol Res 40: 373–381, 1994
10. Glover CV III: On the physiological role of casein kinase II in *Saccharomyces cerevisiae*. Prog Nucleic Acid Res Mol Biol 59: 95–133, 1998
11. Hanna DE, Rethinaswamy A, Glover CVC: Casein kinase II is required for cell cycle progression during G1 and G2/M in *Saccharomyces cerevisiae*. J Biol Chem 270: 25905–25914, 1995
12. Padmanabha R, Chen-Wu JL-P, Hanna DE, Glover CVC: Isolation, sequencing, and distruption of the yeast *CKA2* gene: Casein kinase II is essential for viability in *Saccharomyces cerevisiae*. Mol Cell Biol 10: 4089–4099, 1990
13. Russo GL, Vandenberg MT, Yu IJ, Bae Y-S, Franza BRJ, Marshak DR: Casein kinase II phosphorylates $p34^{cdc2}$ kinase in G1 phase of the HeLa cell division cycle. J Biol Chem 267: 20317–20325, 1992
14. Marshak DR, Russo GL: Regulation of protein kinase CKII during the cell division cycle. Cell Mol Biol Res 40: 513–517, 1994
15. Lorincz AT, Reed SI: Primary structure homology between the product of yeast cell division control gene *CDC28* and vertebrate oncogenes. Nature (London) 307: 183–185, 1984
16. Hartwell LH, Weinert TA: Checkpoints: Controls that ensure the order of cell cycle events. Science 246: 629–634, 1989
17. Forsburg SL, Nurse P: Cell cycle regulation in the yeasts *Saccharomyces cerevisiae* and *Schizosaccharomyces pombe*. Ann Rev Cell Biol 7: 227–256, 1991
18. Nasmyth K: At the heart of the budding yeast cell cycle. Trends Genet 12: 405–412, 1996
19. Deshaies RJ: Phosphorylation and proteolysis: Partners in the regulation of cell division in budding yeast. Curr Opin Genet Dev 7: 7–16, 1997
20. Andrews B, Measday V: The cyclin family of budding yeast: Abundant use of a good idea. Trends Genet 14: 66–72, 1998
21. Yeong FM, Lim HH, Padmashree CG, Surana U: Exit from mitosis in budding yeast: Biphasic inactivation of the Cdc28-Clb2 mitotic kinase and the role of Cdc20. Mol Cell 5: 501–511, 2000
22. Rudner AD, Murray AW: Phosphorylation by Cdc28 activates the Cdc20-dependent activity of the anaphase-promoting complex. J Cell Biol 149: 1377–1390, 2000
23. Elledge SJ: Cell cycle checkpoints: Preventing an identity crisis. Science 274: 1664–1672, 1996
24. Sorger PK, Murray AW: S-phase feedback control in budding yeast independent of tyrosine phosphorylation of p34*cdc28*. Nature (London) 355: 365–368, 1992
25. Amon A, Surana U, Muroff I, Nasmyth K: Regulation of p34 CDC28 tyrosine phoshorylation is not required for entry into mitosis in *S. cerevisiae*. Nature (London) 355: 368–371, 1992
26. Lim HH, Loy CJ, Zaman S, Surana U: Dephosphorylation of threonine 169 of Cdc28 is not required for exit from mitosis but may be necessary for Start in *Saccharomyces cerevisiae*. Mol Cell Biol 16: 4573–4583, 1996
27. Rhind N, Russell P: Tyrosine phosphorylation of Cdc2 is required for the replication checkpoint in *Schizosaccharomyces pombe*. Mol Cell Biol 18: 3782–3787, 1998
28. Ross KE, Kaldis P, Solomon MJ: Activating phosphorylation of the *Saccharomyces cerevisiae* cyclin-dependent kinase, cdc28p, precedes cyclin binding. Mol Biol Cell 11: 1597–1609, 2000
29. Russo GL, Van Den Bos C, Sutton A, Coccetti P, Baroni MD, Alberghina L, Marshak DR: Phosphorylation of Cdc28 and regulation of cell size by the protein kinase CKII in *Saccharomyces cerevisiae*. Biochem J 351: 143–150, 2000
30. Guthrie C, Fink GR: Guide to Yeast Genetics and Molecular Biology, vol 194. Academic Press, San Diego, CA, 1991
31. Sambrook J, Fritsch EF, Maniatis T: Molecular Cloning: A Laboratory Manual. Cold Spring Harbor Laboratory, Cold Spring Harbor, NY, 1989
32. Marshak DR, Vandenberg MT, Bae Y-S, Yu IJ: Characterization of synthetic peptide substrate for $p34^{cdc2}$ protein kinase. J Cell Biochem 45: 391–400, 1991
33. Baroni MD, Martegani E, Monti P, Alberghina L: Cell size modulation by *CDC25* and *RAS2* genes in *Saccharomyces cerevisiae*. Mol Cell Biol 9: 2715–2723, 1989
34. Ducommun B, Brambilla P, Felix M-A, Franza Jr BR, Karsenti E, Draetta G: cdc2 phosphorylation is required for its interaction with cyclin. EMBO J 10: 3311–3319, 1991
35. Marcote MJ, Knighton DR, Basi G, Sowadski JM, Brambilla P, Draetta G, Taylor SS: A three-dimensional model of the Cdc2 protein kinase: Localization of cyclin- and Suc1-binding regions and phosphorylation sites. Mol Cell Biol 13: 5122–5131, 1993
36. Jeffrey PD, Russo AA, Polyak K, Gibbs E, Hurwitz J, Massagué J, Pavletich NP: Mechanism of CDK activation revealed by the structure of a cyclin A-CDK2 complex. Nature (London) 376: 313–320, 1995
37. Tyers M, Tokiwa G, Futcher B: Comparison of the *Saccharomyces cerevisiae* G1 cyclins: Cln3 may be an upstream activator of Cln1, Cln2 and other cyclins. EMBO J 12: 1955–1968, 1993

Molecular and Cellular Biochemistry **227**: 119–127, 2001.

Distinctive features of plant protein kinase CK2

Marta Riera, Giovanna Peracchia and Montserrat Pagès
Departament de Genètica Molecular, Institut de Biologia Molecular de Barcelona, Centre d'Investigació i Desenvolupament, C.S.I.C., Barcelona, Spain

Abstract

In plants, protein kinase CK2 is involved in different processes that control many aspects of metabolism and development. In mammals and yeast the enzyme is a heterotetramer composed of two types of subunits. During years the subunit composition of the maize protein kinase CK2 enzyme has been a source of controversy. We have recently characterized the maize holoenzyme subunits. Our results show that multiple catalytic and regulatory subunits are expressed in maize and are able to specifically interact with other α and β subunits suggesting a high level of heterogeneity in the typical heterotetrameric structure. Here, we summarize data available on plant CK2 enzymes, in order to clarify the distinctive features and functions of plant protein kinase CK2. (Mol Cell Biochem **227**: 119–127, 2001)

Key words: protein kinase CK2, CK2β regulatory subunits, functionality

Introduction

Protein kinase CK2 is a ubiquitous and highly conserved Ser/Thr kinase present in nucleus and cytoplasm of all eukaryotic cells examined to date. The protein kinase CK2 enzyme has been widely studied in animals, where it is involved in different processes such as cell proliferation [1], transcriptional control [2] or cell cycle progression [3].

In plants, the existence of CK2-like activity was first described at the beginning of the 80s [4]; however, during those years, the subunit composition of the plant enzyme was not well defined. During the past decade, both monomeric and oligomeric forms with CK2-like activity were isolated from different plant sources such as maize [5] broccoli [6] or pea [7]; however, the subunit composition of the oligomeric forms was not clear. The maize CK2α subunit was the first catalytic subunit identified in plants [8], and then CK2 catalytic and regulatory subunits were cloned in Arabidopsis [9–11]. It is noteworthy that the recombinant maize CK2α is the only CK2 catalytic subunit crystallized to date [12] and all the structural studies reported are based on the plant subunit. Recently, we have isolated the maize regulatory subunits and clarified the composition of the maize holoenzyme [13]. Information about the role of the enzyme in plants is scarce but, as in the case of other organisms, it seems to be involved in many different processes that are essentials for plant viability such as light-regulated gene expression, plant growth [14], and cell-cycle progression [15]. In this paper, we will summarize the main characteristics of the CK2 enzyme and its possible physiological role in plant development.

Materials and methods

cDNA library screening

A HybriZap two-hybrid vector system library was constructed from poly(A)$^+$ RNA maize stressed leaf of the inbred line W64A according to the manufacturer (Clontech matchmaker™).

Different cDNA library screenings were performed in order to clone the maize CK2β regulatory subunits. The BLAST program was used to screen the NCBI dbest database using the *Arabidopsis thaliana* CK2B1 cDNA sequence (accession number L22563). One candidate maize EST clone (AA979779) of 585 bp was identified, obtained and used as a probe to screen a maize cDNA library. For the cDNA library screening, hybridization with the maize EST clone was carried out at 65°C in a 250 mM sodium phosphate buffer, containing 7% SDS and 1 mM EDTA. For high stringency screening

Address for offprints: M. Pagès, Departament de Genètica Molecular, Institut de Biologia Molecular de Barcelona, Centre d'Investigació i Desenvolupament, C.S.I.C. Jordi Girona 18–26, 08034 Barcelona, Spain (E-mail: mptgmm@cid.csic.es)

filters were washed twice at 65°C in 1 × SSC, 1% SDS and 3 times at 65°C in 0.1 × SSC, 0.1% SDS; as a result, the first CK2β subunit (CK2β-1) was isolated. Moreover, a low stringency screening was performed as described, except that filters were washed twice at 65°C in 1 × SCC, 1% SDS and 3 times at 0.5 × SSC, 0.1% SDS and a second CK2β subunit (CK2β-2) was cloned. Plasmids were excised from a homogeneous population of hybridizing phages into *E. coli* according to the manufacturer (Stratagene). The largest inserts were completely sequenced using the Automated Laser Fluorescent (ALF) system of Pharmacia.

Yeast two-hybrid screening

The maize CK2β-1 (pGBT9-CK2β1) was used as a bait for the two-hybrid screening of the HybriZap two-hybrid vector system library.

For screening, the Saccharomyces cerevisiae strain HF7c (MAT**a**, ura3-52, his3-200, ade2-101, lys2-801, trp1-901, leu2-3,112 gal4-542, gal80-538, LYS2::GAL1$_{UAS}$-GAL1$_{TATA}$-HIS3,URA3::GAL4$_{17MERS(3X)}$-CYC1$_{TATA}$-LacZ) was transformed with the pGBT9-CK2β1 plasmid according to the manufacturer (Clontech matchmaker™). Positive clones were selected in plates lacking triptophan, and transformed with the HybriZap two-hybrid library. Transformants were selected in Leu-, Trp-, His- plates containing 1 mM 3-amino-1,2,4,-triazole. Purified colonies were tested for β-galactosidase activity using filter assays according to the manufacturer (Clontech matchmaker™). Plasmids from His^+ $LacZ^+$ colonies were isolated and electrophorated into *E. coli*, and the DNA sequence of the inserts was determined. Using this method, new maize CK2α subunits (CK2α-3) and CK2β subunits (CK2β-3) were isolated.

Results

CK2α subunits

In mammals and yeast the CK2 enzyme is a heterotetramer composed of two types of catalytic subunits, (α and α′) and regulatory subunits, (β) giving rise to different forms: $\alpha_2\beta_2$, $\alpha\alpha'\beta_2$, $\alpha'_2\beta_2$. The α (42–44 kDa) and α′ (38 kDa) subunits are catalytically active by themselves, and structurally related, although they are encoded by different genes [16]. The CK2α subunits are highly conserved among different species and are closely related to the cdc2 group of protein kinases. This high degree of conservation suggests that some of the main CK2 functions have to be conserved between the different species. CK2 function has been analyzed in *Saccharomyces cerevisiae* by constructing mutants for the different kinase subunits. In this organism, simultaneous disruption of the *CKA1* and *CKA2* genes encoding α and α′ catalytic subunits is lethal for the cell [17].

In plants, the composition of the CK2α family seems to be quite different. In *Arabidopsis thaliana* two cDNA clones have been identified that encode proteins 72% identical to the human CK2α′ catalytic subunit [9], and the same authors, based on the Southern analysis, suggest the existence of a third CK2α subunit. Furthermore, computational analysis of Arabidopsis genome, which has been completely sequenced, indicates the existence of at least four different genes encoding for CK2α subunits. In maize three clones have been described to date [8, 13, 18] that are also highly similar to the human CK2α′ subunit. In Fig. 1, plant CK2α subunits have been aligned with human CK2α and α′ sequences. The different plant CK2α subunits are very similar to each other; they present more than 90% of identity at amino acid level and almost the same length. Most of the different residues are located in the C-terminal region but the structural determinants defined for CK2α catalytic subunits are conserved in all the cases. Both Arabidopsis and maize present no significant homology between their CK2α genes in the 3′ non-coding region at the nucleotide level. According to the data, we can postulate that in plants the CK2α subunits belong to a multigenic family composed at least by three members but only by one type of subunit. The human CK2α′ is 40 amino acids shorter than the α subunit in the carboxyterminal region. It is interesting that the sequences of plant CK2α subunits, due to their length, are more similar to human α′ than α subunit. The C-terminal region of the maize enzyme is 60 amino acids shorter than human CK2α; this fact should explain the unusual high stability and activity of maize CK2α and the reason why maize CK2α is the only CK2 catalytic subunit crystallized to date [12]. Recently, several reports about the maize CK2α structure, have been published [19–21].

The maize CK2α gene corresponding to CK2α-1 is the only CK2 genomic sequence described in plants [18]. The maize genomic clone is 7.5 Kb long and contains 10 exons separated by 9 introns of different sizes. In the promoter region we can find typical elements of eukaryotic promoter motifs, such as TATA boxes, CAAT boxes or GC-boxes. However, further data on plant CK2 genomic sequences will be required to determine if the organization of the maize gene is shared in other plant species.

CK2β subunits

The CK2β regulatory subunits (26–40 kDa) present no homology to regulatory subunits or domains of other protein kinases, except to the *Drosophila melanogaster Stellate* gene product [22]. This CK2β subunit presents three main properties: it is inactive by itself but can stimulate CK2α catalytic

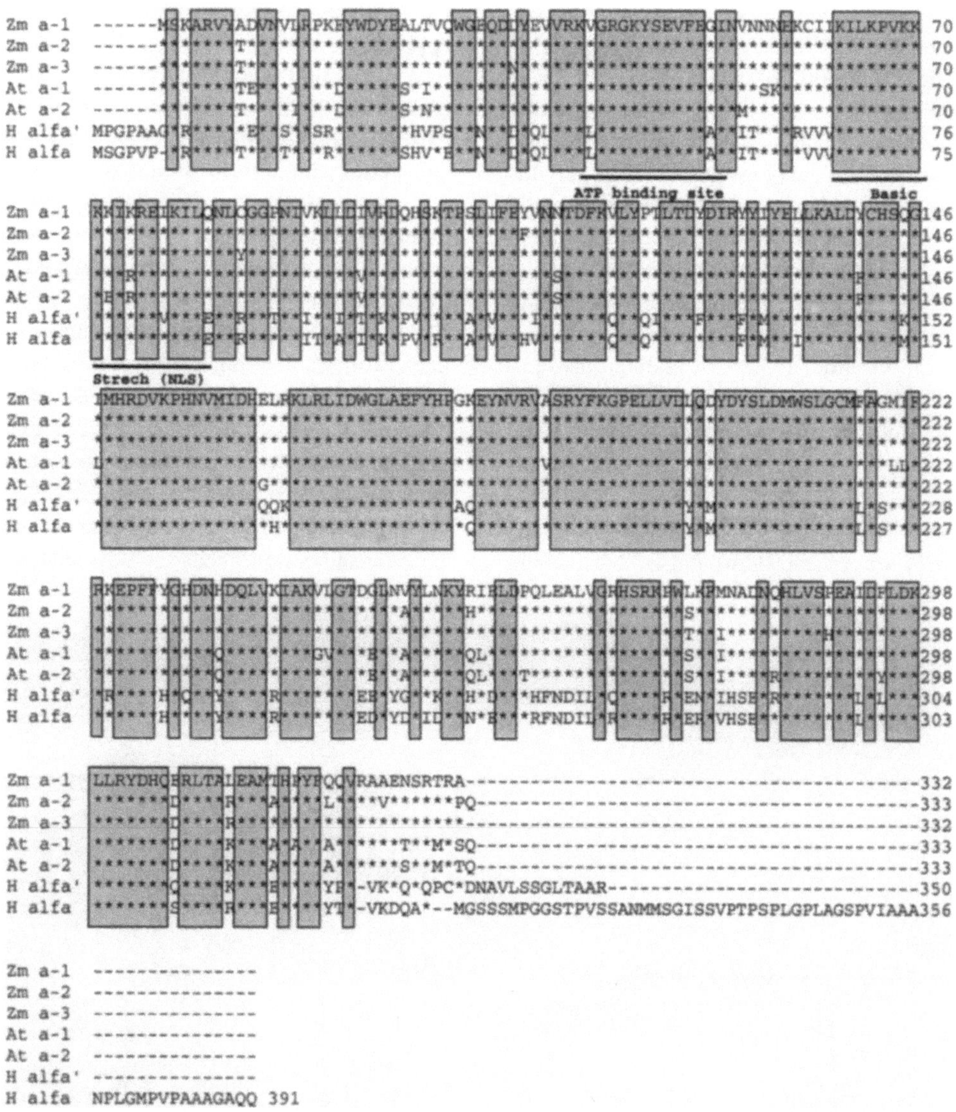

Fig. 1. Alignment of the plant CK2α catalytic subunits with the human CK2α and CK2α′ sequences. The amino acid sequence of the three maize CK2α subunits, Zm a-1 (CAA43659), Zm a-2 (CAA72290), Zm a-3 (AAG36872) and the two Arabidopsis CK2α subunits At a-1 (Q08467) and At a-2 (Q08466) has been aligned with human CK2α′ , H alfa′ (AAA51548) and CK2α, H alfa (AAA355503). Invariant residues are indicated by asterisks and by shaded boxes, and dash indicates a gap introduced to maximize alignment. Characteristics domains of CK2α catalytic subunits, ATP binding site and the basic stretch (NLS) are underlined.

activity [23], it confers stability to the enzyme [24] and it provides specificity for the interaction with substrates and inhibitors [25]. Whereas in most organisms only one gene for CK2β has been described, two genes (*CKB1* and *CKB2*) have been identified in *Saccharomyces cerevisiae* [26]. At least two genes exist in *Drosophila melanogaster* [27, 28], plus the β-like protein *Stellate* [22]. In contrast to the catalytic subunits, deletion of *CKB1* and/or *CKB2* genes coding for regulatory subunits is not lethal for the yeast cells; it does not affect yeast growth under normal conditions but results in a phenotype of hypersensitivity to Na^+ and Li^+ cations [29].

The level of identity between plant, yeast and human CK2β regulatory subunits is not as high as in the case of CK2α subunits; for this reason complementation assays were needed in order to isolate CK2β regulatory subunits (CK2β1 and CK2β2) in *Arabidopsis thaliana* [10]. Recently, a third CK2β regulatory subunit (CK2β3) has been identified in *Arabidopsis* [11]. Data from the computational analysis of the Arabidopsis genome indicates the existence of a fourth CK2β subunit. During several years, there was controversy about the existence of CK2β subunits in maize. For instance, antibodies raised against chicken CK2β failed to recognize the presence of this protein in maize extracts [30]. Furthermore, the resolution of the crystal structure of Zea mays CK2α [12] showed that the enzyme is more stable than recombinant human CK2α. This stability and the high specific activity of the maize catalytic subunit allow speculating that it could exist without the presence of CK2β. On the other hand, two forms of the maize enzyme were originally purified: CK2A, corresponding to the typical heterotetramer, and CK2B, which is a monomeric form related to the catalytic subunit CK2α [30]. However, the properties of the monomeric form CK2B were different from those of the recombinant maize CK2α subunit, because CK2B was unable to assemble with human CK2β, whereas recombinant maize CK2α does [31, 21]. Results obtained in our lab clearly demonstrate that multiple CK2β regulatory subunits do exist and are expressed in maize [13].

The alignment between the plant CK2β regulatory subunits and the human CK2β sequence is shown in Fig. 2. The plant CK2β present an N-terminal extension of about 90 amino acids that shares no homology to other known proteins but it retains a significant level of amino acid identity (55%) between Arabidopsis and maize proteins. This N-terminal extension is not present in any other CK2β from other organisms and its functionality is unknown.

It has been reported [32] that the last 33 residues of the human CK2β regulatory subunit are relevant for oligomerization of the tetramer, as deletion of this region reduced the intensity of its interaction with the CK2α catalytic subunit. It is noteworthy that all Arabidopsis and maize CK2β subunits lack 20 of the mentioned 33 residues. This might imply that, in plants, the interaction between CK2α/β subunits is weaker than in the case of the human CK2 holoenzyme, thus making possible the existence of the two forms of the maize CK2 enzyme that were originally purified, CK2A and CK2B [30].

All the plant CK2β proteins present each of the major conserved features described for CK2β subunits from other organ-

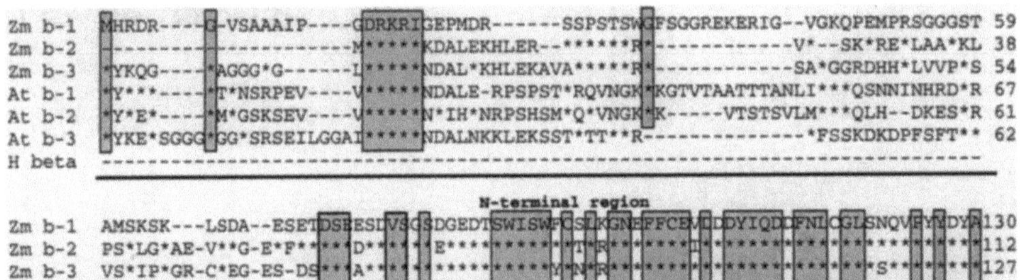

Fig. 2. Alignment of the plant CK2β regulatory subunits with the human CK2β sequence. The amino acid sequence of the three maize CK2β subunits Zm b-1 (AAG36869), Zm b-2 (AAG36870), Zm b-3 (AAG36871) and the three Arabidopsis CK2β subunits, At b-1 (AAA36869), At b-2 (AAA36870) and At b-3 (AF068318) as been aligned with human CK2β , H beta (CAA34379). Invariant residues are indicated by asterisks and by shaded boxes, and dash indicates a gap introduced to maximize alignment. Characteristics domains of CK2β regulatory subunits, the N-terminal region, the destruction box, the acidic stretch and the zinc finger domain are underlined.

isms [33]. They contain the CK2 regulatory subunit signature (C-P-X_3-C-X_{22}–CPXC), a cysteine-rich motif involved in the zinc-finger structure. The characteristic acidic stretch is also present in plants CK2β and, even though its amino acid sequence is not identical to the other CK2β subunits, the acidic residues that have been demonstrated as essentials by site-directed mutagenesis [34] are conserved. Adjacent to the acidic domain is located a region of 9 amino acids, previously reported in other CK2β regulatory subunits [35], that is described as a potential 'destruction box'; however, the functionality of this domain has been not demonstrated in plants.

It has been described that in animals, CK2 should be regulated by $p34^{cdc2}$ phosphorylation. However, the consensus site present in the C-terminal region in human CK2β, which is phosphorylated by $p34^{cdc2}$ [36] is not present in plant CK2β, suggesting a different regulation by phosphorylation of the plant enzyme. Moreover, in addition to the conserved autophosphorylation site described in other organisms, Arabidopsis and maize CK2β contain additional putative CK2 phosphorylation sites (according to the S/T-XX-D/E CK2 consensus), most of them also located in the NH_2-terminal region. The functional significance of autophosphorylation is not well understood but it is suspected to be involved in tuning of the kinase activity [37]. Due to this high number of autophosphorylation sites in plant CK2, it will be interesting to further investigate relevance of CK2β autophosphorylation.

To summarize, previous data in Arabidopsis [38] and our work in maize [13] indicate that plant CK2β subunits are able to interact with the other CK2α and CK2β subunits, allowing the formation of the typical heterotetrameric structure described in all the other organisms examined to date. However, in maize we have detected preferential interactions between α/β isoforms and β/β isoforms, suggesting that a high level of heterogeneity for the CK2β isoforms does exist in plants.

Discussion

Physiological role of CK2 in plants

The plant CK2 has a pleiotropic effect in the cells and is involved in many different processes most of them essentials for plant viability as summarized in Fig. 3. In animals, it has been reported that the protein kinase CK2 is able to phosphorylate more than 160 substrates [39], however a lower number of *in vitro* substrates of CK2 has been described in plants (see Table 1).

It has been clearly demonstrated that plant CK2 is involved in the regulation of the light-signal transduction pathway,

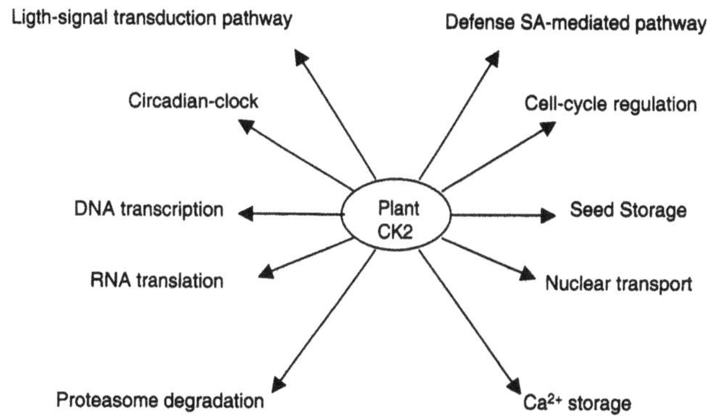

Fig. 3. Schematic representation of the physiological roles of protein kinase CK2 in plants.

through the phosphorylation of several transcription factors. Recently, studies on CK2α antisense Arabidopsis plants confirmed the role of CK2 is in the light-regulated gene expression and plant growth [14]. The CK2 enzyme phosphorylates and therefore affects the DNA binding activity of several transcription factors that bind to elements such as G-box or AT-rich regions which are located in the promoter regions of many light regulated genes. For instance, the GBF factor increases its DNA binding activity to G-box elements when it is phosphorylated by CK2 [6], on the contrary, phosphorylation by CK2 of transcription factors AT-1 [40] and ATBP-1 [41] seems to inhibit their binding to the AT rich regions. The circadian clock-associated (CCA1) and the late elongated hypocotyl (LHY) proteins are two Myb-related transcription factors essentials for the regulation of the circadian rhythms. CK2 is able to interact and phosphorylate both of them affecting their DNA binding activity [11, 42]. In animals, the myb proteins are also regulated by CK2 phosphorylation [43]. Moreover, transgenic plants overexpressing a CK2β subunit (CK2B3) display increased CK2 activity and shorter periods of rhythmic expression of CCA1 and LHY, demonstrating that CK2 is involved in the regulation of circadian rhythms in Arabidopsis [42]. It has been reported recently that the Arabidopsis bZIP transcription factor HY5, which has a role in the promotion of the photomorphogenesis or light-adapted development, is another target of CK2: unphosphorylated HY5 binds DNA of its target promoters, like the G-box in RBCS1a or CHS1 genes stronger than the phosphorylated form [44]. In this case the kinase activity is regulated by light, since a higher number of phosphorylated forms are present in dark conditions, and the CK2 phosphorylation prevents the protein degradation by the proteasome 26A. The authors postulate that this higher activity in darkness might reflect the effect of a specific CK2β regulatory subunit on HY5.

Also the maize bZIP transcription factor the Opaque 2 (O2) is likely to be regulated by CK2 phosphorylation, the phosphorylated forms, which are accumulated in the darkness,

Table 1. *In vitro* substrates of protein kinase CK2 in plants

Name	Type	Source	Role	Reference
Light-signal transduction pathway				
GBF-1	bZIP transcription factor	Arabidopsis	binds to G-box promoter	6
AT-1	transacting factor	Pea	binds to AT-rich promoter	40
ATBP-1	transacting factor	Pea	binds to AT-rich promoter	41
CCA-1	Myb-related transcription factor	Arabidopsis	potential clock gene	43
LHY	Myb-related transcription factor	Arabidopsis	potential clock gene	43
HY5	bZIP transcription factor	Arabidopsis	promotes photomorphogenesis	44
Seed storage				
Opaque2	bZIP transcription factor	Maize	binds to zein promoter	45
β-conglycinin	storage protein	Soybean	storage protein	46
calreticulin	Ca^{2+} binding protein	Spinach	Ca^{2+} metabolism	47
RNA translation				
P-proteins	Ribosomal proteins	Maize	complex with 60S ribosomal subunits	49
p34	ribonucleoprotein	Spinach cloroplasts	mRNA 3′ end processing	51
p36	eIF-2 subunit	Wheat germ	guanine nucleotide exchange	52
DNA transcription				
PDH65	Dna helycase	Pea	opens the duplex DNA during nucleic acid transactions	54
ABA/stress-induced pathway				
Rab 17	LEA protein	Maize	Stress-induced protein	55
TAS-14	LEA protein	Tomato	Stress-induced protein	58
ATP synthesis				
CFOCF1-ATPase	ATPase synthase b subunit	Spinach cloroplasts	ATP synthesis	59
apyrase	Pea	ATP hydrolysis	60	
Lipid synthesis				
gp96	lipoxygenase	Soybean	oxygenation of unsaturated fatty acids	61
Proteasome machinery				
C2	proteasome protein	Rice	protein degradation	62
Nuclear matrix proteins				
lamina-like	lamina matrix protein	Pea	lamina matrix protein	7

have a diminished DNA binding activity compared to the hypophosphorylated forms [45]. In maize endosperm, O2 regulates the expression of the zein genes, which are the primary storage product. In fact, multiple storage protein are *in vitro* substrates of CK2, such as β-conglicin [46] or calreticulin [47]. The plant calreticulin is an abundant ER involved in many different functions in cells, and is specifically phosphorylated by CK2 *in vitro*. On the contrary, animal calreticulin is not a CK2 substrate, even though all the major Era/Ca^{2+}-binding proteins such as calmodulin [25] or calnexin [48] are.

As in the case of animals, proteins involved in the RNA translation are also phosphorylated in plants. In maize, the complex of the 60S ribosomal subunit contains acidic phosphoproteins (P-proteins) involved in the regulation of the protein synthesis which are *in vitro* phosphorylated by CK2 [49] as in the case of the P-proteins from animals and yeast [50]. In spinach chloroplasts, the physiological activity of the ribonucleoprotein p34 which is required for plastid mRNA 3' processing, is regulated by CK2 phosphorylation [51]. Another *in vitro* substrate of CK2 is p36, the small subunit of the wheat germ eukariotic initiation factor, eIF-2α [52]. The functional significance of plant eIF-2α phosphorylation is not clear, whereas in mammals, the eIF-2α phosphorylation by eIF-2 kinases inhibits the protein synthesis [53]. Recently, it has been reported that PDH65, a DNA helicase that may be involved both in rDNA transcription and in early stages of pre-RNA processing, is upregulated by CK2 phosphorylation [54].

The maize Rab17 is a phosphoprotein responsive to abscisic acid and induced under water stress conditions and it has been reported that is strongly phosphorylated by CK2 *in vitro* [55]. It is located both in cytoplasm and nucleus of the maize cells and it is able to interact with synthetic NLS peptides but only in the phosphorylated form, it seems therefore that Rab 17 protein might act regulating transport of proteins from cytoplasm to the nucleus during stress conditions and this nuclear transport may be dependent of CK2 phosphorylation [56, 57]. Recently in animals CK2 re-localization has been described under stress conditions [58]. A homologue of Rab17 in tomato, the TAS-14 protein is also phosphorylated by CK2 [59].

Several enzymes implicated in the ATP metabolism such as an spinach chloroplast ATP synthase [60] and a pea nuclei apyrase [61] are also CK2 *in vitro* substrates. Other plant CK2 targets are a soybean lipoxygenase [62] and a proteasome subunit [63]. In pea nuclei, CK2 has been found in the lamina-matrix and it is able to phosphorylate a lamina-like protein [7].

In yeast and in mammalian cells it has been demonstrated the involvement of CK2 in cell cycle control. Using the synchronizable tobacco BY-2 cell line it has been reported that the Ck2 activity oscillates during the cell cycle, peaking at G1/S and M phases [15]. It is proposed that the CK2β regulatory subunits may play a central role in controlling the level of CK2 activity. Recently, it has been reported that CK2 is involved in the activation of the Salicylic acid-mediated pathway [64], which is one of the most studied pathways in the plant defense reactions. It is the first time in which activation of the CK2 by a plant hormone is reported.

Acknowledgements

This work was supported by grant BIO2000-1562. M.R. was supported by grant 200 1 TDO C000 12 from Generalitat de Catalunya (Spain).

References

1. Seldin DC, Leder P: Casein kinase II alpha transgene-induced murine lymphoma: Relation to theileriosis in cattle. Science 267: 894–896,1995
2. Lüscher B, Kuenzel EA, Krebs EG, Eisenman RN: Myc oncoproteins are phosphorylated by casein kinase II. EMBO J 8: 1111–1119, 1989
3. Hanna D, Rethinaswamy A, Glover CVC: Casein kinase II is required for cell cycle progression during G1 and G2/M in *Saccharomyces cerevisiae*. J Biol Chem 270: 25905–25914, 1995
4. Yan T-FJ, Tao M: Purification and characterization of a wheat germ protein kinase. J Biol Chem 257: 7037–7043, 1982
5. Dobrowolska G, Meggio F, Marchiori F, Pinna LA: Specific determinants of maize casein kinase II-B are related but distinct from those of rat liver casein kinase-2. Biochem Biophys Act 1010: 274–277, 1989
6. Klimczak LJ, Schindler U, Cashmore AR: DNA binding activity of the Arabidopsis G-box binding factor GBF1 is stimulated by phosphorylation by casein kinase II from broccoli. Plant Cell 4: 87–98, 1992
7. Li H, Roux SJ: Casein kinase II protein kinase is bound to lamina-matrix and phosphorylates lamina-like protein in isolated pea nuclei. Proc Natl Acad Sci USA 89: 8434–8438, 1992
8. Dobrowolska G, Boldyreff B, Issinger O-G: Cloning and sequencing of the casein kinase 2 a subunit from Zea mays. Biochem Biophys Acta 1129: 139–140, 1991
9. Mizoguchi T, Yamaguchi-Shinozaki K, Hayashida N, Kamada H Shinozaki K: Cloning and characterization of two cDNAs encoding casein kinase II catalytic subunits in *Arabidopsis thaliana*. Plant Mol Biol 21: 279–289, 1993
10. Collinge MA, Walker JC: Isolation of an *Arabidopsis thaliana* casein kinase II β subunit by complementation in *Saccharomyces cerevisiae*. Plant Mol Biol 25: 649–658, 1994
11. Sugano S, Andronis C, Green RM, Wang Z-Y, Tobin EM: Protein kinase CK2 interacts with and phosphorylates the Arabidopsis circadian clock-associated 1 protein. Proc Natl Acad Sci USA 95: 11020–11025, 1998
12. Niefind K, Guerra B, Pinna LA, Issinger O-G, Schomburg D: Crystal structure of the catalytic subunit of protein kinase CK2 from Zea mays at 2.1 A resolution. EMBO J 17: 2451–2462, 1998
13. Riera M, Peracchia G, de Nadal E, Ariño J, Pagès M: Maize protein kinase CK2: Regulation and functionality of three β regulatory subunits. Plant J 25: 2001 (in press)
14. Lee Y, Lloyd AM, Roux SJ: Antisense expression of the CK2 alpha-subunit gene in Arabidopsis. Effects on light-regulated gene expression and plant growth. Plant Physiol 119: 989–1000, 1999
15. Espunya MC, Combettes B, Dot J, Chaubet-Gigot N, Martinez MC: Cell-cycle modulation of CK2 activity in tobacco BY-2 cells. Plant J 19: 655–666, 1999
16. Litchfield DW, Lozemann FJ, Piening C, Sommercorn J, Takio K, Walsch KA, Krebs, EG: Subunit structure of casein kinase II from bovine testis. J Biol Chem 265: 7638–7644, 1990
17. Padmanabha R, Chen-Wu JLP, Hanna DE, Glover CVC: Isolation, sequencing, and disruption of the CKA2 gene: Casein kinase II is essential for viability in *Saccharomyces cerevisiae*. Mol Cell Biol 10: 4089–4099, 1990
18. Peracchia G, Jensen AB, Culiáñez-Macià FA, Grosset J, Goday A, Issinger O-G, Pagès M: Characterization, subcellular localization and nuclear targeting of casein kinase II from Zea mays. Plant Mol Biol 40: 199–211, 1999
19. Guerra B, Boldyreff B, Sarno S, Cesaro L, Issinger O-G, Pinna LA: CK2: A protein kinase in need of control. Pharmacol Ther 82: 303–313, 1999
20. Niefind K, Putter M, Guerra B, Issinger O-G, Schomburg D: GTP plus water mimic ATP in the active site of protein kinase CK2. Nat Struct Biol 6: 1100–1103, 1999
21. Battistutta R, Sarno S, De Moliner E, Marin O, Issinger O-G, Zanotti G, Pinna LA: The crystal structure of the complex of Zea mays alpha subunit with a fragment of human beta subunit provides the clue to the architecture of protein kinase CK2 holoenzyme. Eur J Biochem 267: 5184–5190, 2000
22. Livak KJ: Detailed structure of the *Drosophila melanogaster* stellate genes and their transcripts. Genetics 124: 303–316, 1990
23. Grankowski N, Boldyreff B, Issinger O-G: Isolation and characterization of recombinant human casein kinase II subunits α and β from bacteria. Eur J Biochem 198: 25–30, 1991
24. Meggio F, Boldyreff B, Marin O, Marchiori F, Perich JW, Issinger O-G, Pinna LA: Role of β subunit of casein kinase-2 on the stability and specificity of the recombinant reconstituted holoenzyme. Eur J Biochem 204: 293–297, 1992
25. Bidwai AP, Reed JC, Glover CVC: Phosphorylation of calmodulin by the catalytic subunit of casein kinase II is inhibited by the regulatory subunit. Arch Biochem Biophys 300: 265–270, 1993
26. Bidwai AP, Reed JC, Glover CVC: Casein kinase II of *Saccharomyces cerevisiae* contains two distinct regulatory subunits, β and β′. Arch Biochem Biophys 309: 348–355, 1994
27. Saxena A, Padmanabha R, Glover CVC: Isolation and sequencing of cDNA clones encoding the alpha and beta subunits of drosophila casein kinase II. Mol Cell Biol 7: 3409–3417, 1987
28. Bidwai AP, Zhao W, Glover CVC: A gene located at 56F1-2 in *Drosophila melanogaster* encodes a novel metazoan beta-like subunit of casein kinase II. Mol Cell Biol Res Commun 1: 21–28, 1999

29. Bidwai AP, Reed JC, Glover CVC: Cloning and disruption of CKB1, the gene encoding the 38-kDa beta subunit of *Saccharomyces cerevisiae* casein kinase II (CKII). Deletion of CKII regulatory subunits elicits a salt-sensitive phenotype. J Biol Chem 270: 10395–10404, 1995
30. Dobrowolska G, Meggio F, Szczegielniak J, Muszynska G, Pinna LA: Purification and characterization of maize seedling casein kinase IIB, a monomeric enzyme immunologically related to the a subunit of animal casein kinase-2. Eur J Biochem 204: 299–303, 1992
31. Boldyreff B, Meggio F, Dobrowolska G, Pinna LA, Issinger O-G: Expression and characterization of a recombinant maize CK-2 α subunit. Biochim Biophys Acta 1173: 32–38, 1993
32. Chantalat L, Leroy D, Filhol O, Nueda A, Benitez MJ, Chambaz EM, Cochet C, Dideberg O: Crystal structure of the human protein kinase CK2 regulatory subunit reveals its zinc finger-mediated dimerization. EMBO J 18: 2930–2940, 1999
33. Reed JC, Bidwai AP, Glover CVC: Cloning and disruption of CKB2, the gene encoding the 32-kDa regulatory beta'-subunit of *Saccharomyces cerevisiae* casein kinase II. J Biol Chem 269: 18192–18200, 1994
34. Krehan A, Lorenz P, Plana-Coll M, Pyerin W: Interaction sites between catalytic and regulatory subunits in human protein kinase CK2 holoenzymes as indicated by chemical cross-linking and immunological investigations. Biochemistry 35: 4966–4975, 1996
35. Allende JE, and Allende CC: Protein kinase CK2: An enzyme with multiple substrates and a puzzling regulation. FASEB J 9: 313–323, 1995
36. Litchfield DW, Lozeman FJ, Cicirelli MF, Harrylock M, Ericsson LH, Piening CJ, Krebs EG: Phosphorylation of the beta subunit of casein kinase II in human A431 cells. Identification of the autophosphorylation site and a site phosphorylated by p34cdc2. J Biol Chem 266: 20380–20389, 1991
37. Lin W-J, Sheu G-T, Traugh JA: Effects of autophosphorylation on casein kinase II activity: Evidence from mutations in the β subunit. Biochemistry 33: 6998–7004, 1994
38. Klimczak LJ, Collinge MA, Farini D, Giuliano G, Walker JC, Cashmore AR: Reconstitution of Arabidopsis casein kinase II from recombinant subunits and phosphorylation of transcription factor GBF1. Plant Cell 7: 105–115, 1995
39. Guerra B, Issinger O-G: Protein kinase CK2 and its role in cellular proliferation, development and pathology. Electrophoresis 20: 2391–406, 1999
40. Datta N, Cashmore AR: Binding of a pea nuclear protein to promoters of certain photoregulated genes is modulated by phosphorylation. Plant Cell 1: 1069–1077, 1989
41. Tjaden G, Coruzzi GM: A novel AT-rich DNA binding protein that combines an HMG-like DNA binding protein with a putative transcription domain. Plant Cell 6: 107–118, 1994
42. Sugano S, Andronis C, Ong MS, Green, Tobin EM: The protein kinase CK2 is involved in regulation of circadian rhythms in Arabidopsis. Proc Natl Acad Sci USA 96: 12362–123626, 1999
43. Lüscher B, Christenson E, Litchfield D, Krebs EG, Eisenman RN: Myb DNA binding inhibited by phosphorylation at site deleted during oncogenic activation. Nature 344: 517–521, 1990
44. Hardtke CS, Gohda K, Osterlund MT, Oyama T, Okada K, Deng XW: HY5 stability and activity in arabidopsis is regulated by phosphorylation in its COP1 binding domain. EMBO J 19: 4997–5006, 2000
45. Ciceri P, Gianazza E, Lazzari B, Lippoli G, Genga A, Hoschek G, Schmidt RJ, Viotti A: Phosphorylation of Opaque 2 changes diurnally and impacts its DNA binding activity. Plant Cell 9: 97–108, 1997
46. Ralet MC, Fouques D, Leonil J, Molle D, Meunier JC: Soybean beta-conglicin alpha subunit is phosphorylated on two distinct serines by protein kinase CK2 *in vitro*. J Protein Chem 18: 315–323, 1999
47. Baldan B, Navazio L, Friso A, Mariani P, Meggio F: Plant calreticulin is specifically and efficiently phosphorylated by protein kinase CK2. Biochem Biophys Res Commun 221: 498–502, 1996
48. Ou WJ, Thomas DY, Bell AW, Bergeron JJ: Protein casein kinase II phosphorylation of signal sequence receptor alpha and the associated membrane chaperone calnexin. J Biol Chem 267: 23789–23796, 1992
49. Bailey-Serres J, Vangala S, Szick K, Lee CH: Acidic phosphoprotein complex of the 60S ribosomal subunit of maize seedling roots. Components and changes in response to flooding. Plant Physiol 114: 1293–305, 1997
50. Hasler P, Brot N, Weissbach H, Parnassa AP, Elkon K: Ribosomal proteins P0, P1 and P2 are phosphorylated by casein kinase II at their conserved carboxy termini. J Biol Chem 266: 13815–13820, 1991
51. Kanekatsu M, Munakata H, Furuzono K, Ohtsuki K: Biochemical characterization of a 34 kDa ribonucleoprotein (p34) purified from the spinach chloroplast fraction as an effective phosphate acceptor for casein kinase II. FEBS Lett 335: 176–180, 1993
52. Janaki N, Krishna VM, Ramaiah KV: Phosphorylation of wheat germ initiation factor 2 (eIF-2) by N-ethylmaleimide-treated wheat germ lysates and by purified casein kinase II does not affect the guanine nucleotide exchange on eIF-2. Arch Biochem Biophys 324: 1–8, 1995
53. Pain VM, Clemens MJ: Assembly and breakdown of mammalian protein synthesis initiation complexes: Regulation by guanine nucleotides and by phosphorylation of initiation factor eIF-2. Biochemistry 22: 726–733, 1983
54. Tuteja N, Beven AF, Shaw PJ, Tuteja R: A pea homologue of human DNA helicase I is localized within the dense fibrillar component of the nucleolus and stimulated by phosphorylation with CK2 and cdc2 protein kinases. Plant J 25: 9–17, 2001
55. Plana M, Itarte E, Eritja R, Goday A, Pagès M, Martinez MC: Phosphorylation of Maize RAB-17 protein by casein kinase 2. J Biol Chem 266: 22510–22514, 1991
56. Goday A, Jensen AB, Culiañez-Macià FB, Albà MM, Figueras M, Serratosa J, Torrent M, Pagès M: The maize abscisic acid-responsive protein rab17 is located in the nucleus and interacts with nuclear localization signals. Plant Cell 6: 351–360, 1994
57. Jensen AB, Goday A, Figueras M, Jessop AC, Pagès M: Phosphorylation mediates the nuclear targeting of the maize Rab17 protein. Plant J 13: 691–697, 1998
58. Gerber DA, Souquere-Besse S, Puvion F, Dubois MF, Bensaude O, Cochet C: Heat-induced relocalization of protein kinase CK2. Implication of CK2 in the context of cellular stress. J Biol Chem 275: 23919–23926, 2000
59. Godoy JA, Lunar R, Torres-Schumann S, Moreno J, Rodrigo RM, Pintor-Toro JA: Expression, tissue distribution and subcellular localization of dehydrin TAS14 in salt-stressed tomato plants. Plant Mol Biol 26: 1921–1934, 1994
60. Kanekatsu M, Saito H, Motohashi K, Hisabori T: The beta subunit of chloroplast ATP synthase (CF0CF1-ATPase) is phosphorylated by casein kinase II. Biochem Mol Biol Int 46: 99–105, 1998
61. Hsieh HL, Song CJ, Roux SJ: Regulation of a recombinant pea nuclear apyrase by calmodulin and casein kinase II. Biochim Biophys Acta 1494: 248–255, 2000

62. Ohtsuki K, Nakamura S, Shimoyama Y, Shibata D, Munakata H, Yoshiki Y, Okubo K: A 96-kDa glycyrrhizin-binding protein (gp96) from soybeans acts as a substrate for casein kinase II, and is highly related to lipoxygenase 3. J Biochem (Tokyo) 118: 1145–1150, 1995
63. Umeda M, Manabe Y, Uchimiya H: Phosphorylation of the C2 subunit of the proteasome in rice (*Oryza sativa* L.). FEBS Lett 403: 313–317, 1997
64. Hidalgo P, Garreton V, Berrios CG, Ojeda H, Jordana X, Holuigue L: A nuclear casein kinase 2 activity is involved in early events of transcriptional activation induced by salicylic acid in tobacco. Plant Physiol 125: 396–405, 2001

Molecular and Cellular Biochemistry **227**: 129–135, 2001.

Cloning and characterization of the cDNA coding for the catalytic α subunit of CK2 from tobacco

Paula Salinas,[1] Brigitte Bantignies,[1] Julio Tapia,[2] Xavier Jordana[1] and Loreto Holuigue[1]

[1]*Departamento de Genética Molecular y Microbiología, Facultad de Ciencias Biológicas, P. Universidad Católica de Chile, Santiago;* [2]*Programa de Biología Celular y Molecular, ICBM, Facultad de Medicina, Universidad de Chile, Santiago, Chile*

Abstract

We have previously reported the participation of the protein kinase CK2 in the mechanism by which salicylic acid activates transcription of genes, such as those coding for glutathion *S*-transferases, in tobacco. With the purpose of further studying the participation of CK2 in this signal transduction pathway, we isolated and sequenced the cDNA from the NtCK2A gene, coding for the catalytic α subunit of CK2 from tobacco. The NtCK2A cDNA was isolated by screening of a tobacco cDNA library with a heterologous probe from *Arabidopsis thaliana*, followed by 3′ RACE to obtain the 3′ region. Sequence analysis of the NtCK2A cDNA showed a high level of identity between this CK2α protein sequence and the corresponding sequences of other plant species such as *Arabidopsis* and maize (92–95% identity), or those of animal species such as human and *Xenopus laevis* (75% identity). The expression of the NtCK2A gene in different tissues from tobacco plants was analyzed by Northern blot. High levels of expression of this gene were observed in proliferating tissues such as shoot and root apical meristems. A recombinant CK2α protein was obtained after expression of the NtCK2A cDNA in *Escherichia coli*. The ability of this recombinant CK2α subunit to phosphorylate casein was inhibited by heparin and stimulated by the CK2β subunit from *Xenopus laevis*. (Mol Cell Biochem **227**: 129–135, 2001)

Key words: protein kinase CK2, alpha subunit cDNA, tobacco, tissue expression

Introduction

In our group we are interested in the study of the signal transduction pathway of salicylic acid (SA) in plants. This hormone plays a crucial role in the defense reactions of plants against pathogens and environmental stressors such as ozone and UV radiation [1–3]. The main role of SA in these reactions is to activate the transcription of a number of defense genes, such as those coding for glutathion *S*-transferases (GSTs) [4, 5]; enzymes that protect plant tissues from oxidative damage [6, 7].

Our interest has focussed on the activation of the SA-responsive element found in the promoter regions of the genes regulated by SA [8, 9]. We have recently reported evidences for the participation of a nuclear CK2 activity in this step of the SA transduction pathway. Interestingly, we found that treatment of plants with SA increases the nuclear CK2 activity [9]. The mechanism by which SA triggers this increase in CK2 activity in the nucleus, as well as the identity of the target substrate(s) for CK2 in this pathway, are still not known. With the purpose of contributing to the further study of the participation of CK2 in this signal transduction pathway, we isolated and sequenced the cDNA from the NtCK2A gene coding for the catalytic CK2α subunit in tobacco. This gene is preferentially expressed in proliferating meristematic tissues in the tobacco plant. Expression of the NtCK2A cDNA in *Escherichia coli* allowed us to obtain a recombinant CK2α subunit (rCK2α), whose catalytic activity is inhibited by heparin and strongly stimulated by the CK2β subunit from *Xenopus laevis*. Characterization of tobacco CK2 subunits should be the starting point for further studies leading to the understanding of the regulation of the CK2 kinase by SA.

Address for offprints: L. Holuigue, P.O. Box 114-D, Santiago, Chile (E-mail: lholuig@genes.bio.puc.cl)

Materials and methods

Screening of the tobacco cDNA library

A heterologous probe from the *Arabidopsis thaliana* AtCK2A cDNA was used for the first two steps of the screening. To obtain this probe, GenBank *Arabidopsis* ESTs were searched with the BLASTN programs, using the nucleotide sequence of the AtCK2A gene (accession number D10246). An EST clone (accession number T45147, clon 77D6T7), showing the highest similarity to the query sequence, was supplied by the *Arabidopsis* Biological Resource Center (Columbus, USA). DNA from this clone was purified, sequenced and used as a template for PCR amplification. A 432 bp-fragment from a conserved region (corresponding to residues 35–178 in the AtCK2A sequence, see Fig. 2), was amplified using the primers 5′ CKA (5′-GAGGTAGTAAGGAAGG TA-GGGCGTGG-3′) and 3′CKA (5′-GTGGTAAAACTCAGC-AAGACCCCAAT-3′). A homologous probe from tobacco was used for the final step of the screening. To obtain this probe, PCR amplification of a 632 bp fragment was performed using as a template the positive clones of the second screening step. The following degenerated primers were used in this amplification: 5′CK2D (5′-AAYGTNNMTNMGNCC-WAANGADTAYTGG-3′) and 3′CK2D (5′-CCWGCRAA CATRCANCCNARNSWCCACAT-3′) (underlined in Fig. 1). Probes were labeled with [α-^{32}P]dCTP by random oligonucleotide-primer synthesis (RTS RadPrime Labeling System, Amersham).

Isolation of cDNA clones was carried out as described in Lambda Library Protocols Handbook (CLONTECH Laboratories). About 4 × 10^5 clones of a shoot tobacco (*Nicotiana tabacum* cv w38) cDNA library in λgt10 (kindly supplied by Dr. Marc Van Montagu, University of Gent, Belgium) were plated onto *Escherichia coli* Y1090r-cells. Screening was performed by hybridization at 37°C in 35% formamide, 5 × SSPE (50 mM NaH_2PO_4, 5 mM EDTA, 750 mM NaCl, pH 7.4), 5 × Denhardt's solution, 0.1% SDS and 100 μg/ml salmon sperm carrier DNA, using the ^{32}P-labeled *Arabidopsis* probe (2–5 × 10^7 cpm). The hybridizing recombinant phages were rescreened at 42°C in 50% formamide, 5 × SSPE, 5 × Denhardt's solution, 0.1% SDS, using the ^{32}P-labeled tobacco probe (2–5 × 10^7 cpm) until pure populations were obtained. Final washing was performed at 42°C in 0.2 × SSC and 0.2% SDS. Phage particles were isolated in the SM buffer (100 mM NaCl, 8 mM $MgSO_4$ × $7H_2O$, 50 mM Tris-HCl, 0.01% gelatin). Inserted cDNAs were obtained from recombinant phages by PCR amplification using the primers 5′gt10 (5′-AGCA-AGTTCAGCCTGGTTAAGT-3′) and 3′gt10 (5′-TTATGA-GTATTTCTTCCAGGG-3′), and ligated to the pGEM-T Easy plasmid (Promega).

```
1    - ATTCCGGCTGTTGATTACGTCACCGGCGGTGGCTTCACATGGAAGATGCTGCCTTCAATT
61   - TCTTCTAGTGTGTGCCGTCCTTGTTGCTTATCGTGCACCTCTTGCCTCACCTCCTGACCT
121  - CCTACAGCTCGTCGCTACTTCTGCCACCGGAAATTTCGTCAGTTTTGGTAAGATCAACAG
181  - CTCTGAACCGGTGATTGACGGTAAAACTATGTCAAAAGCTCGTGTTTACACCGATGTTAA
                                   M  S  K  A  R  V  Y  T  D  V  N
241  - TGTCCTTCGCCCGAGGGAATATTGGGATTATGAAGCCCTTACTGTTCAATGGGGTGATCA
         V  L  R  P  R  E  Y  W  D  Y  E  A  L  T  V  Q  W  G  D  Q
301  - GGATGACTATGAGGTTGTTAGGAAAGTTGGAAGAGGAAAATATAGTGAAGTTTTTGAAGG
         D  D  Y  E  V  V  R  K  V  G  R  G  K  Y  S  E  V  F  E  G
361  - TATAAATGTTAACAGCCATGAAAAGTGCATAATCAAGATCCTGAAACCTGTTAAGAAGAA
         I  N  V  N  S  H  E  K  C  I  I  K  I  L  K  P  V  K  K  K
421  - AAAGATCAAGAGAGAGATAAAAATCTTGCAGAACCTCTGTGGGGGACCAACCATTGTCAA
         K  I  K  R  E  I  K  I  L  Q  N  L  C  G  G  P  T  I  V  K
481  - ACTCCTTGATATTGTCAGAGATCAGCACTCAAAAACTCCAAGCTTAATTTTTGAGTATGT
         L  L  D  I  V  R  D  Q  H  S  K  T  P  S  L  I  F  E  Y  V
541  - GAACAGTACAGATTTCAAAGTACTGTACCCAACATTAACAGATTATGACATCCGGTACTA
         N  S  T  D  F  K  V  L  Y  P  T  L  T  D  Y  D  I  R  Y  Y
601  - CATATATGAGCTTCTCAAGGCACTAGACTATTGTCATTCACAAGGAATAATGCATAGAGA
         I  Y  E  L  L  K  A  L  D  Y  C  H  S  Q  G  I  M  H  R  D
661  - TGTCAAGCCACATAATGTTATGATAGACCATGAATTGCGGAAGCTTCGCTTGATAGACTG
         V  K  P  H  N  V  M  I  D  H  E  L  R  K  L  R  L  I  D  W
721  - GGGTCTTGCTGAATTTTACCATCCAGGAAAGGAATATAATGTTCGTGTGGCTTCAAGATA
         G  L  A  E  F  Y  H  P  G  K  E  Y  N  V  R  V  A  S  R  Y
781  - CTTCAAGGGCCCTGAACTTCTTGTTGACTTGCAAGACTATGACTATTCTTTGGACATGTG
         F  K  G  P  E  L  L  V  D  L  Q  D  Y  D  Y  S  L  D  M  W
841  - GAGTCTTGGATGCATGTTTGCAGGAATGATTTTCCGCAAGGAACCTTTCTTTTATGGTCA
         S  L  G  C  M  F  A  G  M  I  F  R  K  E  P  F  F  Y  G  H
901  - TGATAACCAGGATCAGCTTGTCAAAATTGCTAAGGTACTTGGAACTGATGAGTTGAATGC
         D  N  Q  D  Q  L  V  K  I  A  K  V  L  G  T  D  E  L  N  A
961  - ATATCTGCACAAGTATCAACTAGAACTTGATCCTCAGCTGGAGGCTATGGTTGGGAGACA
         Y  L  H  K  Y  Q  L  E  L  D  P  Q  L  E  A  M  V  G  R  H
1021 - CAGTAGGAAGCCATGGTCCAAATTTATAAATGCAGATAATCAGCATTTAGTGTCGCCAGA
         S  R  K  P  W  S  K  F  I  N  A  D  N  Q  H  L  V  S  P  E
1081 - GGCTATAGATTTTCTTGACAAGCTTCTCCGTTATGACCATCAGGATAGACTTACAGCAAG
         A  I  D  F  L  D  K  L  L  R  Y  D  H  Q  D  R  L  T  A  R
1141 - AGAAGCAATGGCCCATCCTTATTTCCTGCAAGTGAGGGCTGCAGAGAATAGCAGGATGAG
         E  A  M  A  H  P  Y  F  L  Q  V  R  A  A  E  N  S  R  M  R
1201 - GACGCAGTAGCCAGTTGTATTGGTCTTACTGGGAAAGAAGAAAATATGGTGTAATGTAGC
         T  Q  *

1261 - ATACATTGCTCTCTTCATCTATTTGGATTAAAGTACAGGATAAGTTGAACTTAAGAGGGC
1321 - AGAATTTCATCATGGAGCATTTCATGAGCAATTGGTTTCATAAATCTACCTCTCATGACC
1381 - TTGTTGACGATTGCCTATGTTCCCCCTGCCGATTATGAATTCTCTTCATTAGTCTTCATA
1441 - TTTTTGTATCCAAAGCCATCTTCCATGTGGAGTGTATTGTTTGAGAATTGCAATTTACTG
1501 - AAGCTTTTACTGAAAAAAAAAAAAAA
```

Fig. 1. Nucleotide sequence and deduced aminoacid sequence of the NtCK2A cDNA. The DNA sequence contains the coding region and the 5′ and 3′ UTRs. Sequence corresponding to the 15 amino acids from the C-terminal and the 3′UTR were obtained by 3′RACE (indicated by the arrowhead in the figure). The amino acid sequence of the CK2α protein is shown beneath the coding region of the cDNA sequence. The primers used for amplification of the homologous probe (5′ CK2D, **3** and 3′ CK2D, **4**), sequencing (5′ CK2D, **3** and 3′ sec, **7**), 3′RACE (3′ race-2, **5** and 3′ race-1, **6**), and cloning (Bamnt5′ , **1** and Xhont3′ , **2**), are shown.

3′RACE

The 3′ end of the isolated cDNA was obtained using the commercial 3′ RACE system (Rapid Amplification of cDNA Ends, Gibco-BRL), following the supplier's protocol. For this purpose, total RNA was isolated from tobacco leaves as de-

```
NtCK2         MSKARVYTDVNVLRPREYWDYEALTVQWGDQDDYEVVRKVGRGKYSEVFEGINVN  55
ZmCK2         MSKAKVYTDVNVLRPKEYWDYEALTVQWGEQDDYEVVRKVGRGKYSEVFEGINVN  55
AtCK2A        MSKARVYTEVNVIRPKDYWDYESLIVQWGEQDDYEVVRKVGRGKYSEVFEGINVN  55
HsCK2    MSGPVPSRARVYTDVNTHRPREYWDYESHVVEWGNQDDYQLVRKLGRGKYSEVFEAINIT  60
XlCK2    MSGPVPSRARVYTDVNTHRPRDYWDYESHVVEWGNQDDYQLVRKLGRGKYSEVFEAINIT  60
              <---- N-terminal region ---->  <-β 1->   <--β 2-->

NtCK2    SHEKCIIKILKPVKKKKIKREIKILQNLCGGPTIVKLLDIVRDQHSKTPSLIFEYVNSTD  115
ZmCK2    NNEKCIIKILKPVKKKKIKREIKILQNLCGGPNIVKLLDIVRDQHSKTPSLIFEFVNNTD  115
AtCK2A   SKEKCIIKILKPVKKKKIRREIKILQNLCGGPNIVKLLDVVRDQHSKTPSLIFEYVNSTD  115
HsCK2    NNEKVVVKILKPVKKKKIKREIKILENLRGGPNIITLADIVKDPVSRTPALVFEHVNNTD  120
XlCK2    NNEKVVVKILKPVKKKKIKREIKILENLRGGPNIITLADIVKDPVSRTPALVFEHVNNTD  120
          <-β 3->  <--- αC --->         <-β 4->  <-β 5->     <

NtCK2    FKVLYPTLTDYDIRYYIYELLKALDYCHSQGIMHRDVKPHNVMIDHELRKLRLIDWGLAE  175
ZmCK2    FKVLYPTLTDYDIRYYIYELLKALDYCHSQGIMHRDVKPHNVMIDHELRKLRLIDWGLAE  175
AtCK2A   FKVLYPTLTDYDIRYYIYELLKALDFCHSQGLMHRDVKPHNVMIDHELRKLRLIDWGLAE  175
HsCK2    FKQLYQTLTDYDIRFYMYEILKALDYCHSMGIMHRDVKPHNVMIDHEHRKLRLIDWGLAE  180
XlCK2    FKQLYQTLTDYDIRFYMYEILKALDYCHSMGIMHRDVKPHNVMIDHEHRKLRLIDWGLAE  180
    αD->  <---- αE ---->  <β 6>     <β 7>   <β 8>   <β 9
                               <-cat. loop->   <-----------

NtCK2    FYHPGKEYNVRVASRYFKGPELLVDLQDYDYSLDMWSLGCMFAGMIFRKEPFFYGHDNQD  235
ZmCK2    FYHPGKEYNVRVASRYFKGPELLVDLQDYDYSLDMWSLGCMFAGMIFRKEPFFYGHDNHD  235
AtCK2A   FYHPGKEYNVRVVSRYFKGPELLVDLQDYDYSLDMWSLGCMFAGLLFRKEPFFYGHDNQD  235
HsCK2    FYHPGQEYNVRVASRYFKGPELLVDYQMYDYSLDMWSLGCMLASMIFRKEPFFHGHDNYD  240
XlCK2    FYHPGQEYNVRVASRYFKGPELLVDYQMYDYSLDMWSLGCMLASMIFRKEPFFHGHDNYD  240
         ->    <-p+1-loop->            <--- αF --->          <---
activation segment ------>

NtCK2    QLVKIAKVLGTDELNAYLHKYQLELDPQLEAMVGRHSRKPWSKFINADNQHLVSPEAIDF  295
ZmCK2    QLVKIAKVLGTDGLNAYLNKYHIELDPQLEALVGRHSRKPWSKFMNADNQHLVSPEAIDF  295
AtCK2A   QLVKIAKGVGTDELNAYLNKYQLELDPQLEALVGRHSRKPWSKFINADNQHLVSPEAIDF  295
HsCK2    QLVRIAKVLGTEDLYDYIDKYNIELDPRFNDILGRHSRKRWERFVHSENQHLVSPEALDF  300
XlCK2    QLVRIAKVLGTEDLYDYIDKYNIELDPRFNDILGRHSRKRWERFVHSENQHLVSPEALDF  300
          αG ---X--- αH -->  <- αI ->    <-αJ->  <-αK-X---αL

NtCK2    LDKLLRYDHQDRLTAREAMAHPYFLQVRAAENSRMRTQ                       333
ZmCK2    LDKLLRYDHQDRLTAREAMAHPYFLQVRAVENSRTRPQ                       333
AtCK2A   LDKLLRYDHQDRLTAKEAMAHAYFAQVRAAETSRMRSQ                       333
HsCK2    LDKLLRYDHQSRLTAREAMEHPYFYTVVKDQARMGSSSMPGGSTPVSSANMMSGISSVPT 360
XlCK2    LDKLLRYDHQTRLTAREAMDHPYFYPIVKDQSRMAALICPVAAHPSVAPV           350
                   <-αM->    <---αN--->

HsCK2     PSPLGPLAGSPVIAAANPLGMPVPAAAGAQQ                             391
```

Fig. 2. Comparison of the deduced amino acid sequence of NtCK2A with ZmCK2 from *Zea mays* (accession number Y11526), AtCK2A from *Arabidopsis thaliana* (accession number D10246), HsCK2 from human (accession number NM.001895), and XlCK2A from *Xenopus laevis* (accession number X62375). Alignment was performed with CLUSTALW program. Asterisks (*) indicate identical amino acids and points (●) indicate conserved amino acids. β-sheet and α-helix structures, as identified according to Niefind *et al.* [19], are indicated beneath the sequence. Important functional regions such as the N-terminal region, the catalytic loop and the activation segment, are also indicated. Amino acid residues conserved in all kinase families are underlined. The Ser113 residue in NtCK2A and in AtCK2A are indicated in bold.

scribed below, and Poly (A)$^+$ RNA was purified from 1 μg of total RNA, using the Oligotex™ mRNA purification kit according to the manufacturer's instructions (Qiagen). Poly (A)$^+$ RNA was reversely transcribed using a primer containing an oligo dT tail (AP primer, Gibco-BRL). Then, two sequential PCR reactions were carried out using a primer complementary to the AP primer (AUAP primer, Gibco-BRL), and specific sense primers Race3′-2 (5′-GGTCATGATAACCAGGATCAGCTTG-3′) for the first reaction and Race3′-1 (5′-CACAGTAGGAAGCCATGGTCC-3′) for the second reaction. These primers are located at 688 and 810 bp from the ATG codon of the sequenced cDNA, respectively (underlined in Fig. 1).

DNA sequencing

DNA sequencing was performed by the dideoxy chain termination method, using [α-^{33}P]-ddNTPs (Thermosequenase™, Amersham). Primers used for sequencing the NtCK2A cDNA were: 5′ > 10 and 3′ >10 for sequencing the 5′ and 3′ ends; 5′ CK2D and 3′ sec (5′-CAGCAAGAC CCCAGTCTA-3′) for sequencing the internal cDNA region (underlined in Fig. 1). For sequencing the RACE product, the Race3′-1 primer was used. For sequencing the PCR products used as probes for screening, pGEM-T vector primers were used. Nucleotide sequences were analyzed by electrophoresis in urea-polyacrylamide gels. After the electrophoresis, gels were dried and subjected to autoradiography at –70°C with enhancer screens for 3 days.

Expression of the recombinant CK2α protein

To obtain the recombinant CK2α protein encoded by the NtCK2A gene, we cloned the NtCK2A cDNA in the pET-21d(+) vector (Novagen) for expression in *Escherichia coli*. For this purpose, we amplified by PCR the truncated NtCK2A cDNA obtained after screening, using the primers Bamnt5′ (5′-GGATCCTCATGTCAAAAGCTCGTGTTTACACCG-3′) and Xhont3′ (5′-TCTCGAGATAAGGATGGGCCATTG-CTTCTCT-3′) (underlined in Fig. 1). These primers allow the introduction of *Bam* H1 and *Xho* I restriction sites at the 5′ end and the 3′ end of the PCR fragment, respectively. The amplified fragment was cloned in pGEM-T vector and digested with *Bam*H1 and *Xho* I enzymes. Then, the digested fragment containing the NtCK2A cDNA was cloned in pET-21d(+) vector and *Escherichia coli* BL21(DE3) cells were transformed with this recombinant plasmid, essentially as described [10].

To express the NtCK2A cDNA, transformed *E. coli* BL21 (DE3) bacteria were grown overnight at 37°C in LB medium containing 100 μg/ml ampicillin. Erlenmeyer flasks containing 500 ml of LB medium (supplemented with 100 μg/ml ampicillin) were inoculated with 1 ml of the saturated culture medium and grown at 37°C under vigorous agitation. When the OD at 600 nm reached 0.5, protein expression was induced by 1 mM IPTG and the culture was grown for an additional 3 h-period. Bacteria were spun at 6500 × g for 15 min and the pellet was stored at –20°C until use.

Western blot analysis

To detect the recombinant CK2α protein in transformed *E. coli* BL21(DE3) bacteria, a 10 ml-bacterial culture was induced as described above and then centrifuged at 8000 rpm for 10 min at 4°C. The pellet was resuspended in 180 μl of

LB medium. A 10 µl-aliquot of the bacterial suspension was treated with 6 µl of electrophoresis loading buffer, heated at 95°C for 10 min, and loaded onto a 10% polyacrilamide denaturing gel. The Western blot was performed according to the manufacturer's instructions of the Chemiluminescence Reagent Plus kit (New England Nuclear). A monoclonal anti-T7 tag antibody (Novagen), at a dilution of 1/10,000, was used as the primary antibody. A peroxidase conjugated anti-mouse IgG, diluted at 1/10,000 was used as the secondary antibody. Peroxidase activity was revealed using Chemiluminescence Reagent Plus, according to the manufacturer's instructions.

Solubilization of the recombinant CK2α protein from inclusion bodies

To solubilize the recombinant CK2α protein from inclusion bodies, the pellet obtained by centrifugation of 500 ml of bacterial suspension at 8000 rpm was resuspended in 10 ml of 50 mM HEPES-NaOH pH 7.5, 0.5 M NaCl, 1 mM PMSF, 5 mM DTT containing 0.35 mg/ml lysozyme, and then incubated for 30 min at 20°C. After addition of Triton X-100 to a concentration of 1% (v/v) the suspension was sonicated (6 burst of 30 sec) and then cooled in an ice bath. The bacterial extract was then treated with 20 mg/l Dnase I for 1 h at 37°C. The inclusion bodies were sedimented by centrifugation at 30,000 × g for 30 min at 4°C. The pellet was washed twice with PBS containing 1% (v/v) Triton X-100, followed by spinning at 30,000 × g for 30 min at 4°C. The pellet was then solubilized in 2 ml of 50 mM HEPES-NaOH pH 7.5, 6 M guanidine-HCl, 25 mM DTT and left for 1 h at 4°C. Insoluble material was removed by centrifugation at 100,000 × g for 10 min and protein concentration was adjusted to 1 mg/ml using the solubilization buffer.

With the purpose of folding the recombinant CK2α protein, the solubilized proteins were diluted 10 times in cold folding buffer (50 mM HEPES-NaOH pH 7.5, 0.2 M NaCl, 1 mM DTT, 1 M NDSB 201 (non detergent sulfobetaine, Calbiochem)) and left for 1 h at 4°C. Excess guanidine and NDSB were removed by dialysis against 50 mM HEPES-NaOH pH 7.5, 10 mM KCl, 10% glycerol, 1 mM DTT, 7.6 mM $MgCl_2$. After concentrating the proteins through an Amicon cell (membrane OMEGA 10 kDa), the solution was stored at –70°C until use.

CK2 activity assay

CK2 assays were performed in 30 µl of medium containing 50 mM HEPES-NaOH pH 7.9, 8 mM MgCl2, 10 mM KCl, 10 mM NaF, 200 µM ATP, 2.6 × 10^6 cpm [γ-^{32}P]ATP and 5 mg/ml β-casein as a substrate. Heparin (20 µg/ml) was used as inhibitor. Reactions were initiated by the addition of the solubilized recombinant CK2α (rCK2α) protein (6 µg protein). After incubation at 30°C for 25 min, the reaction mixtures (22 µl) were spotted onto Whatman P81 ion-exchange papers. Papers were washed 3 times with 75 mM phosphoric acid, dried, and radioactivity was quantitated in a liquid scintillation counter. Alternatively, reactions were terminated by the addition of 30 µl of ice-cold 2 × SDS-PAGE sample buffer, and the phosphorylated β-casein was separated by PAGE-SDS and detected by autoradiography, following 7 days of exposure. The activity of the rCK2α was assayed in the presence of CK2β, in a reaction medium (30 µl) containing 20 mM HEPES-NaOH pH 7.6, 10 mM $MgCl_2$, 0.5 mM DTT, 100 mM NaCl, 100 µM ATP, 6.6 × 10^6 cpm [γ-^{32}P]ATP, 5 µg of β-casein and 6 µg of rCK2α. Reactions were performed in the absence or presence of 4.3 pmol of GST-XlCK2β and 20 µg/ml of heparin. The reactions were initiated by the addition of ATP. After incubation at 30°C for 15 min, the phosphorylated β-casein was analyzed by PAGE-SDS and autoradiography at –70°C overnight.

RNA extraction and Northern blot analysis

Total RNA was isolated from 100 mg of shoot meristems, young leaves (upper small leaves), older leaves (fully expanded leaves), shoots and roots from *Nicotiana tabacum* plants (cv Xhanti NN) containing 8–12 expanded leaves, as described by Logemann *et al.* [11]. 15 µg of total RNA was resolved on formaldehyde-agarose gels and blotted onto nylon membranes (Hybond N, Amersham). The membranes were hybridized at 65°C in a buffer containing 2.5 × SSPE, 5 × Denhardt's solution, 0.1% SDS, 150 µg/ml salmon sperm carrier DNA and ^{32}P-labeled probe (2–5 × 10^7 cpm). Filters were washed 3 times in 1 × SSC and 0.1% SDS for 15 min at 65°C. The truncated NtCK2A cDNA probe (1,165 bp) was synthesized by PCR using 5′ > 10 and 3′ > 10 primers. The NtCK2A 3′ UTR probe (318 bp) was obtained by digestion of the RACE 3′ product with *Pst* I and *Hind* III restriction enzymes. Both probes were labeled with [α-^{32}P]dCTP by random oligonucleotide-primer synthesis (RTS RadPrime Labeling System, Amersham).

Results and discussion

Isolation of the cDNA for the α subunit of CK2 from tobacco

The cDNA from the NtCK2A gene coding for the α subunit of CK2 was isolated by screening of a tobacco cDNA library cloned in λgt10. For this purpose, a PCR fragment of 432 bp-long, amplified from the cDNA of the AtCK2A gene from *Arabidopsis thaliana*, was used as a probe. This fragment corresponds to a conserved region among plant CK2α subunits (residues 35–178, see Fig. 2). After three steps of purification, nine positive clones were isolated. The inserted cDNAs were obtained by PCR amplification, using primers corresponding to the flanking sequences from the vector. Sequencing of the 3′ end of the nine isolated cDNAs, showed that all corresponded to the same gene. One of these clones was chosen and sequencing of the cDNA was completed using primers from the 5′ end and from internal regions (see Fig. 1). Annealing of the nucleotidic sequence obtained (1165 bp) with the sequence of two maize and two *Arabidopsis* CK2α cDNAs, revealed that the cDNA was truncated 45 bp upstream of the corresponding stop codon. To obtain the sequence of the 3′ end of the cDNA, the 3′ RACE technique was used. For this purpose, Poly $(A)^+$ RNA isolated from tobacco leaves was reversely transcribed, and then two sequential PCR reactions were performed using specific NtCK2A primers (see Fig. 1). After these reactions, a unique 428 pb-fragment corresponding to the 3′ end of the isolated cDNA was obtained and sequenced. This fragment contained (a) a 130 bp-sequence that overlaps with the 3′ end of the cDNA sequence obtained by screening, (b) the TAG stop codon and (c) a 295 bp-sequence from the 3′-untranslated region (UTR). Then, the complete cDNA sequence of the NtCK2A contains 1525 bp, from which 999 correspond to the coding region.

Although after the screening we only found the NtCK2A cDNA, we believe that in tobacco, like in other plant species, more than one CK2α gene must exist. Supporting this idea, in a previous report in which PCR fragments were amplified from CK2α and CK2β gene homologues from BY-2 tobacco cells, two different partial clones for tobacco CK2α, and three for CK2β, were reported [12]. We can not determine yet whether one of these partial CK2α clones corresponds to NtCK2A because the sequences of these fragments have not been published. Furthermore, in *Arabidopsis* and maize – the plant species where CK2α genes have been characterized – two highly homologous genes have been identified [13–15]. Recently, a third CK2α gene was identified in maize [16]. A third putative CK2α gene can be also postulated to exist in *Arabidopsis*. In fact, we used the BLAST programs to search in the GenBank *Arabidopsis* genomic and EST sequences for homologous sequences to the AtCK2A gene (accession number D10246). In addition of the previously characterized AtCK2A and AtCK2B genes found in chromosomes 5 and 3, respectively; a third highly homologous sequence coding for a putative CK2α protein containing a 100-amino acid extension, was found in chromosome 2 (accession number AC007112.5).

The aminoacidic sequence (333 aa) of the CK2α subunit encoded by the NtCK2A gene was deduced from the nucleotidic sequence of the cDNA and annealed with the sequence of other plant and animal CK2α subunits (Fig. 2). The aminoacidic sequence of NtCK2A was highly homologous to the other plant CK2α sequences reported up to date: 92.5 and 93.1% identical to AtCK2A and AtCK2B from *Arabidopsis thaliana*, respectively [14]; 93.4 and 94.6% identical to ZmACK2 and ZmCK2 from maize, respectively [13, 15]. The homology of NtCK2A with other CK2α subunits from animal species albeit lower is still considerable: 75.4% identical to the human HsCK2 [17]; and 74.8% identical to XlCK2 from *Xenopus laevis* [18]. The main difference in the sequence of CK2α between plants and animals, is that CK2αs from plants have a shorter carboxy terminal region (see Fig. 2).

It is interesting to note that a Ser residue occupies the position 113 (corresponding to position 118 in HsCK2, in bold in Fig. 2) in the aminoacidic sequence of the tobacco CK2α. In most CK2αs from different species, including maize, an Asn is in this position. It has been demonstrated in maize that the Asn 113 helps to the correct alignment of the Val 111 carbonyl group, a key residue in the binding of GTP [19]. Furthermore, studies using site-directed mutagenesis of the XlCK2 gene, have corroborated the relevance of this Asn to the ability of the enzyme to use GTP as a substrate [20]. Whether the presence of the Ser113 in the tobacco CK2α, and also in the *Arabidopsis* CK2α, affects the capacity of these enzymes to use GTP as a substrate, remains to be established. We have previously detected a CK2 activity in nuclear extracts from tobacco leaves that is able to use GTP as substrate [9]. However, we are not certain whether this activity involves the participation of the α subunit encoded by the NtCK2A gene.

Expression of the NtCK2A gene in different plant tissues

To evaluate whether the NtCK2A gene is differentially expressed in the tissues of tobacco plants, we measured the levels of the corresponding mRNA by Northern blot analysis, in samples of total RNA obtained from shoot meristems, young leaves, old leaves, shoots and roots. To detect the NtCK2A transcript we used as probes a 1,165 bp-DNA fragment corresponding to the truncated NtCK2A cDNA sequence isolated from screening, and a 318 bp-DNA fragment corresponding to the NtCK2A 3′UTR. We thought that the first probe should be able to recognize transcripts from different CK2α genes, while the second probe should be specific for the NtCK2A transcript. Surprisingly, similar expression pattern was obtained with both probes. This indicates that either only the NtCK2A transcript was detected with both probes, or that other putative CK2α genes have the same expression pattern than the NtCK2A gene.

Results shown in Fig. 3 indicate that, although NtCK2A gene was expressed in all tissues, it was preferentially expressed in proliferating tissues, such as shoot meristems and roots. This result is consistent with previous reports show-

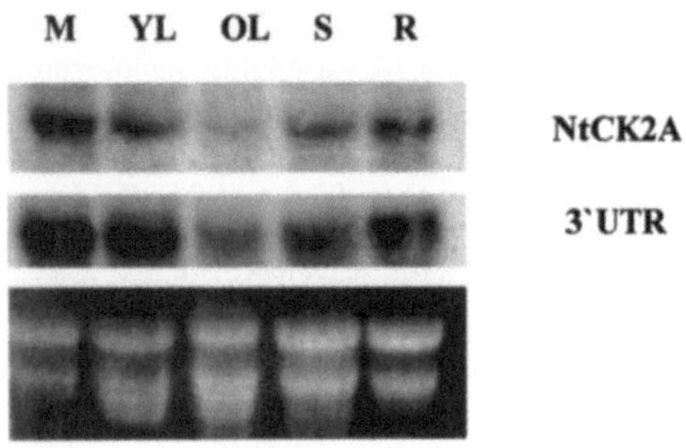

Fig. 3. Northern analysis of NtCK2A gene expression in different tissues of tobacco. Levels of NtCK2A mRNA were detected using total RNA (15 μg per lane) isolated from the apical shoot meristem (M), young leaves (YL), old leaves (OL), shoots (S) and roots (R). ^{32}P-labeled probes used for hibridization were a 1,165 pb-sequence of the NtCK2A cDNA isolated by screening (upper panel, NtCK2A), and a 318 pb-sequence from the 3′ UTR of the NtCK2A cDNA (middle panel, 3′ UTR). Staining the gels with ethidium bromide controlled for an equal RNA loading (lower panel).

ing a high expression and activity of CK2 in actively proliferating animal and plant cells [12, 21–23].

Expression of the NtCK2A cDNA in Escherichia coli

In order to obtain the tobacco CK2α protein, the NTCK2A cDNA was expressed in *E. coli*. For this purpose, we cloned the cDNA isolated by screening in the pET-21d(+) vector. This cloning allowed us to obtain a recombinant CK2α protein (rCK2α) which contains 318 residues from the tobacco CK2α protein (lacking 15 residues from the C-terminal) fused to a 14 amino acid-N-terminal extension (11 residues from the T7 tag followed by Arg-Ile-Leu) and to a 8 amino acid-C-terminal extension (Leu-Glu followed by 6 His residues). Bacteria (*E. coli*, BL21 DE3 strain) was transformed with the recombinant vector and, after induction with IPTG, the rCK2α protein was detected in the inclusion body fraction using a monoclonal anti T7 tag antibody (Fig. 4A).

The rCK2α protein was solubilized from the inclusion bodies by denaturation, and then refolded for the recovery of the catalytic activity. The activity of the rCK2α protein obtained after this process was assayed using β-casein and [γ-^{32}P]-ATP as substrates. Figs 4B and 4C show that the rCK2α protein was able to phosphorylate β-casein, reaction that was inhibited by heparin (20 μg/ml). Furthermore, the

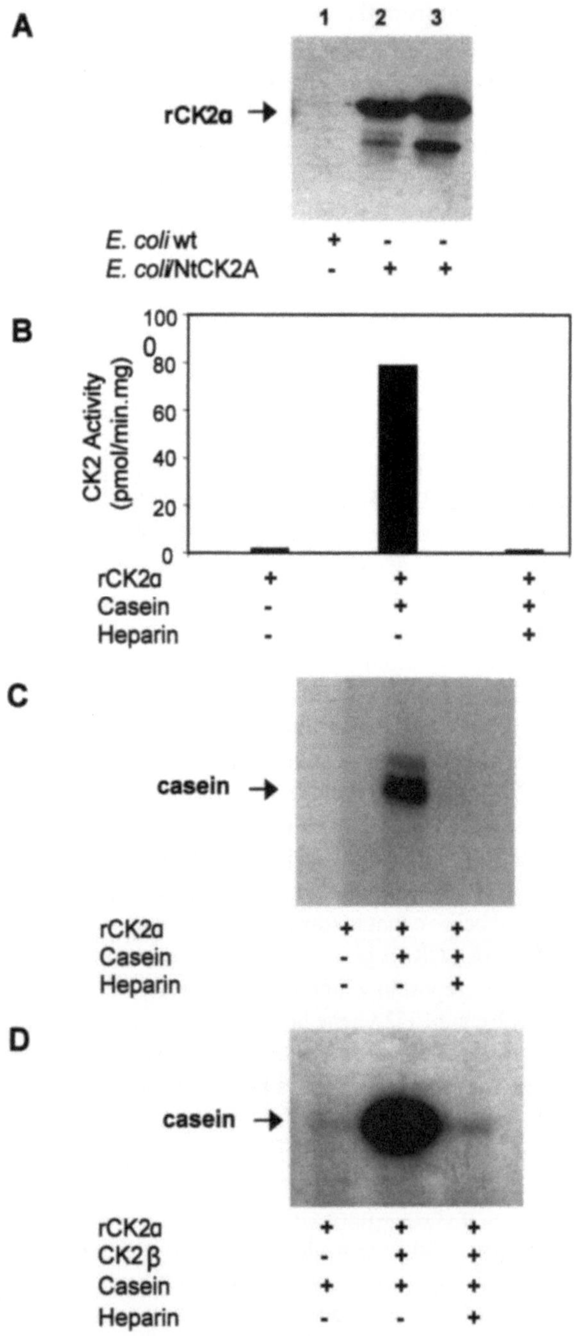

Fig. 4. Expression and activity assay of the recombinant tobacco CK2α subunit. The tobacco NtCK2A cDNA was cloned in the vector pET21d, wich contains a N-terminal T7 tag. (A) Expression of the tobacco NtCK2A cDNA in *E. coli*. *E. coli* extracts were assayed for Western blot with an anti-T7 tag antibody. The figure shows the expression of the recombinant CK2α (rCK2α) in total cell protein sample (lane 2) and in the insoluble inclusion body fraction (lane 3). Extract from untransformed *E. coli* was used as a control (lane 1). (B) rCK2α catalytic activity detected by adsorption on phosphocellulose papers. The rCK2α was solubilized from *E. coli* inclusion bodies and renatured with NDSB. The phosphorylation reaction contained 0.2 mg/ml of rCK2α protein, 200 μM [γ-^{32}P] ATP, 5 mg/ml of β-casein and 20 μg/ml of heparin, as indicated in the figure. (C) rCK2α catalytic activity detected by SDS-PAGE and autoradiography. The rCK2α was used for the phosphorylation of β-casein under the same condition described in (B). Phosphorylated β-casein was detected after SDS-PAGE and 7 days of autoradiography at –70°C. (D) Effect of CK2β on the rCK2α activity. The phosphorylation reaction mixture contained 0.2 mg/ml of rCK2α solution, 100 μM [γ-^{32}P] ATP, 0.16 mg/ml of β-casein, 20 μg/ml of heparin and 4.3 pmol GST-XLCK2β, as indicated in the figure. The phosphorylated β-casein was detected as in (C), but after 16 h of exposure.

catalytic activity of the rCK2α protein was strongly stimulated by the *Xenopus laevis* regulatory β subunit (Fig. 4D).

It was surprising that the rCK2α protein, lacking 15 residues from the C-terminal region, was catalytically active and able to bind functionally an heterologous b subunit. Studies carried out with the CK2α subunit from *Xenopus laevis* (expressed from the XlCK2 gene) indicated that a truncated form of the protein (lacking a portion equivalent to the last 11 residues from the C-terminal region of the NtCK2A protein) while able to interact with the CK2β regulatory subunit, was not active (Tapia and Allende, pers. comm.). In this case, the lack of activity could be explained by the instability of the protein due to exposition of a hydrophobic region located in the near αE-helix. It is possible that in the rCK2α protein from tobacco the additional LEHHHHHH sequence present in the C-terminal region plays a role in the stabilization of this hydrophobic αE-helix, thus conferring activity to the protein. We have to consider that – in keeping with the previous reasoning – under the conditions used to refold and assay the rCK2α protein (pH:7.8–8.0), the C-terminal LEHHHHHH sequence has a hydrophobic character, due to the uncharged state of the imidazol groups of the H residues (pK:6.8).

Characterization of the tobacco CK2 subunits should be the starting point for further studies leading to the understanding of the mechanism by which SA increases the nuclear CK2 activity, and to the identification of the target protein(s) for this enzyme in the SA-activated pathway.

Acknowledgements

We thank Dr. Marc Van Montagu for kindly providing the tobacco cDNA library; Virginia Garretón and Perla Hidalgo for helpful discussions, and Dr. Marcela Bitran for improving the manuscript. J. Tapia is the recipient of a doctoral thesis grant of FONDECYT. This work was supported by a grant from FONDECYT/CONICYT, Chile (N° 8980005).

References

1. Delaney TP, Uknes S, Vernooij B, Friedrich L, Weymann K, Negrotto D, Gaffney T, Gut-Rella M, Kessmann H, Ward E, Ryals J: A central role of salicylic acid in plant disease resistance. Science 266: 1247–1250, 1994
2. Yalpani N, Enyedi AJ, León J, Raskin I: Ultraviolet light and ozone stimulate accumulation of salicylic acid, pathogenesis-related proteins and virus resistance in tobacco. Planta 193: 372–376, 1994
3. Sharma YK, León J, Raskin I, Davis KR: Ozone-induced responses in *Arabidopsis thaliana*: The role of salicylic acid in the accumulation of defense-related transcripts and induced resistance. Proc Natl Acad Sci USA 93: 5099–5104, 1996
4. Schenk PM, Kazan K, Wilson I, Anderson JP, Richmond T, Somerville SC, Manners JM: Coordinated plant defense responses in Arabidopsis revealed by microarray analysis. Proc Natl Acad Sci USA 97: 11655–11660, 2000
5. Maleck K, Levine A, Eulgem T, Morgan A, Schmid J, Lawton KA, Dangl JL, Dietrich RA: The transcriptome of *Arabidopsis thaliana* during systemic acquired resistance. Nat Genet 26: 403–410, 2000
6. Marrs KA: The functions and regulation of glutathione *S*-transferases in plants. Annu Rev Plant Physiol Plant Mol Biol 47: 127–158, 1996
7. Alvarez ME, Pennell RI, Meijer P-J, Ishikawa A, Dixon RA, Lamb C: Reactive oxygen intermediates mediate a systemic signal network in the establishment of plant immunity. Cell 92: 773–784, 1998
8. Stange C, Ramírez I, Gómez I, Jordana X, Holuigue L: Phosphorylation of nuclear proteins directs binding to salicylic acid-responsive elements. Plant J 11: 1315–1324, 1997
9. Hidalgo P, Garretón V, Berríos CG, Ojeda H, Jordana X, Holuigue L: A nuclear casein kinase 2 activity is involved in early events of transcriptional activation induced by salicylic acid in tobacco. Plant Physiol 125: 396–405, 2001
10. Ausubel FM: Current Protocols in Molecular Biology. John Wiley & Sons, 1997, pp 1–3
11. Logemann J, Schell J, Willmitzer L: Improved method for the isolation of RNA from plant tissues. Anal Biochem 163: 16–20, 1987
12. Espunya MC, Combettes B, Dot J, Chaubet-Gigot N, Martínez MC: Cell-cycle modulation of CK2 activity in tobacco BY-2 cells. Plant J 19: 655–666, 1999
13. Dobrowolska G, Boldyreff B, Issinger OG: Cloning and sequencing of the casein kinase 2 alpha subunit from *Zea mays*. Biochem Biophys Acta 1129: 139–140, 1991
14. Mizoguchi T, Yamaguchi-Shinozaki K, Hayashida N, Kamada H, Shinozaki K: Cloning and characterization of two cDNAs encoding casein kinase II catalytic subunits in *Arabidopsis thaliana*. Plant Mol Biol 21: 279–289, 1993
15. Peracchia G, Jensen AB, Culianez-Macia FA, Grosset J, Goday A, Issinger OG, Pages M: Characterization, subcellular localization and nuclear targeting of casein kinase 2 from *Zea mays*. Plant Mol Biol 40: 199–211, 1999
16. Riera M, Peracchia G, de Nadal E, Ariño J, Pagès M: Maize protein kinase CK2: Regulation and functionality of three β regulatory subunits. Plant J 25: 365–374, 2001
17. Meisner H, Heller-Harrison R, Buxton J, Czech MP: Molecular cloning of the human casein kinase II alpha subunit. Biochemistry 28: 4072–4076, 1989
18. Jedlicki A, Hinrichs MV, Allende CC, Allende JE: The cDNAs coding for the alpha- and beta-subunits of *Xenopus laevis* casein kinase II. FEBS Lett 297: 280–284, 1992
19. Niefind K, Putter M, Guerra B, Issinger OG, Schomburg D: GTP plus water mimic ATP in the active site of protein kinase CK2. Nat Struct Biol 12:1100–1103, 1999
20. Jacob G, Neckelman G, Gimenez M, Allende CC, Allende J: Involvement of asparagine 118 in the nucleotide specificity of the catalytic subunit of protein kinase CK2. FEBS Lett 466: 363–366, 2000
21. Litchfield D, Lüscher B: Casein kinase II in signal transduction and cell cycle regulation. Mol Cell Biochem 127–128: 187–199, 1993
22. Pinna L, Meggio F: Protein kinase CK2 ('casein kinase-2') and its implication in cell division and proliferation. Cell Cycle Res 3: 77–96, 1997
23. Bosc DG, Lüscher B, Litchfield DW: Expression and regulation of protein kinase CK2 during the cell cycle. Mol Cell Biochem 191: 213–222, 1999

Molecular and Cellular Biochemistry **227**: 137–144, 2001.

Regulation of the human papillomavirus oncoproteins by differential phosphorylation

Paola Massimi, David Pim, Christian Kühne and Lawrence Banks
International Centre for Genetic Engineering and Biotechnology, Trieste, Italy

Abstract

Human papillomaviruses (HPVs) are intimately associated with the development of cervical cancer. The virus encodes two oncoproteins, E6 and E7, that are primarily responsible for inducing malignant transformation. The last few years have seen significant progress in elucidating the mechanisms by which these two viral proteins bring about cell transformation. Both proteins interact with a large number of cellular targets, many of which are involved in regulating diverse functions such as cell cycle regulation, transcription, differentiation and apoptosis. However both E6 and E7 are normally present at low levels within the virally infected cell, and how all these interactions are achieved and regulated has, until recently, been unclear. We have found that both E6 and E7 are subject to differential phosphorylation, the net results of which regulate their abilities to interact with some of their respective target proteins. In the case of E6, phosphorylation by Protein Kinase A (PKA) negatively regulates its ability to interact with the Discs Large (Dlg) tumour suppressor. In the case of E7, phosphorylation by Casein Kinase II (CKII) significantly increases its ability to interact with the TATA Box Binding Protein (TBP). Further, CKII regulation of E7 appears to vary during the cell cycle, therefore this provides a means of specifically targeting E7 to a given substrate at a given point within the cell cycle. This differential regulation of E6 and E7 by phosphorylation thus provides specificity to a diverse set of protein–protein interactions. (Mol Cell Biochem **227**: 137–144, 2001)

Key words: human papillomavirus, E6, E7, protein kinase A, phosphorylation, casein kinase II

Introduction

Human papillomaviruses (HPVs) are small double stranded DNA viruses which are associated with the development of a number of clinically important diseases. The most important of these is cervical cancer. This is the third most common cancer in women worldwide with over 370,000 new cases per annum [1]. It is now widely accepted that HPV is a causative agent in over 99% of these cases. Over a 100 different HPV types have so far been identified but only a small subset of these types are associated with the development of cervical cancer [2]. Based on this association, the viruses are classified as either 'high' or 'low' risk depending on whether they are found associated with malignant or benign lesions respectively. The most frequently found types in cervical cancer are HPV-16 and HPV-18, although other types such as HPV-31 and HPV-33 are also found [3].

A striking feature of cervical cancers, and their derived cell lines, is the continued retention and expression of viral sequences many years after the initial transforming events. In many such cases the viral DNA is integrated into the host genome, with concomitant deletions of large sections of the viral genome [4, 5]. However, the viral upstream regulatory region (URR), which is responsible for regulating viral gene expression, and the viral E6 and E7 open reading frames are invariably retained [6–8]. It is now known that E6 and E7 are the major viral oncoproteins and their continued function is essential for maintenance of the transformed phenotype. Numerous studies have blocked E6 or E7 functions, either by antisense, ribozyme or inhibitory peptide approaches [9–11]. These blocks result in either growth arrest or apoptosis, depending on whether E7 or E6 functions respectively are being inhibited. These data therefore suggest that the two viral oncoproteins are valid targets for therapeutic intervention in

Address for offprints: L. Banks, International Centre for Genetic Engineering and Biotechnology, Padriciano 99, I-34012 Trieste, Italy (E-mail: banks@icgeb.trieste.it)

HPV induced disease and a large amount of work has now been done to attempt to elucidate the mechanism of action of these two viral proteins.

Biological activities of E6 and E7

The first evidence that HPV harboured oncogenic activity was observed in established rodent cells, where the HPV-16 E7 gene was found to efficiently induce malignant transformation [12, 13]. Numerous other studies have now shown that both E6 and E7 have intrinsic transforming activity. Both will cooperate with an activated ras oncogene in the transformation of primary rodent cells [14, 15] and both exhibit transforming potential in primary mammary epithelial cells [16]. However the most relevant assay system, with respect to the natural target cells of the virus, is immortalisation of primary human cervical keratinocytes. These cells are notoriously difficult to immortalise, yet E6 and E7 will cooperate to bring about the immortalisation of this cell type [17, 18]. It should be emphasised that a key feature of these assays is that both E6 and E7 are required for this to occur, implying that the two viral oncoproteins normally cooperate in the processes of cell immortalisation *in vivo*. Another aspect of these assays is that these cells are not fully transformed. Additional genetic insults, achieved either through extended passage in culture, or by addition of activated oncogenes [19, 20], are required before these cells become fully transformed, thus highlighting the multistep nature of the disease process. An understanding of how these two viral proteins cooperate to bring about cell transformation can be gained from an analysis of the cellular proteins with which they interact.

Cellular targets of E6 and E7

Both viral oncoproteins interact with a number of key cellular proteins involved in the regulation of cell growth and differentiation. Perhaps the two most important targets of E6 and E7 are the cellular tumour suppressor proteins, p53 and pRb respectively [21, 22], the interactions with which give a vital clue as to how these viral proteins are working. Thus, E7 interferes with pathways regulating cell cycle and promotes a G1/S phase transition in the normally quiescent infected cell [23, 24]. On the other hand, E6 targets proteins involved in the protection of the cell from unscheduled DNA synthesis, as typified by p53, and it thereby blocks the cell's normal response, which is to undergo apoptosis [25, 26]. The net result of these interactions is the survival of an unnaturally replicating cell, and thence the initiation of the steps leading towards malignancy.

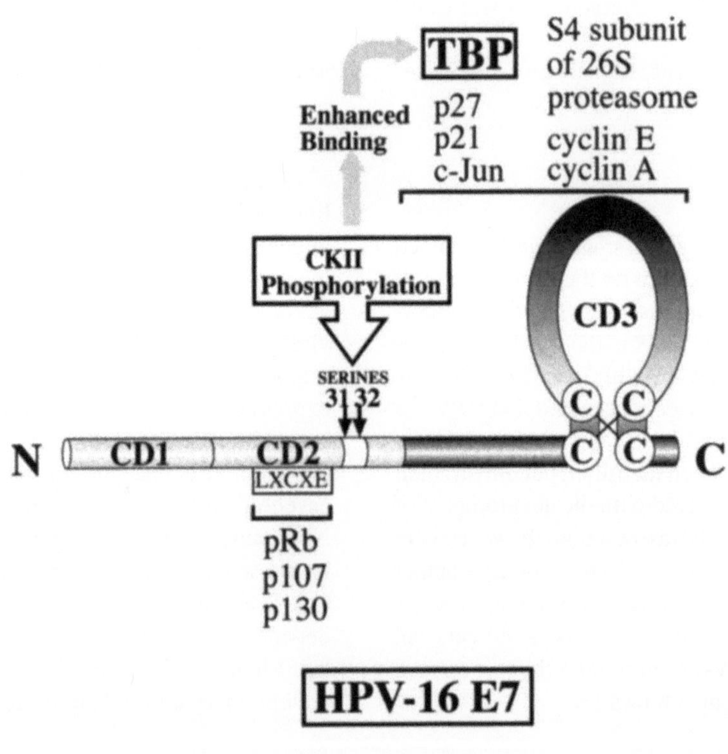

Fig. 1. Schematic diagram showing the three conserved domains (CD1–3) of the HPV-16 E7 protein, together with some of its known cellular target proteins and their sites of interaction. Also highlighted is the site of CKII recognition and the enhanced binding with TBP that occurs following CKII phosphorylation.

It is clear however that there are many other protein interactions which contribute to the ability of these two viral proteins to bring about cell transformation. As can be seen from Fig. 1, E7 interacts with a large number of cellular proteins. Interestingly the majority of these are linked to processes regulating cell cycle and gene expression. Thus, in addition to pRb, E7 also interacts with the two other members of the pocket protein family, p130 and p107 [27, 28]. This, coupled with E7's associations with cyclins E [29] and A [30], plus its interactions with the cyclin-dependent kinase inhibitors, p27 [31] and p21 [32, 33], ultimately results in the constitutive activation of cell cycle progression by interfering with the cyclin/cdk complexes, which, together with the release of free E2F family members, then upregulates cellular genes required for the G1/S phase transition [34].

E7 has also been found to associate with a number of cellular transcription factors, including TBP [35] and members of the AP-1 family [36]. Although the precise cellular genes which are upregulated and/or repressed as a consequence of these interactions remains to be determined, it is clear that both appear to be required for the efficient transforming activity of E7 [37, 38].

Similarly, E6 has been shown to have a large number of cellular targets including Mcm7, a component of DNA replication licensing factor [39], and the transcriptional co-activator p300 [40] (Fig. 2). As well as p53, the pro-apoptotic member of the Bcl-2 family of proteins, Bak, has also been shown to be a target of E6 [41, 42]. As a consequence of the Bak and p53 interactions, E6 has potent anti-apoptotic activity. This has obvious and important implications for the continued survival of cells that would normally be cleared either during a viral infection, or in response to the subsequent accumulation of DNA damage in an immortalised cell. This therefore represents one of the means by which E6 contributes to malignant progression.

Recent studies have identified a new class of proteins, containing PDZ domains, that bind to the extreme C-terminus of only the high risk E6 proteins. This is particularly important, since this interaction occurs through a highly defined domain encompassing just four amino acids [43, 44]. The site that is recognised on the target proteins is a PDZ domain; these are sites of protein–protein interaction [45, 46]. The proteins so far shown to be recognised in this manner, Dlg, MAGI-1 and hScrib [43, 47, 48], are all expressed at sites of cell–cell contact, and are believed to be involved in assembling multi-protein complexes that regulate the processes of cell division and polarity [49–51]. Certainly both Dlg and hScrib have all the attributes of tumour suppressor proteins

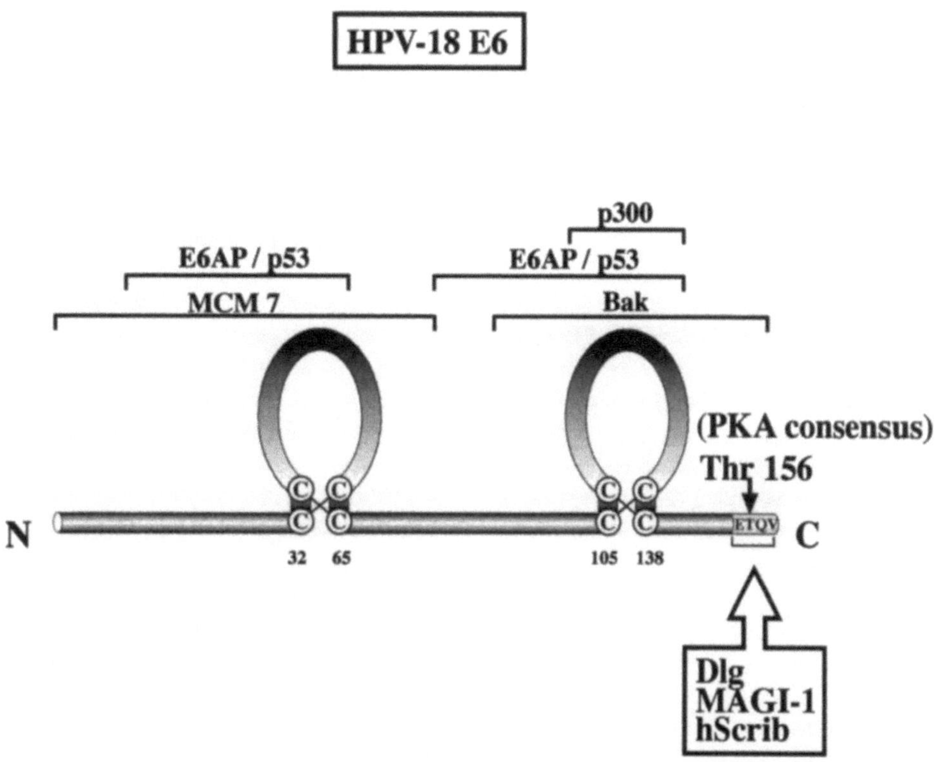

Fig. 2. Schematic diagram showing the zinc finger structure of the HPV-18 E6 protein together with some of its known target proteins and their sites of interaction. Also highlighted is the PKA recognition site overlapping the consensus PDZ binding domain, phosphorylation of which reduces binding to Dlg and possibly to MAGI-1 and hScrib.

based on studies performed in Drosophila where the knockouts result in uncontrolled cell proliferation and loss of cell polarity, characteristic of a metastatic phenotype [52, 53].

Based on the functions of the cellular proteins with which the viral oncoproteins interact, it is not surprising that it is generally thought that E7 induces cell proliferation, whereas E6 contributes more towards cell survival and metastasis. Indeed, in recent studies where the two viral oncoproteins were expressed separately in the skin of transgenic mice, E7 was found to induce largely benign cellular proliferations whereas E6 gave rise to fewer lesions with a more aggressive behaviour and with a higher metastatic potential [54]. It is very tempting to suggest that this effect of E6 is related, in part, to its ability to target the PDZ domain containing proteins, particularly since these regulate cell polarity and cell attachment; processes which are grossly perturbed during malignant progression.

Materials and methods

Cells

Baby Rat Kidney (BRK) and HaCat cells were grown in DMEM supplemented with 10% fetal calf serum. Transformation assays in primary BRK cells were performed by cotransfecting HPV E7 and EJ-ras together with a neomycin resistance marker and cells selected over a period of 2 weeks after which time the dishes were fixed and stained with Giemsa and colonies counted. Transfections were performed using standard calcium phosphate precipitation.

In vivo *labeling and immunoprecipitation*

Cells were placed in 0.2% fetal calf serum for 48 h and then placed in phosphate free medium for a further 2 h. After this time 1 mCi [^{32}P] orthophosphate was added to the time zero cells for 1 h. The remaining cells were induced with 10% fetal calf serum and labeled with orthophosphate for 1 h at different times after serum addition. Cells were harvested at 2 h intervals and immunoprecipitated with a rabbit polyclonal anti-E7 antibody. Immunoprecipitates were run on SDS-PAGE and analysed by autoradiography.

Cell cycle analysis

Cells were harvested at different times post addition of serum to serum arrested cells. The DNA content was then assessed by propidium iodide staining and FACS analysis.

In vitro *kinase assays*

Cells were placed in 0.2% fetal calf serum for 48 h, after which time 10% fetal calf serum was added back to the cells which were then extracted at different time points post addition of serum. Equal amounts of purified GST fusion proteins were then incubated with the cell extracts for 20 min in the presence of 56 nM [^{32}P] ATP. After extensive washing the phosphorylated proteins were monitored by SDS-PAGE and autoradiography. The level of phosphorylation was quantitated using a Packard Instant Phosphor Imager.

Results and discussion

The specificity problem

The E6 and E7 proteins from the high risk HPV types are present in very low amounts and detection of these oncoproteins in virally infected cells has often been difficult. From the above discussion it is clear that these viral proteins interact with many different cellular targets which are often present in much larger stoichiometric amounts. This suggests that the function of the viral proteins cannot be explained by simple static complex formation with their respective cellular targets, and implies specific selective regulation of these associations, together with a catalytic aspect to viral protein function.

Thus, the virus has evolved a number of interesting ways of dealing with this. In the case of E6 it has been shown to also interact with another cellular protein, E6AP, which is the prototype HECT domain containing protein and which acts as a ubiquitin ligase. Thus, in the case of p53, E6 catalytically conjugates ubiquitin to p53 via its association with E6AP and thereby targets it for proteasome mediated degradation [55–57]. A similar process occurs with the PDZ domain containing proteins, although in this case the ubiquitin ligase involved appears to be distinct from E6AP and remains yet to be identified [48, 58].

Similarly, E7 also induces a high turnover of pRb by targeting it for ubiquitin mediated degradation [59]. This proteolytic destruction of pRb in the presence of E7 thereby irreversibly inactivates pRb and promotes G1/S progression. The ubiquitin ligase employed by E7 remains to be identified and as yet, no other cellular proteins have been found to be targeted for degradation by E7. However, it is intriguing that E7 itself has recently been shown to be subject to proteasome mediated degradation [60], as well as being able to associate with the S4 subunit of the 26S proteasome [61]. This suggests that these activities may be part of a regulated, connected pathway of E7 function.

As well as using an enzymatic approach for overcoming the stoichiometric difficulties of their respective interactions, E6 and E7 have also been shown to be regulated by phosphorylation. In the case of E6 this involves phosphorylation by PKA within the extreme C-terminal PDZ binding domain [62]. As a consequence of this, the ability of E6 to interact with its PDZ domain containing targets is inhibited, and thus their degradation is blocked. Therefore, this suggests that there are specific points during the viral life cycle where E6 will be able to target this class of proteins. At other times, however, it will be phosphorylated by PKA and free to perform other functions. Likewise E7 has also recently been shown to be regulated by CKII phosphorylation and this will be discussed in more detail below.

CKII regulation of E7

Sequence alignment of HPV-16 E7 with other viral transforming proteins reveals marked sequence conservation, within the conserved domains 1 and 2 (CD1 and CD2) of E7, with Adenovirus E1a and SV40 LT antigen [63]. Further analysis revealed that a key element within CD2 is a conserved LXCXE motif (see Fig. 1), which serves as the recognition site for the pocket protein interactions [64]. In addition, it was noted that there is a highly conserved potential CKII recognition motif on Ser 31 and Ser 32. Indeed, further studies showed that HPV-16 E7 is an excellent substrate for phosphorylation by CKII [65]. Perhaps most interesting, however, were studies demonstrating that mutants of E7 defective in phosphorylation by CKII had a greatly reduced transforming potential. Moreover, replacement of the mutant residues with two acidic Asp residues restored the full transforming activity of the E7 protein [65, 66]. This strongly suggests that phosphorylation of E7 is required for optimal E7 transforming activity. However, the molecular basis behind these observations, other than that E7 required an acidic charge at this site for optimal activity, remained obscure.

Recent studies, however, have shown that the ability of E7 to interact with TBP is regulated by CKII phosphorylation [35]. Thus, TBP was found to be an excellent binding partner of E7 both *in vitro* and *in vivo*. Mutational analysis of TBP identified a region within the highly conserved C-terminal domain of the protein as being the region recognised by E7. Interestingly this also corresponds to the same region of TBP recognised by p53, HPV E6 and HPV E2, suggesting a conservation of function in interacting with this region of TBP [37]. Most interesting, however, was the observation that the phosphorylation of E7 by CKII greatly increased its affinity for TBP. Similarly, the substitution of two acidic Asp residues at the CKII site also increased the affinity with which E7 bound TBP. Therefore this represents the first molecular evidence of a function for the phosphorylation of E7. Interestingly, the TBP recognition site in Adenovirus E1a is also immediately downstream of its putative CKII recognition sequence, and phosphorylation of E1a also stimulates its ability to bind TBP [35]. This is a particularly important series of observations, since it implies that this mechanism of association, and subsequent regulation of the interaction between these viral proteins and TBP is conserved across widely different viral types.

Obviously the next question was whether this interaction might account for the reduced levels of transformation seen with E7 when the CKII site is mutated. One way in which this was addressed was by performing a mutational analysis of E7 in order to identify the basic region of interaction with TBP. This identified residues at 79–83 as being required for the association and, most interestingly, this mutant also displayed a greatly reduced level of transforming activity, very similar to that observed with the CKII recognition site mutants of E7 [37, 65]. Therefore these studies suggest that CKII regulation of the E7-TBP association contributes to the ability of E7 to induce cell transformation.

As mentioned above, a key element of the investigations into the functions of these viral proteins has been the question of stoichiometry and specificity. An obvious question was whether or not the phosphorylation status of E7 might change during the cell cycle, thereby defining a time when the ability to associate with TBP was optimal. Indeed, a series of cell cycle analyses demonstrated that E7 is differentially phosphorylated during the cell cycle. Thus, E7 is highly phosphorylated by CKII during the G1/S phase transition of the cell cycle. Thereafter there is a decrease in the level of E7 phosphorylation, followed by an additional phosphorylation of E7 at a CKII independent site later in the cell cycle [67]. However, the main finding of this analysis was the observation that E7 was most highly phosphorylated by CKII at the time when E7 function is optimal; i.e. in stimulating the progression through G1 into S phase. Therefore this suggests that the time when E7 is most likely to associate with TBP is during the late stages of G1, and that this contributes to the ability of E7 to stimulate cell cycle progression. At present, it is still not clear which cellular promoters are affected by this particular association. However, it is intriguing that there are now several reports of E7 influencing the transcription, both positively and negatively, of a growing number of different cellular promoters [36, 68–70]. It will be extremely interesting in the coming months to determine which of these activities are associated with the TBP/CKII aspect of E7 function.

Summary

The identification of many different cellular targets of the viral oncoproteins is fundamental to our understanding of the molecular pathogenesis of these viruses. However the issues of specificity and stoichiometry remain one of the most important questions in DNA Tumour Virus biology. How do, often small, highly multi-functional proteins perform so many different tasks? Targeting their cellular substrates for ubiquitin mediated degradation is clearly one answer to the problem. However, regulation of the viral proteins by differential phosphorylation by cellular kinases provides another strategy. In this case, CKII can be expected to play an important function in the activity of many viral proteins. Perhaps the most relevant viral target to human disease, however, is the HPV E7 protein. As we have seen, it is intimately involved in the processes which lead to malignancy of the cervix, and its recognition by CKII appears to play an integral part in the ability of E7 to contribute towards that process.

Acknowledgements

We are most grateful to Miranda Thomas for valuable comments on the manuscript. The work described here was supported in part by a research grant from the Associazione Italiana per la Ricerca sul Cancro.

References

1. Parkin DM, Pisani P, Ferlay J: Estimates of the worldwide incidence of 25 major cancers in 1990. Int J Cancer 80: 827–841, 1999
2. de Villiers EM: Human pathogenic papillomavirus types: An update. Curr Top Microbiol Immunol 186: 1–12, 1994
3. zur Hausen H, Schneider A: The role of papillomaviruses in human anogenital cancers. 1987 In: N. Salzman, P.M. Howley (eds). The Papovaviridae, vol 2. New York, Plenum Press, pp 245–263
4. Schwarz E, Freese U, Gissmann L, Mayer W, Roggenbuck B, Stremlau A, zur Hausen H: Structure and transcription of human papillomavirus sequences in cervical carcinoma cells. Nature 314: 111–114, 1985
5. Yee C, Krishnan-Hewlett I, Baker CC, Schlegel R, Howley PM: Presence and expression of human papillomavirus sequences in human cervical carcinoma cell lines. Am J Pathol 119: 361–366, 1985
6. Smotkin D, Wettstein F: Transcription of human papillomavirus type 16 early genes in a cervical cancer and a cancer-derived cell line, and identification of the E7 protein. Proc Natl Acad Sci USA 83: 4680–4684, 1986
7. Androphy EJ, Hubbert N, Schiller JT, Lowy DR: Identification of the HPV 16 E6 protein from transformed mouse cells and human cervical carcinoma cells. EMBO J 6: 989–992, 1987
8. Banks L, Spence P, Androphy E, Hubbert N, Matlashewski G, Murray A, Crawford L: Identification of human papillomavirus type 18 E6 polypeptide in cells derived from human cervical carcinomas. J Gen Virol 68: 1351–1359, 1987
9. Alvarez-Salas LM, Cullinan AE, Siwkowski A, Hampel A, DiPaolo JA: Inhibition of HPV-16 E6/E7 immortalisation of normal keratinocytes by hairpin ribozymes. Proc Natl Acad Sci USA 95: 1189–1194, 1998
10. von Knebel Doeberitz M, Rittmuller C, zur Hausen H, Durst M: Inhibition of tumorigenicity of cervical cancer cells in nude mice by HPV E6–E7 anti-sense RNA. Int J Cancer 51: 831–834, 1992
11. Butz K, Denk C, Ullmann A, Scheffner M, Hoppe-Seyler F: Induction of apoptosis in human papillomavirus-positive cancer cells by peptide aptamers targeting the viral E6 oncoprotein. Proc Natl Acad Sci USA 97: 6693–6697, 2000
12. Kanda T, Furuno A, Yoshiike K: Human papillomavirus type 16 open reading frame E7 encodes a transforming gene for rat 3Y1 cells. J. Virol 62: 610–613, 1988
13. Vousden KH, Doniger J, DiPaolo JA, Lowy DR: The E7 open reading frame of human papillomavirus type 16 encodes a transforming gene. Oncogene Res 3: 167–175, 1988
14. Storey A, Pim D, Murray A, Osborn K, Banks L, Crawford L: Comparison of the *in vitro* transforming activities of human papillomavirus types. EMBO J 7: 1815–1820, 1988
15. Storey A, Banks L: Human papillomavirus type 16 E6 gene cooperates with EJ-ras to immortalise primary mouse cells. Oncogene 8: 919–924, 1993
16. Wazer DE, Liu XL, Chu Q, Gao Q, Band V: Immortalisation of distinct human mammary epithelial cell types by human papillomavirus 16 E6 or E7. Proc Natl Acad Sci USA 92: 3687–3691, 1995
17. Münger K, Phelps WC, Bubb V, Howley PM, Schlegel R: The E6 and E7 genes of human papillomavirus type 16 are necessary and sufficient for transformation of primary human keratinocytes. J Virol 63: 4417–4423, 1989
18. Hawley-Nelson P, Vousden KH, Hubbert NL, Lowy DR, Schiller JT: HPV 16 E6 and E7 proteins cooperate to immortalise human foreskin keratinocytes. EMBO J 8: 3905–3910, 1989
19. Dürst M, Gallahan D, Gilbert J, Rhim JS: Glucocorticoid enhanced neoplastic transformation of human keratinocytes by human papillomavirus type 16 and an activated ras oncogene. Virology 173: 767–771, 1989
20. DiPaolo J, Woodworth C, Popescu MC, Notario V, Doniger J: Induction of human cervical squamous cell carcinoma by sequential transfection with human papillomavirus 16 DNA and viral Harvey ras. Oncogene 4: 395–399, 1989
21. Werness B, Levine A, Howley PM: Association of human papillomavirus types 16 and 18 E6 proteins with p53. Science 248: 76–79, 1990
22. Dyson N, Howley PM, Münger K, Harlow E: The human papillomavirus-16 E7 oncoprotein is able to bind to the retinoblastoma gene product. Science 243: 934–936, 1989
23. Sato H, Furuno A, Yoshiike K: Expression of human papillomavirus type 16 E7 gene induces DNA synthesis of rat 3Y1 cells. Virology 168: 195–199, 1989
24. Banks L, Edmonds C, Vousden KH: Ability of the HPV16 E7 protein to bind RB and induce DNA synthesis is not sufficient for efficient transforming activity in NIH3T3 cells. Oncogene 5: 1383–1389, 1990
25. Kessis TD, Slebos RJ, Nelson WG, Kastan MB, Plunkett BS, Han SM, Lorincz AT, Hedrick L, Cho KR: Human papillomavirus 16 E6 expression disrupts the p53-mediated cellular response to DNA damage. Proc Natl Acad Sci USA 90: 3988–3992, 1993
26. Foster SA, Demers GW, Etscheid BG, Galloway DA: The ability of human papillomavirus E6 proteins to target p53 for degradation *in vivo* correlates with their ability to abrogate actinomycin D-induced growth arrest. J Virol 68: 5698–5705, 1994
27. Davies R, Hicks R, Crook T, Morris J, Vousden KH: Human papillomavirus type 16 E7 associates with a histone H1 kinase and with p107 through sequences necessary for transformation. J Virol 67: 2521–2528, 1993
28. Dyson N, Guida P, Münger, K, Harlow E: Homologous sequences in adenovirus E1A and human papillomavirus E7 proteins mediate interaction with the same set of cellular proteins. J Virol 66: 6893–6902, 1992
29. McIntyre MC, Ruesch M, Laimins L: Human papillomavirus E7 oncoproteins bind a single form of cyclin E in a complex with cdk2 and p107. Virology 215: 73–82, 1996
30. Tommasino M, Adamczewski J P, Carlotti F, Barth CF, Manetti R, Contorni M, Cavalieri F, Hunt T, Crawford L: HPV16 E7 protein associates with the protein kinase p33CDK2 and cyclin A. Onco-

gene 8: 195–202, 1993
31. Zerfass-Thome K, Zwerschke W, Mannhardt B, Tindle R, Botz JW, Jansen-Dürr P: Inactivation of the cdk inhibitor $p27^{KIP1}$ by the human papillomavirus type 16 E7 oncoprotein. Oncogene 13: 2323–2330, 1996
32. Funk JO, Waga S, Harry JB, Espling E, Stillman B, Galloway DA: Inhibition of CDK activity and PCNA-dependent DNA replication by p21 is blocked by interaction with the HPV-16 E7 oncoprotein. Genes Dev 11: 2090–2100, 1997
33. Jones DL, Alani RM, Münger K: The human papillomavirus E7 oncoprotein can uncouple cellular differentiation and proliferation in human keratinocytes by abrogating $p21^{cip1}$-mediated inhibition of cdk2. Genes Dev 11: 2101–2111, 1997
34. Nevins JR: E2F: A link between the Rb tumor suppressor protein and viral oncoproteins. Science 258: 424–429, 1992
35. Massimi P, Pim D, Storey A, and Banks L: HPV-16 E7 and adenovirus E1a complex formation with TATA box binding protein is enhanced by casein kinase II phosphorylation. Oncogene 12: 2325–2330, 1996
36. Antinore M, Birrer M, Patel D, Nader L, McCance J: The human papillomavirus type 16 E7 gene product interacts with and trans-activates the AP-1 family of transcription factors. EMBO J 15: 1950–1960, 1996
37. Massimi P, Pim D, Banks L: Human papillomavirus type 16 E7 binds to the conserved carboxy-terminal region of the TATA box binding protein and this contributes to E7 transforming activity. J Gen Virol 78: 2607–2613, 1997
38. Li JJ, Rhim JS, Schlegel R, Vousden KH, Colburn NH: Expression of dominant negative Jun inhibits elevated AP-1 and NF-kappaB transcription and suppresses anchorage independent growth of HPV immortalised human keratinocytes. Oncogene 16: 2711–2721, 1998
39. Kühne C, Banks L: E3-ubiquitin ligase/E6-AP links multicopy maintenance protein 7 to the ubiquitination pathway by a novel motif, the L2G box. J Biol Chem 273: 34302–34309, 1998
40. Patel D, Huang SM, Baglia LA, McCance DJ: The E6 protein of human papillomavirus type 16 binds to and inhibits co-activation by CBP and p300. EMBO J 18: 5061–5072, 1998
41. Thomas M, Banks L: Inhibition of Bak-induced apoptosis by HPV-18 E6. Oncogene 17: 2943–2954, 1998
42. Jackson S, Harwood C, Thomas M, Banks L, Storey A: Role of bak in UV-induced apoptosis in skin cancer and abrogation by HPV E6 proteins. Genes Dev 14: 3065–3073, 2000
43. Lee S, Weiss R, Javier R: Binding of human virus oncoproteins to hDlg/SAP97, a mammalian homologue of the Drosophila discs large tumour suppressor protein. Proc Natl Acad Sci USA 94: 6670–6675, 1997
44. Kiyono T, Hiraiwa A, Ishii S, Takahashi T, Ishibashi M: Binding of high-risk human papillomavirus E6 oncoproteins to a human homologue of the Drosophila discs large tumour suppressor protein. Proc Natl Acad Sci USA 94: 11612–11616, 1997
45. Ponting C, Philips C: DHR domains in syntrophins, neuronal NO synthases and other intracellular proteins. Trends Biochem Sci 20: 102–103, 1995
46. Saras J, Heldin CH: PDZ domains bind carboxy-terminal sequences of target proteins. Trends Biochem Sci 21: 455–458, 1996
47. Nakagawa S, Huibregtse J: Human scribble (Vartul) is targeted for ubiquitin-mediated degradation by the high-risk papillomavirus E6 proteins and the E6-AP ubiquitin-protein ligase. Mol Cell Biol 20: 8244–8253, 2000
48. Glaunsinger B, Lee S, Thomas M, Banks L, Javier R: Interactions of the PDZ-protein MAGI-1 with adenovirus E4-ORF1 and high-risk papillomavirus E6 oncoproteins. Oncogene 19: 1093–1098, 2000
49. Kim E, Niethammer M, Rothschild A, Nung Jan Y, Sheng M: Clustering of Shaker-type K^+ channels by interaction with a family of membrane-associated guanylate kinases. Nature 378: 85–88, 1995
50. Kim S: Polarised signalling: Basolateral receptor localisation in epithelial cells by PDZ-containing proteins. Curr Opin Cell Biol 9: 853–859, 1997
51. Marfatia J, Lin L, Hough C, Bryant P, Stolz L, Chishti A: Modular organisation of the PDZ domains in the human discs-large protein suggests a mechanism for coupling PDZ domain-binding proteins to ATP and membrane cytoskeleton. J Cell Biol 135: 753–766, 1996
52. Goode S, Perrimon N: Inhibition of patterned cell shape change and cell invasion by discs large during Drosophila oogenesis. Genes Dev 11: 2532–2544, 1997
53. Bilder D, Li M, Perrimon N: Cooperative regulation of cell polarity and growth by *Drosophila* tumor suppressors. Science 289: 113–116, 2000
54. Song S, Liem A, Miller JA, Lambert PF: Human papillomavirus types 16 E6 and E7 contribute differently to carcinogenesis. Virology 267: 141–150, 2000
55. Scheffner M, Werness B, Huibregtse J, Levine A, Howley PM: The E6 oncoprotein encoded by human papillomavirus types 16 and 18 promotes the degradation of p53. Cell 63: 1129–1136, 1990
56. Huibregtse J, Scheffner M, Howley PM: Cloning and expression of the cDNA for E6-AP, a protein that mediates the interaction of the human papillomavirus E6 oncoprotein with p53. Mol Cell Biol 13: 775–784, 1993
57. Scheffner M, Huibregtse J, Vierstra R, Howley PM: The HPV-16 E6 and E6-AP complex functions as a ubiquitin-protein ligase in the ubiquitination of p53. Cell 75: 495–505, 1993
58. Gardiol D, Kühne C, Glaunsinger B, Lee SS, Javier R, Banks L: Oncogenic human papillomavirus E6 proteins target the discs large tumour suppressor for proteasome-mediated degradation. Oncogene 18: 5487–5496, 1999
59. Boyer SN, Wazer DE, Band V: E7 protein of human papilloma virus-16 induces degradation of retinoblastoma protein through the ubiquitin-proteasome pathway. Cancer Res 56: 4620–4624, 1996
60. Reinstein E, Scheffner M, Oren M, Ciechanover A, Schwartz A: Degradation of the E7 human papillomavirus oncoprotein by the ubiquitin-proteasome system: Targeting via ubiquitination of the N-terminal residue. Oncogene 19: 5944–5950, 2000
61. Berezutskaya E, Bagchi S: The human papillomavirus E7 oncoprotein functionally interacts with the S4 subunit of the 26S proteasome. J Biol Chem 272: 30135–30140, 1997
62. Kühne C, Gardiol D, Guarnaccia C, Amenitsch H, Banks L: Differential regulation of human papillomavirus E6 by protein kinase A: Conditional degradation of human discs large protein by oncogenic E6. Oncogene 19: 5884–5891, 2000
63. Phelps WC, Yee CL, Münger K, Howley PM: The human papillomavirus type 16 E7 gene encodes transactivation and transformation functions similar to adenovirus E1a. Cell 53: 539–547, 1988
64. Münger K, Werness BA, Dyson N, Phelps WC, Harlow E, Howley PM: Complex formation of human papillomavirus E7 proteins with the retinoblastoma tumor gene suppressor product. EMBO J 8: 4099–4105, 1989
65. Barbosa MS, Edmonds C, Fisher C, Schiller JT, Lowy DR, Vousden KH: The region of the HPV E7 oncoprotein homologous to adenovirus E1a and SV40 large T antigen contains separate domains for Rb binding and casein kinase II phosphorylation. EMBO J 9: 153–160, 1990
66. Firzlaff JM, Luscher B, Eisenman RN: Negative charge at the casein kinase II phosphorylation site is important for transformation but not for Rb binding by the E7 protein of human papillomavirus type 16. Proc Natl Acad Sci USA 88: 5187–5191, 1991
67. Massimi, P, Banks L: Differential phosphorylation of the HPV-16 E7 oncoprotein during the cell cycle. Virology 276: 388–394, 2000
68. Zerfass K, Schulze A, Spitkovsky D, Friedman V, Henglein B, Jansen-

Dürr P: Sequential activation of cyclin E and cyclin A gene expression by human papillomavirus type 16 E7 through sequences necessary for transformation. J Virol 69: 6389–6399, 1995
69. Zwerschke W, Joswig S, Jansen-Dürr P: Identification of domains required for transcriptional activation and protein dimerisation in the human papillomavirus type-16 E7 protein. Oncogene 12: 213–220, 1996
70. Rey O, Lee S, Park N-H: Human papillomavirus type 16 E7 oncoprotein represses transcription of human fibronectin. J Virol 74: 4912–4918, 2000

Molecular and Cellular Biochemistry **227**: 145–151, 2001.

HIV-1 Rev transactivator: A β-subunit directed substrate and effector of protein kinase CK2

Flavio Meggio, Oriano Marin, Marco Boschetti, Stefania Sarno and Lorenzo A. Pinna

[1]*Dipartimento di Chimica Biologica, Centro di Studio delle Biomembrane del CNR, and CRIBI, Università di Padova, Padova, Italy*

Abstract

The phosphorylation of HIV-1 Rev by protein kinase CK2 is strictly dependent on the regulatory β subunit of the kinase and is deeply affected by conformational changes of the substrate outside the phosphorylation site [12]. Here we show that Rev modulates a variety of CK2 properties, including autophosphorylation, catalytic activity toward calmodulin, and susceptibility to polycationic effectors, whose common denominator is the involvement of the β subunit. Rev's two major CK2 sites are located at its N-terminus, immediately adjacent to a helix-loop-helix motif. By comparing the behaviour of full-size Rev with that of synthetic peptides reproducing, with suitable modifications, its N-terminal 26 amino acids including the phosphoacceptor site (Ser 5, Ser 8) and amphipathic helix-1, it appears that the functional interaction of the N-terminal portion of Rev with the N-terminal domain of the β subunit must rely on both electrostatic and hydrophobic interactions. The former mainly involve Rev's arginine-rich domain (residues 35–50) in helix-2, while the latter are mostly mediated by residues 12–24 of helix-1. These data disclose the possibility that, besides displaying protective, regulatory and targeting properties with respect to the catalytic subunit, the CK2 β subunit also plays a role as a docking site for a subset of CK2 substrates. (Mol Cell Biochem **227**: 145–151, 2001)

Key words: CK2, CKII, casein kinase, protein phosphorylation, Rev

Introduction

Almost all intracellular signaling pathways use protein phosphorylation as a propagation and regulatory device. Among the members of the large family of protein kinases implicated in this complicated network, protein kinase CK2 is probably the most pleiotropic one as it affects more than 200 different targets variably committed in the regulation of many cellular functions, in primis gene expression and cell proliferation [1–3]. Such a large spectrum of engagement probably reflects a number of unique features of CK2 notably its ubiquity and high constitutive activity. The activity of CK2 is invariably elevated in proliferating tissues [3], and cooperates with other proto-oncogenes to promote cell transformation [4, 5]. It is expectable, therefore, that CK2 might also be involved in biochemical events leading to important pathologies such as infectious and neuro-degenerative diseases. The abundance of this kinase as well as the absence of a tight mechanism of regulation makes CK2 the first-choice tool for performing the rapid and unscheduled phosphorylation of proteins expressed by host cell and infectious agents. Indeed CK2 is exploited by many viruses during the early phases of their life cycle in the infected host cell. At least 40 viral proteins (more or less one-fifth of the total number of CK2 substrates identified to date) have been reported to be directly phosphorylated by CK2. It should be noted, moreover, that these proteins are encoded by 25 different viruses belonging to nearly all the main classes of both DNA and RNA viruses, namely Adenoviruses, Herpesviruses (with 13 phosphorylated viral proteins) and Retroviruses, whose best known example is HIV-1. This latter alone accounts for at least 9 different CK2 targets representing either structural or regulatory viral products and

Address for offprints: F. Meggio, Dipartimento di Chimica Biologica, viale G. Colombo 3, 35121 Padova, Italy (E-mail: meggio@civ.bio.unipd.it)

Table 1. Viral products of HIV-1 as substrates of protein kinase CK2

Substrate	Reference
Vpu	[6]
Rev	[7]
Gp120	[8]
Reverse transcriptase p51	[8]
Reverse transcriptase p66	[8]
Envelope protein p41	[8]
Capside protein p27	[8]
Capside protein p17	[8]

including, among others, gp120, Vpu, Rev and reverse transcriptase [6–8] (see also Table 1).

Among the HIV-1 proteins, we have recently investigated the CK2-mediated phosphorylation of the Rev transactivator, a 116-amino-acid RNA-binding protein that, through its activity as RNA chaperone, allows expression of incompletely spliced HIV-1 RNAs necessary for completion of the viral replication cycle [9, 10]. A secondary structure theoretical prediction indicates that Rev consists of an unstructured region spanning residues 1–8, which includes two CK2 phosphorylation sites, followed by a helix-loop-helix region, with the two helices spanning residues 9–24 and 35–61 and a C-terminal unstructured tail [11]. Especially important for Rev's biological activity appears to be its arginine-rich region spanning residues 35 and 50 of helix-2, which serves as both a nuclear localisation signal (NLS) and an RNA binding domain (RBD) [9, 10]. Phosphorylation of Rev by CK2 was demonstrated to occur at a canonical site close to the N-terminal end of the molecule that fulfils the consensus sequence of this acidophilic protein kinase [7]. The phosphorylation mechanism is a 'hierarchical' one with Ser 5 becoming phosphorylated only after a phosphate has been incorporated by the same kinase into Ser 8. Later it was shown that CK2-mediated phosphorylation of Rev is unique in that it is absolutely dependent on the presence of the non catalytic β subunit [12]. Three lines of evidence suggested that an electrostatic interaction between the N-terminal 55–64 acidic domain of the regulatory β subunit and the 35–50 basic stretch of Rev is a prerequisite for CK2-mediated Rev phosphorylation: firstly, Rev phosphorylation is inhibited by polybasic compounds (spermine, polylysine etc.) which bind to the acidic N-terminal domain of the β subunit and usually behave as stimulators of CK2 activity [13, 14]; secondly, phosphorylation of Rev is reduced instead of being increased if wild type β subunit was replaced by the hyperactive mutant β55,57 in which two acidic residues of the 55–64 domain have been substituted with alanines; thirdly, Rev phosphorylation is lost upon alteration of its 35–50 basic stretch by replacing Arg 38 and Arg 39 with leucines. The overall architecture of the N-terminal helix-loop-helix motif of Rev also appears to be functionally important, since the deletion of the loop gives rise to a mutant of Rev (RevΔloop) which is both refractory to CK2 phosphorylation and functionally inactive, due in part to its inability to bind to its RNA target [11, 12]. Here we present data suggesting that the binding of Rev to the N-terminal 55–64 domain of the CK2 β subunit, which is the prerequisite for its own efficient phosphorylation, relies on both electrostatic and hydrophobic interactions mediated by its helix-loop-helix motif. The identification of this interaction could provide a tool for tuning the activity of CK2 toward a subset of targets exploiting the β subunit as a docking platform.

Material and methods

Materials

Purified GST-Rev [12] and recombinant Rev [15] were kindly provided by T. Ferro (University of Padova) and B.K. Felber (NCI-FCRDC), respectively. Human recombinant calmodulin was expressed in *E. coli* and purified by anion exchange and hydrophobic interaction chromatography as described in [16]. Recombinant α and β subunits of wild type human CK2 were obtained as detailed in [17, 18]. Native CK2 was isolated from rat liver cytosol [19].

A synthetic peptide corresponding to the 26 amino-terminal residues of wild type Rev (Rev[1–26]) as well as its derivative Rev[1–26]AAA with Arg14, Arg17 and Lys20 replaced by alanine residues were prepared as previously described [12].

Phosphorylation assays

Phosphorylation experiments were performed by incubating substrates in a medium containing 50 mM Tris-HCl pH 7.5, 12 mM $MgCl_2$, 100 mM NaCl and 20 μM [γ^{32}P]-ATP (specific radioactivity 500–1000 cpm/pmol) for 10 min at 37°C. The reaction was started with the addition of the kinase and was stopped by cooling in ice. Recombinant CK2 holoenzyme was reconstituted *in vitro* by mixing at 1:1 molar ratio α and β subunits, followed by incubation in ice for 10 min prior to use. Phosphate incorporated into the protein substrate was evaluated by subjecting samples to SDS-PAGE according to Laemmli [20]; the phosphocellulose paper procedure [21] was adopted in the case of synthetic peptide substrates.

Results and discussion

Effect of Rev on the autophosphorylation of CK2 holoenzyme

Efficient phosphorylation of Rev by CK2 is strictly dependent on the presence of the β subunit. Mutational data support the view that electrostatic interactions between the N-terminal acidic 55–64 domain of β and the basic 35–50 stretch located in the helix-loop-helix motif of Rev are required for Rev phosphorylation [11, 12]. The acidic domain of the β subunit would therefore play a role as a docking site allowing the proper positioning of Rev, with its N-terminal phosphoacceptor site entering the catalytic site of the kinase. This docking event also permits Rev to promote de-inhibition of CK2 activity, as it displaces the pseudosubstrate N-terminal acidic region of β, which has been shown to interact with basic residues recognizing the phosphoacceptor peptide [22–24], from the active site. The occurrence of such a molecular alteration is documented by the effect of Rev on the autophosphorylation pattern of CK2 kinase. This normally occurs at the N-terminal Ser 2/Ser 3 of the β subunit and depends on the integrity of its 55–64 region, which provides additional acidic specificity determinants [25]. Addition of polybasic compounds like histones and polylysine, which bind to the acidic stretch of the β subunit, promotes a reduced phosphorylation of the β subunit accompanied by increased phosphorylation of the catalytic α subunit. Figure 1 shows that Rev alters the autophosphorylation pattern of CK2 holoenzyme in the same way by decreasing phosphorylation of β while triggering that of α. The hypothesis that both effects are mediated by the β subunit is corroborated by the observation that no autophosphorylation of α is induced by Rev if the β subunit is absent.

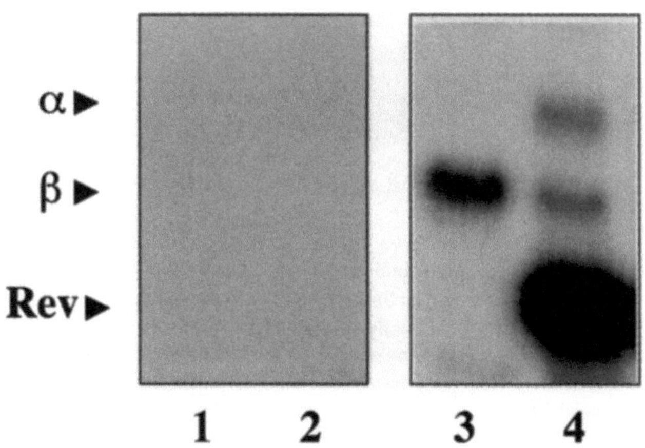

Fig. 1. Alteration of the autophosphorylation pattern of CK2 holoenzyme by Rev. Recombinant wild type Rev protein (2.25 μg, i.e. 150 pmoles) was added to 5 pmoles of recombinant CK2 α subunit either alone or previously mixed to equimolar amounts of β subunit and phosphorylated under conditions described in the experimental section. The autoradiogram relative to a 15% SDS-PAGE is shown. Bands corresponding to α, β and Rev are indicated by arrows. Lane 1 = α subunit alone; lane 2 = α subunit plus Rev; lane 3 = *in vitro* reconstituted CK2 holoenzyme; lane 4 = as lane 3 plus Rev.

Table 2. Inhibitory effect of polylysine on full length and N-terminal synthetic fragments of Rev phosphorylated by CK2 holoenzyme

Substrate	Phosphorylation rate (cpm) control	+ polylysine
Rev	956	145
Rev[1–26]	14,861	6,390
Rev[1–26]AAA	n.d.	n.d.
RRRAADSDDDDD	53,724	67,155

Phosphorylation conditions are detailed under Material and methods. The concentrations of the phosphorylatable substrates were 7 and 40 μM for Rev and synthetic peptides, respectively; n.d. = not detectable.

A simplistic interpretation of these results would presuppose the implication of Rev as a polycationic effector, analogous to polylysine and histones, by virtue of the polyarginine sequence present in its helix-2. Consistent with this possibility, Rev has recently been shown to stimulate CK2 activity toward a number of protein substrates including, among others, reverse transcriptase; a similar stimulation was exerted by a peptide reproducing the Arg-rich basic 35–50 stretch located in Rev's helix 2 region [8]. This interpretation would also be consistent with the observed inhibition (instead of stimulation) of Rev phosphorylation by polybasic effectors [12]. However, the additional finding that the phosphorylation of Rev 1–26 peptide, lacking the polybasic domain of helix-2, is also inhibited by polylysine (Table 2) would indicate that the mode of binding of Rev to the β subunit does not simply rely on electrostatic interactions between the acidic N-terminal domain of β and the polybasic sequence in helix-2. This prompted us to re-evaluate the possible Rev-CK2 β interactions by taking advantage of the recently solved crystal structure of the β subunit [26] and computer-aided prediction of Rev secondary structure [27]. As schematically illustrated in Fig. 2, the amino acid content of both the two N-terminal helices of Rev display an amphipathic motif with a predominance of basic residues within helix 2 and of hydrophobic side chains within helix 1. Intriguingly the so-called 'acidic' N-terminal domain of the β subunit also displays the features of an amphipathic region, with all the acidic side chains protruding on one side and non-polar and hydrophobic residues segregated on the opposite side. This latter observation is based on molecular modeling from the crystal structure of the β subunit [26]. These structural features, taken together with the observed functional importance of Rev's flexible loop [12], lead us to envision a 'sandwich' double molecular mode of interaction with CK2 β in the middle and two helical domains of Rev (H1 and H2 of Fig. 2) properly oriented on the outside.

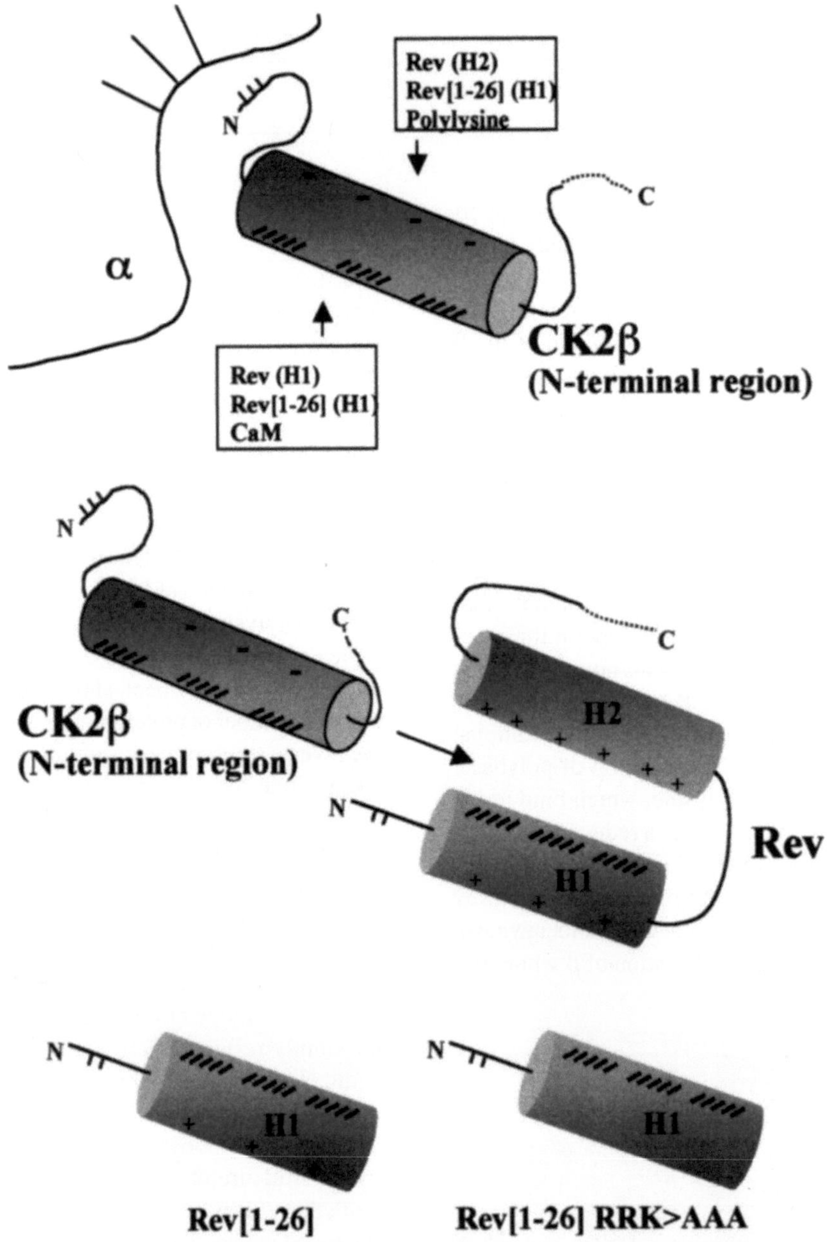

Fig. 2. Schematic representation of functional domains involved during the molecular interaction of Rev with the CK2 β subunit. The N-terminal 55–64 domain of the CK2 β subunit is schematically illustrated by a cylinder in the upper part of the scheme, highlighting the amphipathic distribution of the amino acids on the opposite sides of its external surface resulting from the crystal structure of the β dimer [26]. The lower part of the scheme shows the predicted interaction of the helix-loop-helix structural motif of Rev with the amphipathic domain of β in a sandwich-like structure through electrostatic interactions of positively charged H2 on one side and non-polar interactions of the mainly hydrophobic surface of H1 on the other. Due to the residual presence of three basic residues, the truncated derivative of Rev (Rev[1–26]) is able to flank either side of the β domain, although only the ionic contact would lead to its correct orientation and successful phosphorylation. The Rev[1–26]AAA derivative, which obviously can interact only at the hydrophobic side, is able to bind β thus competing against CaM phosphorylation (see Fig. 4) but fails to be phosphorylated by the catalytic subunit (see Table 2).

To test this model we took advantage of a unique property of calmodulin (CaM) phosphorylation by CK2 holoenzyme, i.e. its total dependence on polybasic compounds.

HIV-1 Rev as a multifunctional modulator of CK2 activity

Despite their opposite behaviour as far as their phosphorylation by CK2 α vs CK2 holoenzyme is concerned [12, 13, 28], Rev and CaM share the same dependency on the β subunit

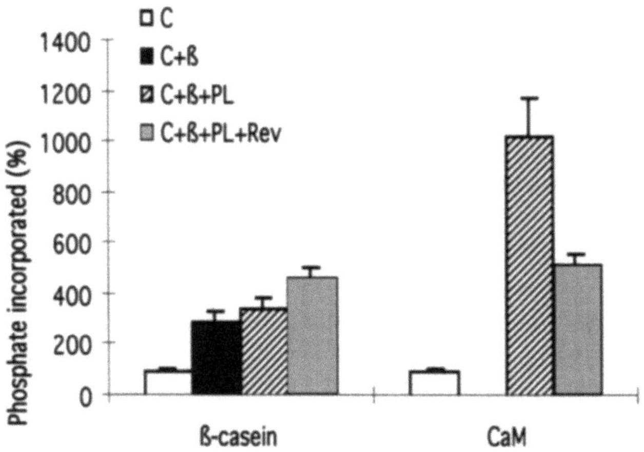

Fig. 3. Mutually exclusive effect of Rev and polylysine on CK2-mediated phosphorylation of calmodulin. Two micrograms of β-casein or calmodulin were separately phosphorylated under conditions described in Material and methods by 5 pmoles of CK2 α subunit alone (C) or previously combined with equimolar amounts of β subunit (C + β). Where indicated, 336 nM polylysine in the absence or presence of 10 μM GST-Rev were added to CK2 reconstituted holoenzyme. Identical results were obtained if GST-Rev was replaced by recombinant Rev protein.

in order to undergo efficient phosphorylation by CK2. However, in the case of calmodulin, this requires the presence of polybasic effectors, which, conversely, are detrimental to the phosphorylation of Rev by CK2 holoenzyme, probably because the mechanism of Rev phosphorylation is self-activatory, and polybasic effectors compete against self-activation [12]. This raised the possibility that the polycation-dependent phosphorylation of CaM and the auto-induced phosphorylation of Rev might be mutually exclusive. That this is indeed the case is confirmed by experiments shown in Fig. 3, in which β-casein, representing a 'canonical' CK2 substrate, and

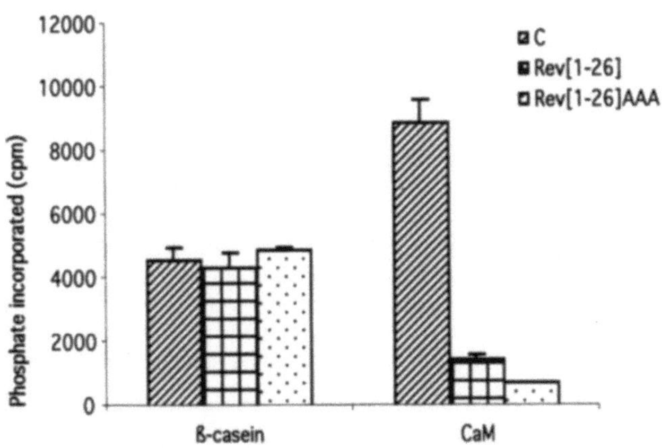

Fig. 4. Inhibition of polylysine-induced, CK2-mediated calmodulin phosphorylation by Rev-derived synthetic peptides. Phosphorylation conditions for β-casein and calmodulin were as detailed in Fig. 3. CK2 holoenzyme, either native or recombinant, was incubated with the substrates in the presence of 336 nM polylysine in the absence (C) or presence of 200 μM indicated peptide.

calmodulin were subjected to phosphorylation by CK2 α alone and by CK2 holoenzyme in the absence and presence of 336 nM polylysine. As previously shown, stimulation of phosphorylation by polylysine is modest with β-casein but dramatic in the case of calmodulin. The simultaneous addition of both polylysine and 10 μM Rev gives rise to an opposite effect: while the phosphorylation of β-casein, which is not crucially dependent on CK2 β subunit, is slightly but significantly stimulated by Rev, in agreement with its role as a polycationic activator of CK2 [8], a 50% reduction in phosphorylation is observed in the case of calmodulin. However, a synthetic derivative (Rev[1–26]) reproducing the N-terminal 1–26 amino acids of Rev including, besides the phosphorylation sites, only helix 1 and therefore lacking the polybasic region of helix 2, inhibits the CK2 holoenzyme-catalyzed, polylysine-dependent phosphorylation of CaM, without any appreciable effect on the phosphorylation of β-casein under the same conditions (Fig. 4). As previously outlined (Table 2), phosphorylation of this fragment of Rev is not strictly dependent on the β-subunit, but is nevertheless susceptible to polylysine inhibition, similar to full length Rev. The possibility that the inhibitory potential of the peptide is accounted for by the presence of three basic amino acids in its sequence (Arg14, Arg17 and Lys20), capable of acting as a surrogate for the 35–50 basic stretch of Rev, was ruled out by showing that Rev[1–26]AAA, in which these basic residues were replaced by alanines, inhibits calmodulin phosphorylation as efficiently as the parent peptide. This would indicate that hydrophobic interactions are responsible for the observed effects and might, in general, be important for binding of full length Rev to the regulatory β subunit of CK2. On the other hand, Arg14, Arg17 and Lys20 do play a role in the phosphorylation of the Rev peptide, since their simultaneous substitution causes the complete loss of phosphorylation by CK2 (see Table 2). These data disclose novel perspectives regarding the vexing question concerning the molecular mechanism of calmodulin phosphorylation by CK2. In fact, in order to account for the adverse effect of Rev[1–26]AAA peptide on CaM phosphorylation, it would be necessary to assume that calmodulin also binds to the β subunit through hydrophobic interactions, whose formation apparently requires the neutralization of the acidic side of the amphipathic 55–64 domain of β (see Fig. 2). The implication of hydrophobic interactions with both Rev and CaM is also corroborated by results of phosphorylation assays performed in the presence of Triton X-100. As shown in Fig. 5, the addition of 0.08% Triton X-100 to the assay negatively affects the phosphorylation of the Rev[1–26] peptide (60% inhibition) and, even more efficiently, the polylysine-induced phosphorylation of calmodulin (87% inhibition) while, under identical conditions, it is not inhibitory but instead significantly increases the phosphorylation of the standard peptide substrate RRRADDSDDDDD and β-casein.

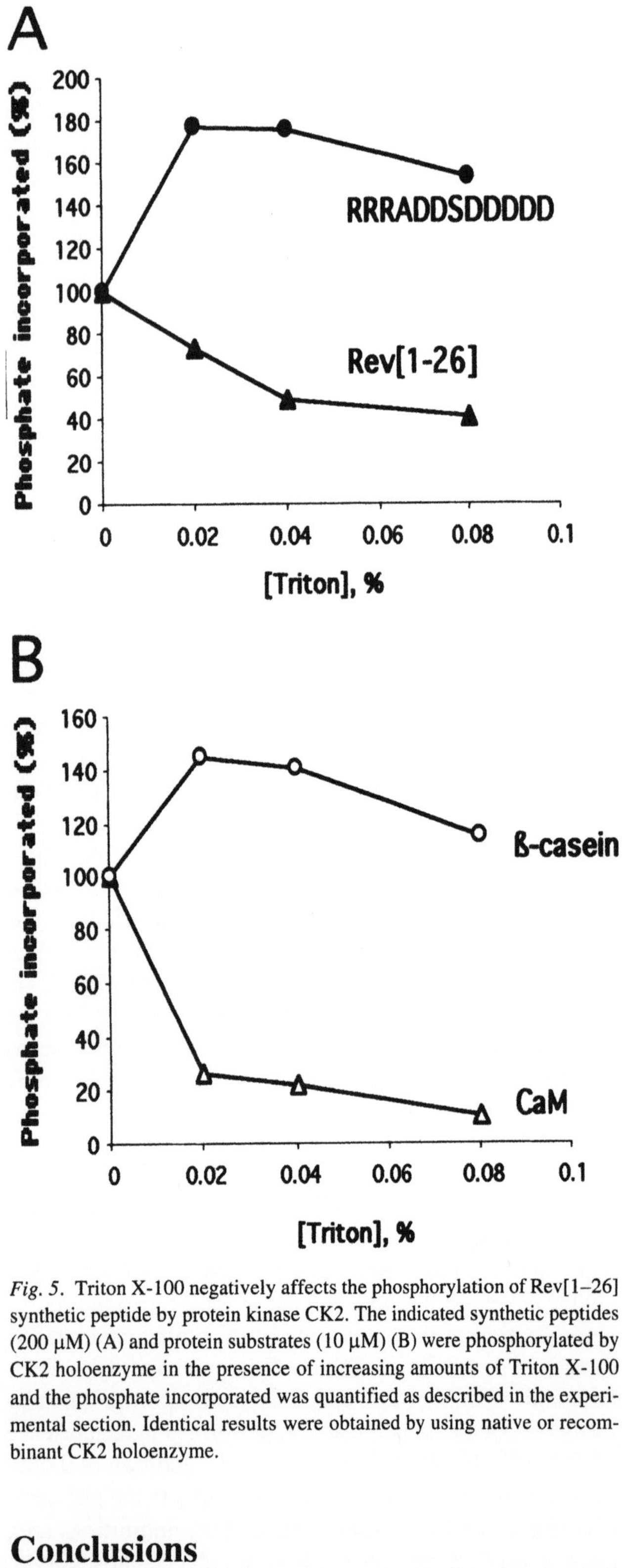

Fig. 5. Triton X-100 negatively affects the phosphorylation of Rev[1–26] synthetic peptide by protein kinase CK2. The indicated synthetic peptides (200 μM) (A) and protein substrates (10 μM) (B) were phosphorylated by CK2 holoenzyme in the presence of increasing amounts of Triton X-100 and the phosphate incorporated was quantified as described in the experimental section. Identical results were obtained by using native or recombinant CK2 holoenzyme.

Conclusions

The data presented corroborate the concept that, besides conferring stability to the catalytic α subunit through a protection against denaturing conditions, the regulatory β subunit of protein kinase CK2 also profoundly influences the biological activity of the enzyme. This reflects not only in modulation of catalytic activity but also in providing a multifunctional docking site for the recruitment of effectors and substrates, these latter otherwise completely refractory to CK2. This highlights the special relevance of the N-terminal 55–64 domain of the β subunit: while already known to play a role as a pseudosubstrate by interacting with the network of basic residues located around the catalytic site, this region also shows other unexpected features as a hydrophobic docking site, e.g. with respect to calmodulin and Rev protein. This property reinforces the need for further studies in order to fully understand the molecular mechanism(s) of interactions of CK2 with its biological targets, especially those whose recruitment is mediated by the β subunit. Clear-cut indications will certainly be provided by the crystal definition of molecular complexes of CK2 bound to its substrates.

Acknowledgements

We thank Vincenzo Ciminale and Donna M. D'Agostino for discussions, and Tiziana Ferro (University of Padova) and Barbara K. Felber (NCI-FCRDC) for providing purified Rev proteins. Financial support was provided by the Italian Ministero della Sanità (Progetto AIDS), The Armenise-Harvard Foundation, AIRC, MURST (PRIN 2000) and CNR (00.00369.ST74 and T.P. on Biotechnology).

References

1. Pinna LA, Meggio F: Protein kinase CK2 ('casein kinase-2') and its implication in cell division and proliferation. Prog Cell Cycle Res 3: 77–97, 1997
2. Allende CC, Allende JE: Promiscuous subunit interactions: A possible mechanism for the regulation of protein kinase CK2. J Cell Biochem 30/31 (suppl): 129–136, 1998
3. Guerra B, Issinger OG: Protein kinase CK2 and its role in cellular proliferation, development and pathology. Electrophoresis 20: 391–408, 1999
4. Seldin DC, Leder P: Casein kinase II α transgene-induced murine lymphoma: Relation to teileriosis in cattle. Science 267: 894–897, 1995
5. Orlandini M, Semplici F, Ferruzzi R, Meggio F, Pinna LA, Oliviero S: Protein kinase CK2 α' is induced by serum as a delayed early gene and cooperates with Ha-ras in cell transformation of primary embryo fibroblasts. J Biol Chem 273: 21291–21297, 1998
6. Schubert U, Schneider T, Henklein P, Hoffmann K, Berthold E, Hauser H, Pauli G, Portsmann T: HIV-1 encoded Vpu protein is phosphorylated by casein kinase II. Eur J Biochem 204: 875–883, 1992
7. Meggio F, D'Agostino DM, Ciminale V, Chieco-Bianchi L, Pinna LA: Phosphorylation of HIV Rev protein: Implications of protein kinase CK2 and Pro-directed kinases. Biochem Biophys Res Commun 226: 547–554, 1996
8. Ohtsuki K, Maekawa T, Harada S, Karino A, Morikawa Y, Ito M: Biochemical characterization of HIV-1 Rev as a potent activator of casein kinase II *in vitro*. FEBS Lett 428: 235–240, 1998

9. Hope TJ: The ins and outs of HIV-1 Rev. Arch Biochem Biophys 365: 186–191, 1999
10. Pollard VW, Malim MH: HIV-1 Rev protein. Annu Rev Microbiol 52: 491–532, 1998
11. D'Agostino DM, Ferro T, Zotti L, Meggio F, Pinna LA, Chieco-Bianchi L, Ciminale V: Identification of a domain in human immunodeficiency virus type 1 Rev that is required for functional activity and modulates association with subnuclear compartments containing splicing factor SC35. J Virol 74: 11899–11910, 2000
12. Marin O, Sarno S, Boschetti M, Pagano MA, Meggio F, Ciminale V, D'Agostino DM, Pinna LA: Unique features of HIV-1 Rev protein phosphorylation by protein kinase CK2 ('casein kinase-2'). FEBS Lett 481: 63–67, 2000
13. Meggio F, Boldyreff B, Issinger OG, Pinna LA: Casein kinase-2 down-regulation and activation by polybasic peptides are mediated by acidic residues in the 55–64 region of the β subunit. A study with calmodulin as phosphorylatable substrate. Biochemistry 33: 4336–4342, 1994
14. Leroy D, Filhol O, Delcros JG, Pares S, Chambaz EM, Cochet C: Chemical features of the protein kinase CK2 polyamine binding site. Biochemistry 36: 1242–1250, 1997
15. Benko DM, Robinson R, Solomin L, Mellini M, Felber BK, Pavlakis GN: Binding of trans-dominant mutant Rev protein of human immunodeficiency virus type 1 to the cis-acting Rev-responsive element does not affect the fate of viral mRNA. New Biologist 2: 1111–1122, 1990
16. Rhyner JA, Koller M, Drussel-Gerber I, Cox JA, Strehler, EE: Characterization of the human calmodulin-like protein expressed in *Escherichia coli*. Biochemistry 31: 12826–12832, 1992
17. Sarno S, Vaglio P, Meggio F, Issinger OG, Pinna LA: Protein kinase CK2 mutants defective in substrate recognition. Purification and kinetic analysis. J Biol Chem 271: 10595–10601, 1996
18. Boldyreff B, Meggio F, Pinna LA, Issinger OG: Casein kinase-2 structure-function relationship: Creation of a set of mutants of the β subunit that variably surrogate the wild type β subunit function. Biochem Biophys Res Commun 188: 228–234, 1992
19. Meggio F, Donella Deana A, Pinna LA: Endogenous phosphate acceptor proteins for rat liver cytosolic casein kinases. J Biol Chem 256: 11958–11961, 1981
20. Laemli UK: Cleavage of the structural proteins during the assembly of the head of bacteriophage T4. Nature 227: 680–685, 1970
21. Ruzzene M, Pinna LA: Assay of protein kinases and phosphatases using specific peptide substrates. In: D.G. Hardie (ed). Protein Phosphorylation. A Practical Approach. Oxford University Press, pp 221–253
22. Sarno S, Vaglio P, Cesaro L, Marin O, Pinna LA: A multifunctional network of basic residues confers unique properties to protein kinase CK2. Mol Cell Biochem 191: 13–19, 1999
23. Guerra B, Boldyreff B, Sarno S, Cesaro L, Issinger OG, Pinna LA: CK2: A protein kinase in need of control. Pharmacol Ther 82: 303–313, 1999
24. Sarno S, Vaglio P, Marin O, Meggio F, Issinger OG, Pinna LA: Basic residues in the 74–83 and 191–198 segments of protein kinase CK2 catalytic subunit are implicated in negative but not in positive regulation by the β subunit. Eur J Biochem 248: 290–295, 1997
25. Boldyreff B, Meggio F, Pinna OG, Issinger OG: Efficient autophosphorylation and phosphorylation of the β subunit by casein kinase-2 require the integrity of an acidic cluster 50 residues downstream from the phosphoacceptor site. J Biol Chem 269: 4827–4831, 1994
26. Chantalat L, Leroy D, Filhol O, Nueda A, Benitez MJ, Chambaz EM, Cochet C, Dideberg O: Crystal structure of the human protein kinase CK2 regulatory subunit reveals its zinc finger-mediated dimerization. EMBO J 18: 2930–2940, 1999
27. Rost B, Sander C: Prediction of protein structure at better than 70% accuracy. J Mol Biol 232: 584–599, 1993
28. Meggio F, Boldyreff B, Marin O, Marchiori F, Perich JW, Issinger OG, Pinna LA: The effect of polylysine on casein kinase-2 activity is influenced by both the structure of the protein/peptide substrates and the subunit composition of the enzyme. Eur J Biochem 205: 939–945, 1992

Molecular and Cellular Biochemistry **227**: 153–165, 2001.

Protein kinase CK2: Signaling and tumorigenesis in the mammary gland

Esther Landesman-Bollag,[1]* Diane H. Song,[2]* Raphaelle Romieu-Mourez,[3] Daniel J. Sussman,[4] Robert D. Cardiff,[5] Gail E. Sonenshein[3] and David C. Seldin[6]

Departments of [1]Pathology, [3]Biochemistry and [6]Medicine and Microbiology, Boston University School of Medicine; [2]Department of Chemistry, Boston University, Boston, MA; [4]Department of Pharmaceutical Sciences, University of Maryland School of Pharmacy, College Park, MD; [5]Center for Comparative Medicine, U.C. Davis School of Medicine, Davis, CA, USA

Abstract

Breast cancer is a major cause of cancer death in women, and the genetic abnormalities leading to the common sporadic forms of the disease are still under active investigation. CK2 has been reported to be upregulated in human breast cancer, which these studies confirm; CK2 is also upregulated in rat carcinogen-induced breast tumors. Transgenic mice overexpressing CK2α in the mammary gland develop mammary hyperplasia, dysplasia, and eventually adenocarcinomas, demonstrating that dysregulated expression of CK2 can contribute to transformation of the mammary epithelium. These mammary tumors have evidence of activation of the Wnt and NFκB pathways and upregulation of c-Myc. CK2 is capable of phosphorylating the key signaling molecule in the Wnt pathway, the transcriptional cofactor β-catenin, and regulating its turnover. CK2 is known to phosphorylate IκB and thereby regulate basal NFκB levels; in the mammary cell lines and tumors, CK2 activity correlates with NFκB levels and inhibition of CK2 downregulates NFκB. Thus, CK2 may promote breast cancer through dysregulation of key pathways of transcriptional control in the mammary epithelium, and inhibition of CK2 has a potential role in the treatment of breast and other cancers. (Mol Cell Biochem **227**: 153–165, 2001)

Key words: protein kinase CK2, dishevelled, β-catenin, NFκB, breast cancer, transgenic mice

Introduction

CK2 is highly conserved and has many essential roles in eukaryotic cells. Studies from our laboratory have demonstrated that lack of one of two homologous catalytic isoenzymes leads to male infertility [1], and preliminary results from Dr. Boldyreff and colleagues indicate the sole mammalian regulatory subunit is required for normal embryonic development (see elsewhere in this volume). Overexpression of CK2 appears to be nearly as deleterious as underexpression, as a series of observations in the literature have found that CK2 activity is enhanced in many human solid tumors, transformed cell lines, and rapidly proliferating tissues. Our laboratory has used transgenic mice overexpressing CK2α in lymphocytes to demonstrate that CK2 has a causative, oncogenic role in hematopoietic malignancies, as these mice develop T cell lymphomas with an incidence that is potentiated by overexpression of the transcription factors myc [2] or tal-1 [3] or by loss of p53 [4].

Epithelial malignancies including lung cancer, breast cancer, and colon cancer are much more common causes of cancer mortality than lymphoid neoplasms. For these diseases, it has been difficult to identify unique chromosomal translocations leading to cancer. Rather, studies from many laboratories have demonstrated that a series of gains and losses of a menu of oncogenes and anti-oncogenes leads to complete transformation. Many of these oncogenes and anti-oncogenes

*These authors contributed equally to these studies.

Address for offprints: D.C. Seldin, Departments of Medicine and Microbiology, Boston University School of Medicine, EBRC4, 650 Albany Street, Boston, MA 02118, USA (E-mail: dseldin@medicine.bu.edu)

Article No. ND-01-5610 cp, copyprep, typeset, disc, bf MCBI PIPS No. 383877

regulate gene transcription, and large scale analyses of arrays of expressed genes in tumors indicate that there may be hundreds of genes whose expression is altered in a cancer cell.

Familial breast cancer is frequently caused by hereditary heterozygous deficiency of the genes encoding the BRCA1 or BRCA2 transcription factors, but somatic chromosomal loss of these genes is rarely seen in the more common sporadic breast cancers. Thus, transcriptional alterations in sporadic breast cancer remains poorly understood. Many transcription factors have been reported to be phosphorylated and regulated by CK2 (see reviews such as [5, 6] and others). In this set of studies, we demonstrate that CK2 is highly expressed in human breast cancer specimens and in carcinogen-induced mammary tumors in rats, and that enforced expression in the mammary gland of transgenic mice leads to mammary tumors. Alterations in the Wnt/TCF/LEF and NFκB transcription factor pathways were found that suggest these pathways may be targets of CK2 in transformation.

Materials and methods

Cell culture

pMV7-Wnt-1 (kindly provided by Dr. Anthony M. C. Brown) as well as control pMV7 empty vector were transfected into C57MG cells via electroporation. Stable clones were generated by selection in 400 μg/mL G418 (GibcoBRL, Gaithersburg, MD, USA). Wnt positive clones, as determined by Northern and RT-PCR, were compared to those for neomycin-resistant clones transfected with the empty vector in each experiment. In addition to the stably transfected clones, we also analyzed pools of cells transiently infected with retrovirus expressing Wnt-1 (Kindly provided by Dr. Daniel J. Sussman) to confirm our results. Transgenic breast tumor cell lines were generated by mincing the tumor in DMEM under sterile conditions followed by culture in DMEM supplemented with 10% fetal bovine serum, 4 mM glutamine, 50 U/ml penicillin, 50 μg/ml streptomycin (Mediatech Inc. Cellgro, Herndon, VA, USA) in a 5% CO_2 incubator at 37°C to establish immortalized cell lines. In some cases, the tumorigenic nature of these cell lines was confirmed by transplantation of approximately 10^6 cells subcutaneously into syngeneic recipients.

Immunoprecipitation and Western blotting

To extract cellular proteins, Wnt-1 cells were lysed in buffer supplemented with protease and phosphatase inhibitors (50 mM Tris-HCl pH 8.0, 1% NP-40, 125 mM NaCl, 1 mM NaF, 1 mM PMSF, 1 μg/mL aprotinin, 1 μg/mL pepstatin, 1 μg/mL leupeptin, 1 mM Na_3VO_4, and 10 mM sodium pyrophosphate). The BCA protein assay (Pierce, Rockford, IL, USA) was used to determine the protein concentrations of the whole cell lysates. Equal amounts of protein were loaded onto polyacrylamide gels, electrophoresed, and transferred onto nitrocellulose membranes (Schleicher & Schuell, Keene, NH, USA) using a semi-dry electroblotter (Owl Scientific, Woburn, MA, USA). After transfer, the membranes were blocked overnight in 5% dry milk and probed for protein expressions using antibodies against CK2α (UBI, Lake Placid, NY, USA), GSK3β (Transduction Laboratories, Lexington, KY, USA), β-catenin (Transduction Laboratories) or c-myc (UBI). Densitometry analysis was performed using Quantity One software (BioRad, Hercules, CA, USA). Dvl proteins were detected with monospecific monoclonal antibodies. Immunoreactive bands using appropriate secondary antibodies coupled to HRP (Santa Cruz, Santa Cruz, CA, USA) were visualized by chemiluminescence in Signaling Solution (Pierce). Ponceau S (Sigma, St. Louis, MO, USA) staining and/or a monoclonal β-actin antibody (Sigma) were used to confirm equal loading of all Western blot membranes. For immunoprecipitation, 300 μg of protein extracts were first pre-cleared with protein A-agarose beads (Sigma). 100 μL of Dvl-1, 2, or 3 hybridoma supernatant or 1 μL of 68 mg/mL β-catenin antibody (Sigma) were added to precleared lysates and incubated overnight at 4°C with constant rotation. 40 μL of protein A-agarose beads were added to each of the incubations for 2 h at 4°C. Control immunoprecipitations with irrelevant antibody or beads alone were also performed. The immunoprecipitated complexes were washed 5 times in PBS and Western blotting was performed as described.

In vitro *kinase assay*

Following immunoprecipitation of the whole cell lysates with Dvl-1, 2, and 3 antibodies as described above, the bead-bound proteins were incubated with 5 μCi [γ-$_{32}$P]GTP at 37°C for 20 min alone or in the presence of apigenin (80 μM) or Wortmannin (25 nM). The kinase reactions were stopped by the addition of 2 × sample loading buffer. The samples were boiled, centrifuged, and subsequently loaded onto a denaturing SDS-polyacrylamide gel. The gel was transferred onto a nitrocellulose membrane and autoradiography of the membrane was performed to visualize [γ-$_{32}$P]GTP labeled proteins. Western blotting was performed to identify the phosphorylated target using specific antibodies.

Human and rat mammary tumor tissue analysis

Breast tumors were induced in rats by a single gavage with 15 mg/kg of DMBA at 8 weeks of age. The rats were sacrificed 17 weeks later. Tumors and pooled normal mammary glands were snap frozen at the time of necropsy; portions of these tissues were used for other experiments on diet and breast cancer [7]. All animal experimentation was carried out in accordance with the regulations of the American Association for the Accreditation of Laboratory Animal Care and with approval of the Boston University School of Medicine I.A.C.U.C committee. Unneeded portions of anonymous frozen human breast tissue samples and breast tumor specimens were obtained with approval of the Institutional Review Board, Boston Medical Center.

Transgenic mice

Mice from the FVB/N strain were utilized to generate transgenic animals. The murine CK2α catalytic subunit cDNA was subcloned into a vector in which the mouse mammary tumor virus (MMTV) long terminal repeat (LTR) directs expression chiefly to the mammary epithelium, with ras 5′ untranslated sequences provided upstream of the cDNA and an SV40 intron and polyadenylation signal downstream [8]. Plasmid sequences were removed by restriction digestion at the SalI and SpeI sites, and the excised transgene construct was gel purified and microinjected into pronuclei of fertilized one-cell zygotes. These were reimplanted into pseudopregnant foster mothers and the offspring were screened for presence of the transgene. Carriers were bred to establish 4 independent transgenic lines. Female transgenic mice were continuously bred to induce transgene expression through activation of the hormone-dependent MMTV-LTR. Mice were monitored weekly for the appearance of tumors. Ill mice were sacrificed and tissues collected for histopathological analysis, cell culture, RNA and protein analyses. To assess expression levels of the transgene in the mouse organs, tissues were collected from 6–8 week old virgin females, from females that were pregnant, and from females at day 7 of mammary gland involution. Mice were kept in a 2-way barrier at the Boston University School of Medicine Core Transgenic Mouse Facility in accordance with the regulations of the American Association for the Accreditation of Laboratory Animal Care.

Histology

Upon necropsy, tumors and organs were removed and immediately fixed in Optimal Fix (American Histology Reagent Company) and shipped in alcohol. The tissues were processed, embedded in paraffin and sectioned at 7 microns. The sections were mounted on glass slides and stained with Hematoxylin and eosin using routine laboratory procedures in the Transgenic Core Pathology Laboratory at the University of California, Davis. Sections were compared with other specimens in the extensive mouse mammary tumor database [9].

CK2 peptide kinase assay

Samples (2–5 μg) of protein lysate were incubated in a 1 mM solution of the specific protein kinase CK2 substrate peptide RRREEETEEE (Sigma-Genosys, The Woodlands, TX, USA) in CK2 kinase buffer (100 mM Tris pH 8.0, 20 mM $MgCl_2$, 100 mM NaCl, 50 mM KCl, 0.1 μg/μl BSA and 100 μM Na_3VO_4) and 5 μCi of [γ-$_{32}$P]-GTP (6000 Ci/mmol) at 30°C for 10 min [10]. The reaction was stopped by adding 25 μl of 100 mM ATP in 0.4 N HCl. Samples were spotted onto a P81 Whatman filter and washed in 150 mM H_3PO_4, 4 times, 5 min each, to remove unincorporated [γ-^{32}P]-GTP. Phosphorylated peptides were quantified in an automatic scintillation counter. The samples were assayed in duplicate and background kinase activity in the absence of the specific peptide substrate was subtracted. The statistical significance of differences in activity between pairs was determined by the t-Test for Paired Two Sample for Means.

NFκB reporter assay

To evaluate NFκB transcriptional activity, wild type (E8-CAT) and mutant (dmE8-CAT) NF-κB element-thymidine kinase (TK) promoter-CAT reporter vectors, containing 2 copies of either the wild type or mutant NFκB element from upstream of the c-*myc* promoter were employed [11]. Cells were transfected, in duplicate, using Fugene reagent (Boehringer Mannheim) according to the manufacturer's directions. CAT assays were performed, and the pSVβ-galactosidase (SV40-βgal) reporter vector was used to normalize transfection efficiency, as previously described [12]. The mean and standard deviation of duplicates were calculated.

Results

CK2 is overexpressed in rat and human breast tumors

To evaluate the role of CK2 role in the pathogenesis of mammary tumors, we measured CK2 catalytic protein subunit expression and activity in rat carcinogen-induced breast tumors [13, 14] and human breast tumors. For rat tissues, tumors and control pooled normal mammary tissue from the same rats were analyzed as matched pairs of samples. In most pairs, CK2 protein is upregulated in the tumors; den-

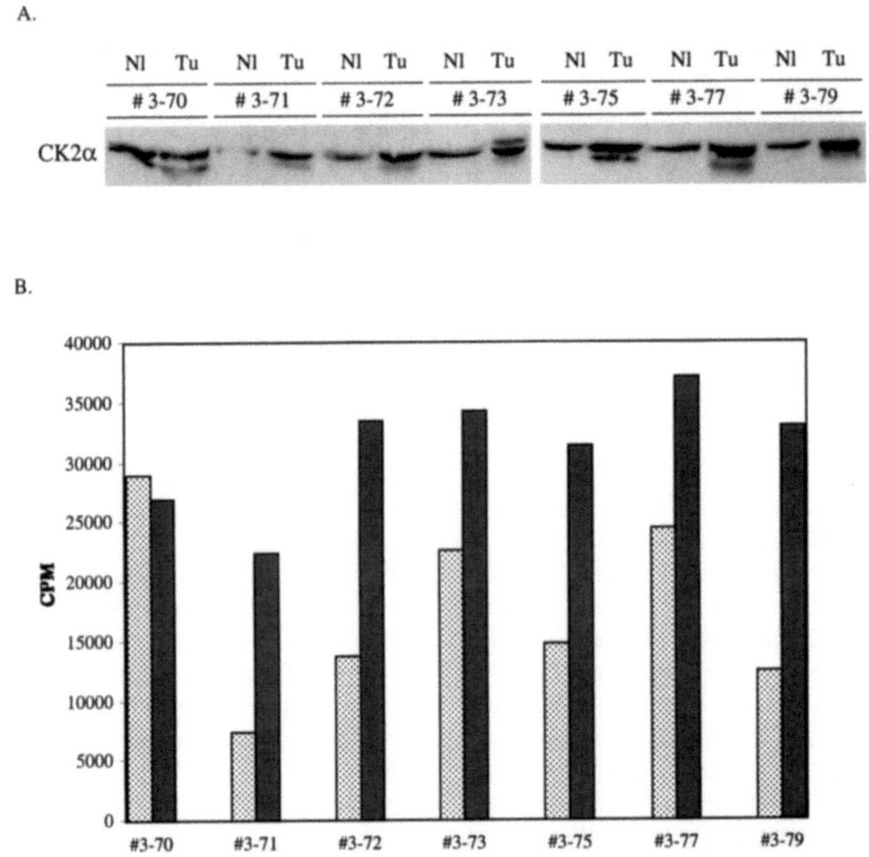

Fig. 1. Protein kinase CK2 expression and activity in carcinogen-induced mammary tumors in rats. (A) 100 µg of protein extracted from paired normal mammary glands (Nl) and mammary tumors (Tu) from rats treated with DMBA were subjected to Western blotting for CK2; equal loading was confirmed by Ponceau S staining (not shown). CK2 expression is elevated in 6 out of 7 tumor samples. (B) CK2 kinase activity: equal amounts of protein lysates were assayed in duplicate for phosphorylation of a specific CK2 substrate peptide [10]. Data is expressed as the difference of the mean counts of the kinase reaction carried out without the specific peptide subtracted from the mean with peptide; error bars cannot be displayed for the difference of means but were generally <10% for the means themselves. In 6 out of 7 pairs, a marked increase in CK2 kinase activity in tumors (dark bars) is seen compared to normal mammary tissue (light bars). The overall difference in kinase activity between mammary glands and tumors was statistically significant ($p < 0.003$).

sitometry analysis showed an average increase of 2.7-fold (Fig. 1A). Both the 42 kDa CK2α and the 38 kDa CK2α′ subunits appear to be recognized by the polyclonal rabbit antibody. No significant change in CK2β protein levels were seen (not shown). To correlate CK2 activity with protein, samples were assayed in duplicate using the peptide substrate RRREEETEEE [10], and [γ-^{32}P]-GTP was employed as a phosphate donor to further ensure specificity of the assay. CK2 activity was higher in the tumors (mean 2-fold, Fig. 1B).

Human breast tumor samples and a non-malignant control reduction mammoplasty were obtained from eleven anonymous specimens, with approval of the Institutional Review Board at Boston University School of Medicine. All breast tumor specimens exhibited abundant expression of the catalytic subunits compared to the non-malignant mammoplasty specimen (Fig. 2A). CK2 kinase activity was determined in 7 samples in which there was adequate material for duplicate measurements. All of these samples had more CK2 activity than the control, and on average the increase was more than 10-fold (Fig. 2B). Thus, in both carcinogen-induced breast tumors in a rodent model system and in human breast tumors, CK2 protein levels and activity are elevated, suggesting that CK2 may play a role in mammary tumorigenesis.

Transgenic MMTV-CK2α mice develop hyperplasia and dysplasia in the post-lactating mammary gland

To determine whether CK2 overexpression plays a causative role in mammary tumorigenesis, we engineered transgenic mice to overexpress the α catalytic subunit of CK2 in the mammary gland using an MMTV-LTR transgene vector [15]. Expression of the hormone-dependent MMTV-LTR transgene was promoted by continuous breeding of the females.

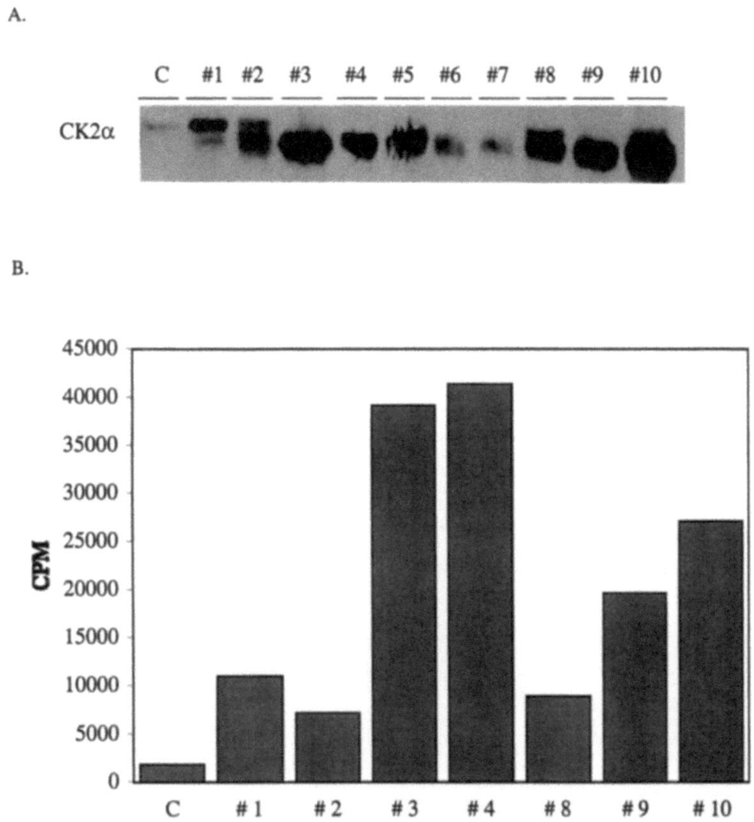

Fig. 2. Protein kinase CK2 expression and activity in human breast tumors. (A) Representative Western blot of tissue extracts with 50 μg of protein lysate in each lane. Breast tissue obtained from a reduction mammoplasty was used as a control and compared with 10 breast cancer specimens; increased CK2 protein was seen in all 10 specimens. Equal loading was obtained as in Fig. 1. (B) CK2 kinase activity: equal amounts of protein lysates were assayed for phosphorylation of the specific CK2 peptide as described in Fig. 1B; insufficient material was available for analysis from samples 5–7. All breast tumor specimens assayed have increased levels of CK2 activity.

Mammary glands from female mice derived from four different founder lines were harvested at several stages of mammary gland development. Histological abnormalities were observed in half of the MMTV-CK2α female transgenic mice (14 out of 27 mice examined). In 3 mice, pubertal mammary gland development was incomplete or retarded, with less than 20% of the mammary fat pad being filled with glandular tissue, although ultimately the females were able to produce adequate milk. Incomplete mammary gland regression was observed in 8 mice whose mammary glands were harvested 2–12 months after the last pregnancy. Importantly, 7 mice developed a variety of dysplastic lesions that are rarely seen in wild type FVB/N females, and are considered to be premalignant [16]. These included areas of dysplastic alveoli and squamous metaplasia, commonly seen with areas of inflammation in the stroma and the epithelium (Fig. 3). Thus, mice with dysregulated expression of CK2α in the mammary gland frequently exhibit abnormalities of mammary gland development and regression, and inflammatory and preneoplastic lesions.

MMTV-CK2α mice develop late-onset mammary adenocarcinomas

A cohort of 56 transgenic female mice was observed for over 2 years. 30% of the mice developed mammary tumors at a median age of 23 months. In addition, over 35% of the mice developed non-mammary malignancies, including lung tumors and lymphomas. While the non-mammary tumor incidence is similar to that reported in wild type FVB mice, the incidence of spontaneous mammary tumors in this strain is very low [17]. The tumors generally arose as solitary masses in single mammary glands, although multiple tumors developed synchronously in separate glands in 6 mice. All the tumors were adenocarcinomas; the most common histologic subtypes were glandular and papillary adenocarcinomas, in 7 tumors each (Fig. 4). Six tumors from 4 mice had histologic features of spindle cell carcinomas. Epithelial origin of the tumors was confirmed by immunohistochemistry analysis of epithelial and mesenchymal markers (not shown). Four were classified as adenosquamous carcinomas (keratoacanthomas)

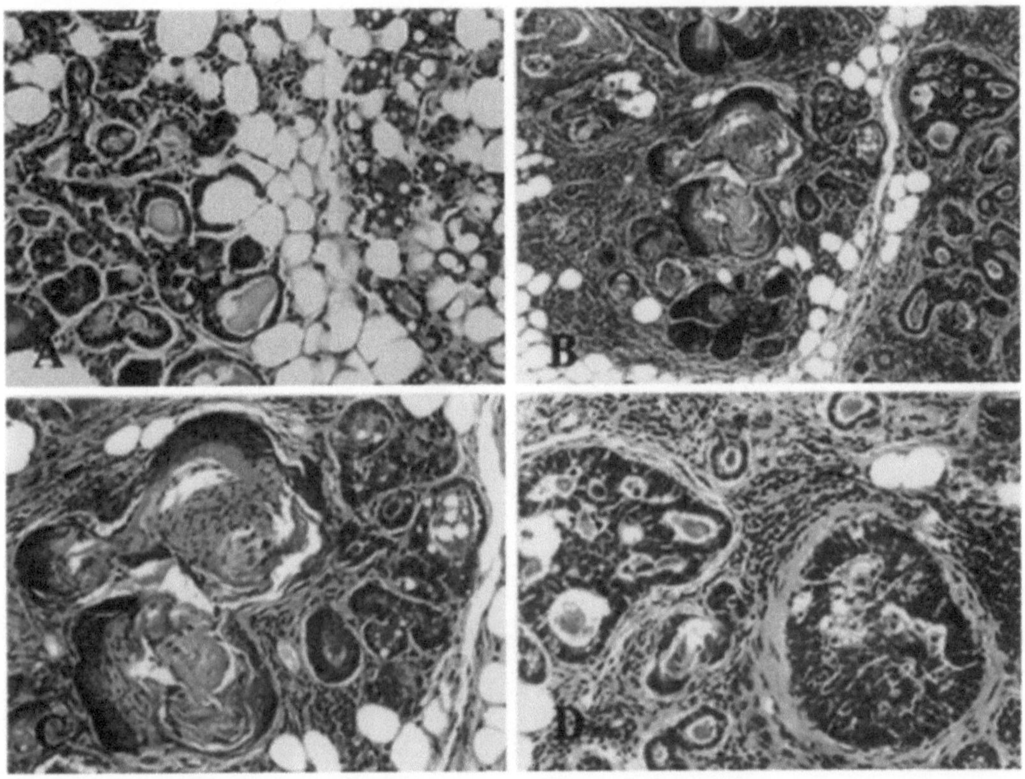

Fig. 3. Mammary gland dysplasia in regressed, multiparous MMTV-CK2α transgenic mice. Hematoxylin and eosin-stained sections reveal that mammary glands from female transgenics frequently present with foci of dysplasia. (A) This field contains a dysplastic region on the left hand side and normal-appearing regressing mammary gland on the right hand side of the image. Compare the size and density of the nuclei in the dysplastic alveolar nodules on the left with the normal epithelium on the right. (B) This field includes both alveolar and squamous nodules, with the squamous nodules enlarged in panel C and the alveolar nodules enlarged in panel D. Note that all fields of dysplasia are associated with extensive fibrosis and inflammation.

and 2 as scirrhous carcinomas. One tumor consisted of large cells similar to those seen with expression of a c-*myc* transgene in the mammary gland, and one had the solid cord-like pattern of tumors as seen arising through transgenic expression of *int-2* [18]. Some tumors exhibited mixed patterns with classical adenocarcinoma in some regions and the cellular connective tissue stroma characteristic of scirrhous carcinomas in others.

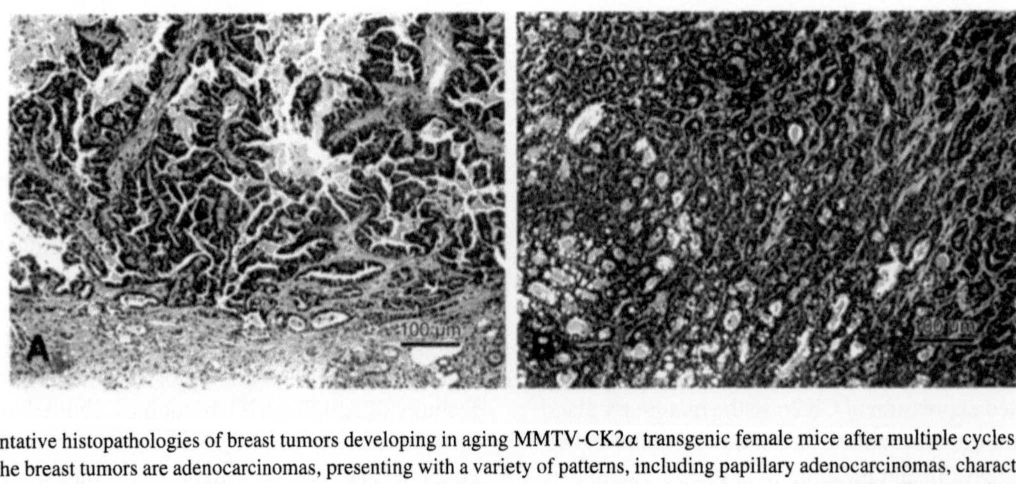

Fig. 4. Representative histopathologies of breast tumors developing in aging MMTV-CK2α transgenic female mice after multiple cycles of pregnancy and regression. All the breast tumors are adenocarcinomas, presenting with a variety of patterns, including papillary adenocarcinomas, characterized by a frond-like pattern with central fibrovascular stalks covered by neoplastic epithelium (A), or pure glandular adenocarcinomas (B). Hematoxylin and eosin staining (100×).

Candidate downstream targets in transgenic mammary tumors

CK2 protein expression and kinase activity in the tumors was determined as previously done in the rat and human breast tumor specimens. Twelve pairs of tumors and normal looking adjacent glands were analyzed, and all exhibit higher levels of CK2α protein in the tumors. CK2 kinase activity was also elevated in the tumors, averaging 6-fold in 6 pairs assayed. Similarly to rat tissues, no significant change in CK2β expression was detected (not shown).

The mammary tumors in the transgenic mice occurred in older mice, suggesting a requirement for multiple additional molecular events for complete mammary epithelial cell transformation. Because of a recent report associating CK2 with the Wnt pathway [19], which is mutated in many human cancers, we looked for activation of the Wnt pathway in these tumors. Levels of β-catenin, the critical transcriptional co-factor for TCF/LEF signaling, vary with Wnt pathway activation. β-catenin levels were assessed by Western blot analysis of tumor tissue extracts; a representative Western blot is shown (Fig. 5A). Overall, 6 of 11 breast tumors tested exhibited elevated levels of β-catenin compared to non-malignant transgenic mammary tissues.

CK2 phosphorylates IκB in its C-terminus and promotes its degradation, regulating basal and inducible levels of nuclear NFκB [20–23]. N-terminal phosphorylation of IκB by the IKKs regulate inducible NFκB activation only. We and others have demonstrated constitutive high levels of NFκB in rat and human breast cancer [7, 24–27]. To determine whether this is also the case in the CK2α transgenic mammary tumors, we first examined nuclear NFκB in primary CK2α transgenic tumors. While all 4 specimens examined contained constitutive nuclear NF-κB (data not shown), some of the specimens derived from the primary tumors also proved to be contaminated with lymphocytes. Thus, to determine whether there is transcriptionally active NFκB in the malignant mammary epithelial cells, 3 cell lines derived from the primary tumors were transfected *in vitro* with an NFκB-dependent chloramphenicol acetyl transferase (CAT) reporter. As a control, we used C57MG normal mouse mammary epithelial cell line [28]. To normalize for transfection efficiency, CMV-βgal was transfected in parallel. All 3 cell lines exhibited higher CAT activity than the C57MG control cell line, with an average 4-fold activation, indicating that the transgenic tumor lines contain transcriptionally active NFκB (Fig. 5B).

Overexpression of the c-*myc* oncogene [2] or loss of *p53* [4] are collaborating events with CK2α in lymphomagenesis, and we hypothesized c-*myc* overexpression or *p53* loss might occur in CK2α transgenic mammary tumorigenesis. In particular, we expected to find elevated c-Myc expression, as c-*myc* is reported to be a transcriptional target of both NF-κB and β-catenin [11, 29, 30]. Levels of c-Myc were determined by Western blot analysis of tumor tissue extracts. Three out of 4 tumors analyzed had evidence of elevated c-Myc protein levels (Fig. 5C). On the other hand, in the majority of primary tumors, *p53* mRNA was expressed and appeared

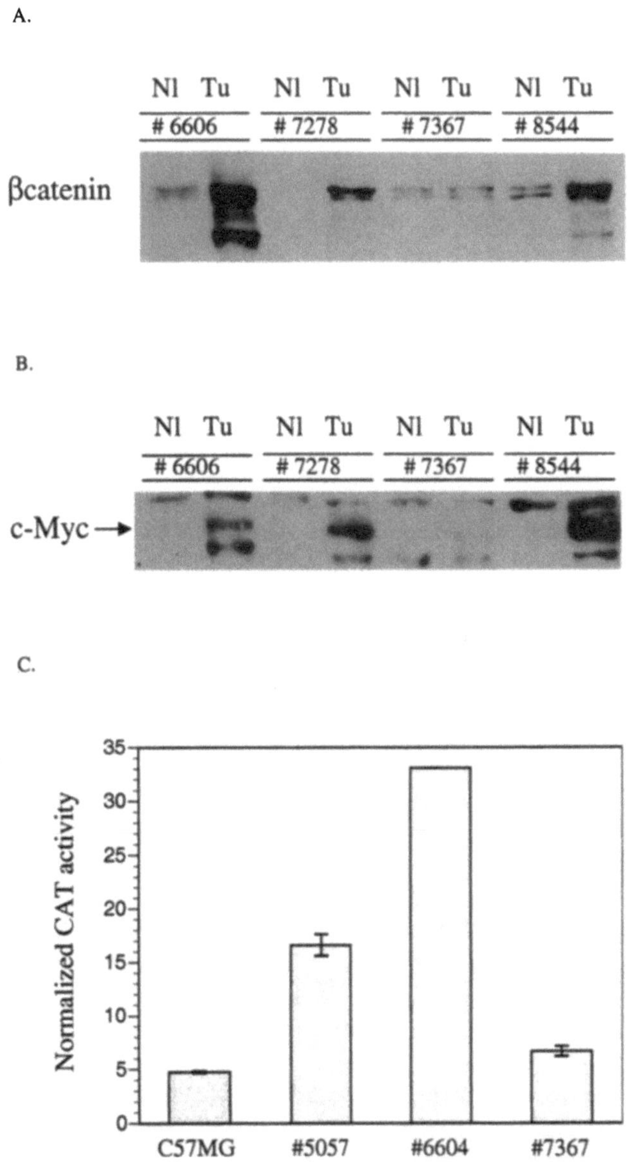

Fig. 5. Expression and activity of β-catenin, NFκB and c-myc in transgenic female MMTV-CK2α breast tumors. 40 μg of nuclear extracts obtained from paired normal mammary glands (Nl) and breast tumors (Tu) of transgenic mice were subjected to Western blotting with β-catenin (A) and c-Myc antibodies (B). Note the increase in β-catenin and c-myc levels in 3 out of 4 breast tumors. Equal loading was verified with Ponceau S staining. (C) Transgenic breast tumor and C57MG cell lines were transfected with an NFκB-CAT reporter construct to evaluate NFκB transcriptional activity. Transfection efficiency was normalized using an SV40βgal reporter construct; data is expressed as mean and standard deviation of duplicate transfections. The increase in NFκB transcriptional activity in the tumor cell lines averages 4-fold over the control cell line.

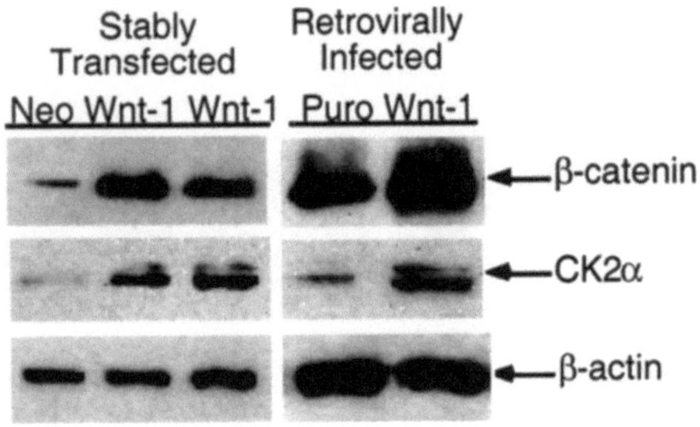

Fig. 6. β-catenin and CK2α proteins are upregulated in Wnt-1 overexpressing C57MG cells. Protein extracts were prepared from clones stably transfected with Wnt-1 or neomycin plasmid vector alone (left panel) or from cells transiently infected with a Wnt-1 retrovirus or puromycin retrovirus control (right panel). 30 μg of protein was transferred onto a nitrocellulose membrane and Western blotting was performed with antibodies against β-catenin, CK2α, or β-actin as a loading control.

normal (not shown). Thus, *p53* loss does not appear to be a frequent event in these transgenic mammary tumors.

Wnt-1 upregulates protein levels of CK2 and β-catenin

While CK2 was reported to associate with components of the Wnt pathway in insect cells [19], its functional role had not been elucidated. Since we found evidence of Wnt pathway activation in CK2-dependent mammary tumors, we carried out studies of the role of CK2 in Wnt signaling. Wnt-1 expressing subclones of C57MG cells were generated and these clones were initially screened by Northern and RT-PCR analyses for Wnt-1 expression. Stabilization of β-catenin is a hallmark of activated Wnt signaling, and as expected β-catenin protein level was upregulated in our Wnt-1-overexpressing cells. CK2 protein level was also upregulated in both the pMV7-Wnt-1 transfected and pBABE-Wnt-1 infected cells (Fig. 6). Northern analysis further showed that CK2 mRNA expression was elevated in the Wnt-1 expressing clones and exhibited 2-fold higher CK2 activity than the vector controls (Data not shown).

Endogenous CK2 and β-catenin associate with Dvl

To determine whether CK2 associates with Dvl in Wnt-1 expressing mammary epithelial cells, as suggested in the insect cells, immunoprecipitations with specific monoclonal antibodies raised against Dvl-1, 2, and 3 were performed, followed by Western blotting. Each of the Dvl proteins was detected in Western blots of immunoprecipitates performed with the cognate antibody (Fig. 7). We also detected immunoreactive CK2 and β-catenin protein in the complexes precipitated by each of the Dvl antibodies. We attempted to perform the reciprocal immunoprecipitates, however neither the commercial CK2 antibody nor the β-catenin antibody proved to be efficient at immunoprecipitation.

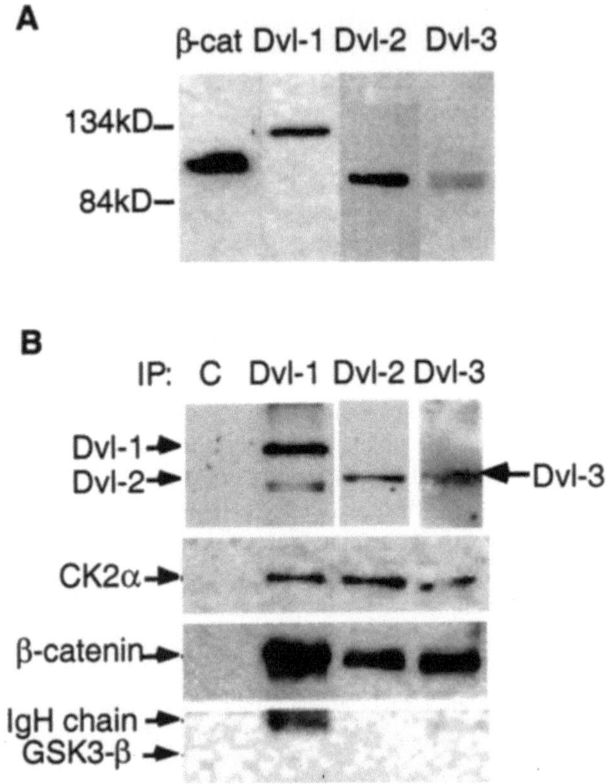

Fig. 7. CK2 and β-catenin associate with Dvl-1, 2, and 3. (A) 100 μg of protein extracted from Wnt-1 expressing cells were run on a denaturing polyacrylamide gel and Western blotted with antibodies to β-catenin and Dvl 1, 2, and 3 to ascertain the electrophoretic mobilities of the corresponding proteins. (B) 300 μg of precleared protein lysates from Wnt-1-transfected cells were incubated overnight with the specific Dvl monoclonal antibodies followed by a 2 h incubation with protein A agarose beads which were washed throroughly, boiled in 2× sample buffer, and applied to a polyacrylamide gel. A control immunoprecipitation (IP:C) was performed with an unrelated mouse monoclonal antibody. By Western blotting, each Dvl antibody immunoprecipitated its cognate Dvl. β-catenin and CK2 were found to co-immunoprecipitate with the Dvl proteins. GSK3β was not detected in the Dvl immunocomplexes; the band migrating slightly more slowly than the GSK3β protein detected in the whole cell lysate is the immunoglobulin heavy chain (Ig H chain).

CK2 phosphorylates β-catenin in vitro

To determine whether the CK2 in the Dvl immunoprecipitates was capable of phosphorylating its partners, an *in vitro* kinase reaction was performed on the immunoprecipitates. Consistently, we found a single major phosphoprotein of approximately 97 kd appearing in all three Dvl immunoprecipitates

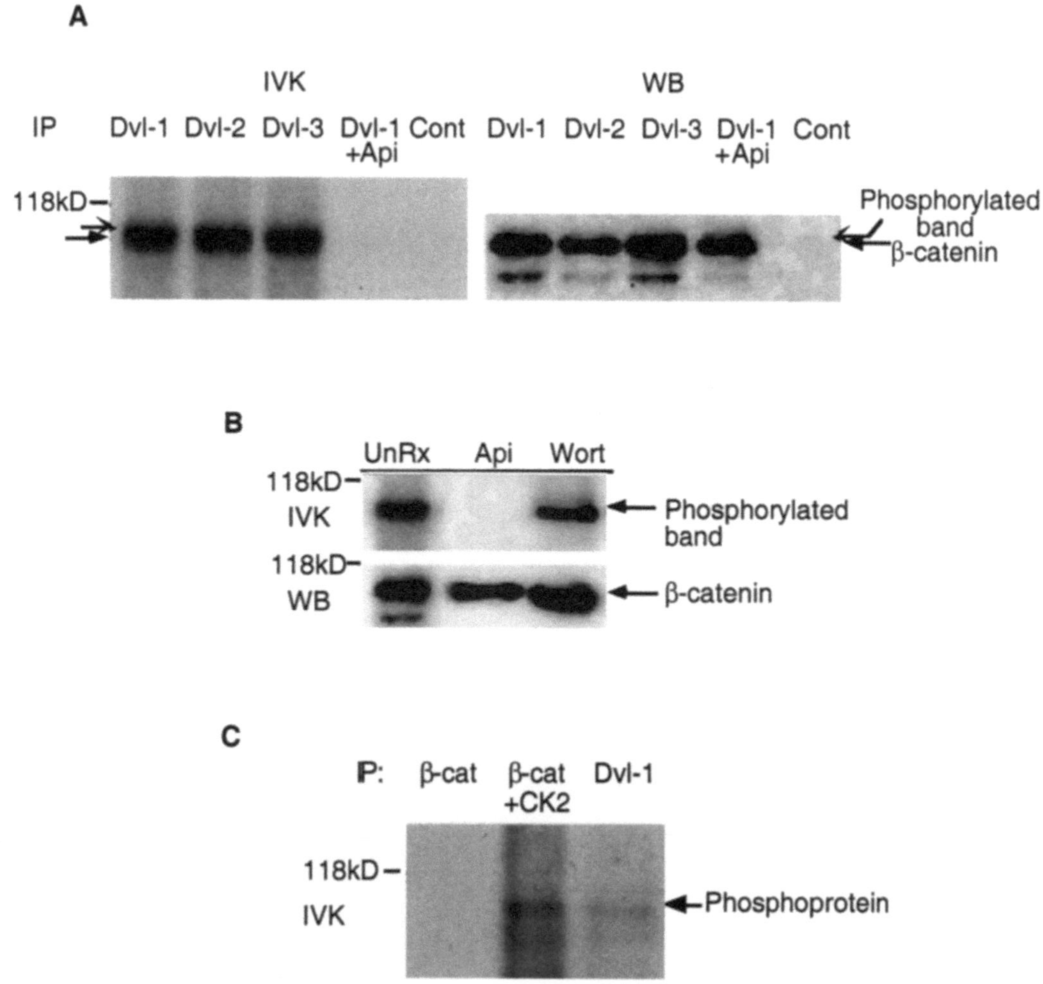

Fig. 8. CK2 phosphorylates β-catenin. (A) An *in vitro* kinase reaction was performed on thoroughly washed Dvl-1, 2, or 3 immunoprecipitates using [γ-^{32}P]GTP as a phosphate donor. The reaction was stopped by the addition of 2× sample buffer, electrophoresed, and transferred onto a nitrocellulose membrane. The [γ-^{32}P]GTP labeled proteins were identified by autoradiography (left panel). Western blotting of the same membrane was subsequently performed with the β-catenin antibody (right panel). A single major phosphoprotein (regular arrow) migrating slightly more slowly than β-catenin (triangular arrow) was seen in the *in vitro* kinase reaction. Phosphorylation of this protein was completely abolished in the presence of 80 μM apigenin (Dvl-1 + Api). Control immunoprecipitates (Cont) lacked both the phosphoprotein and β-catenin. (B) Since CK2 is one of the few kinases capable of utilizing GTP as a phosphate donor, the apigenin-inhibitable enzyme active in the *in vitro* kinase reaction was likely to be CK2. Since apigenin can also inhibit PI3-kinase *in vitro*, we used an authentic PI3-kinase inhibitor, Wortmannin, and found no decrease in the phosphoprotein band. Western blot with β-catenin antibody (bottom panel) documents the presence of β-catenin protein in each immunoprecipitate. (C) When recombinant CK2 was added to the β-catenin immunoprecipitate, a phosphoprotein having same mobility as endogenously phosphorylated phosphoprotein seen in the Dvl immunoprecipitate was seen.

(Fig. 8). The CK2 inhibitor apigenin completely eliminated phosphorylation of the 97 kd band in the *in vitro* kinase reaction suggesting that the endogenous CK2 in the immunoprecipitate was responsible for generation of 97 kd phosphoprotein. This phosphoprotein migrated in polyacrylamide gels slightly slower than immunoreactive β-catenin detected on Western blotting of the same gel, consistent with the phosphorylated form of β-catenin [31]. Dvl-1 migrates slightly slower than the 97 kd phosphoprotein while Dvl-2 and Dvl-3 run faster. Phosphorylation of each of the Dvl proteins should yield three distinct band sizes, rather single sized phosphoprotein bands were detected all three Dvl immunoprecipitates. These data suggested that the phosphoprotein might be β-catenin. One report suggests that apigenin may also inhibit the lipid kinase PI3K *in vitro* [32], however Wortmannin, an authentic PI3K inhibitor, did not affect phosphorylation of the 97 kd phosphoprotein band, nor is there any evidence that PI3K can utilize GTP as a phosphate donor. As previously stated, immunoprecipitate of β-catenin by polyclonal β-catenin antibody did not bring down CK2 nor Dvl proteins. Thus, *in vitro* kinase reaction of β-catenin immunoprecipitate alone could not generate a phosphoprotein as expected. While the β-catenin immunoprecipitates do not contain CK2 or Dvl, treatment of the immunoprecipitates

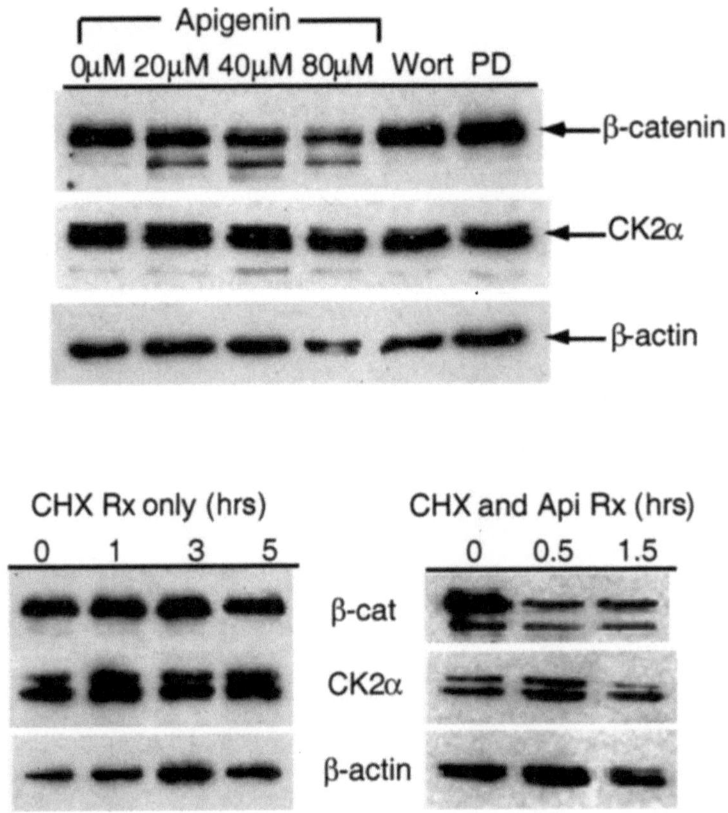

Fig. 9. Apigenin treatment accelerates the rate of degradation of β-catenin. Wnt-1-expressing cells were treated for 2 h with increasing doses of apigenin and whole cell lysates were prepared. 30 μg of protein were loaded onto a polyacrylamide gel and Western blotted with specific primary antibodies. Increasing doses of apigenin progressively decreased the protein levels of β-catenin, while levels of CK2 and β-actin remained constant. The PI3-kinase inhibitor Wortmannin at 25 nM and the MAP kinase inhibitor PD98059 at 20 μM did not have any effect on the levels of these proteins. To measure half life of β-catenin, cycloheximide was employed to block new protein synthesis. Western blotting was used to measure changes in β-catenin protein level over time. In the absence of apigenin (left panels), β-catenin protein was quite stable, with half life of more than 5 h. In the presence of apigenin (right panel), β-catenin protein level rapidly declined, with half life of less than 30 min.

with exogenous recombinant CK2 results in a phosphoprotein of identical mobility.

Apigenin causes destabilization of β-catenin

Having shown that apigenin treatment blocks phosphorylation of the 97 kd phosphoprotein in an *in vitro* kinase reaction, we treated cells with apigenin to determine whether this would affect the stability of β-catenin protein *in vivo*. In a dose-dependent manner, we found a dramatic decrease in the levels of β-catenin, while CK2 and β-actin protein levels were not affected (Fig. 9). Protein levels of β-catenin were not altered by thr PI3 kinase inhibitor Wortmannin or the MAPK inhibitor PD98059. To determine whether the reduction in β-catenin occurred through a decreased rate of synthesis or increased rate of degradation, we measured the half life of β-catenin in the presence of cycloheximide that blocked new protein synthesis. We observed that β-catenin is quite stable in Wnt-1 expressing cells, with a half life of more than 5 h (Fig. 9). However, in the presence of apigenin, β-catenin protein level rapidly declined to less than 50% of baseline and then plateau, suggesting that not all pools of β-catenin are regulated via apigenin-sensitive phosphorylation.

Discussion

Since the discovery of this tetrameric serine-threonine kinase capable of phosphorylating casein, a plethora of studies have demonstrated the remarkable conservation of CK2, its ubiquitous expression, and its ability to phosphorylate many important substrates in cells. The essential functions of CK2 in normal cell physiology and in development are being elucidated in many laboratories through a variety of genetic and biochemical approaches, as described in other papers in this volume. In addition to the physiologic importance of this regulatory enzyme, the consistent overexpression of CK2 seen in human and animal tumors suggests that CK2 may have a role in cancer pathogenesis. In previous and ongoing studies, we have studied CK2 in lymphocytes and identified pathways and collaborating events in lymphocyte transformation, based upon transgenic and *in vitro* studies. Here we review a series of experiments on CK2 in mammary tumorigenesis. A previous study suggested that CK2 is highly expressed in human breast cancer, as immunohistochemical staining for CK2 was much greater in tumors than in normal mammary tissue [33]. We have confirmed these results with a more quantitative approach using Western blotting and kinase activity measurements. Furthermore, in a standard rat model in which mammary tumors are produced by the administration of the polycyclic aromatic hydrocarbon DMBA [7, 34], upregulation of CK2 protein and activity was again evident.

To test the hypothesis that the overexpression of CK2 may directly contribute to mammary tumorigenesis, we used the MMTV-LTR to misexpress CK2α in the mammary epithelium of transgenic mice. Females from 4 independent transgenic lines exhibited histological abnormalities of the mammary gland, such as defects in involution and dysplasia. Dysplasic lesions are uncommon in wild type FVB/N mice and are thought to be precancerous [18]. Thirty percent of the female transgenics went on to develop mammary adenocarcinomas; in FVB/N mice these are rare as spontaneous malignancies [17]. Furthermore, the bulk of the histological subtypes in our transgenics are glandular and papillary adenocarcinomas, and spindle cell carcinomas, while the rare spontaneous tumors in FVB/N are keratoacanthomas or squamous cell carcinomas. Thus, mammary hyperplasia, dysplasia, and transformation were likely caused by transgenic expression of CK2α.

The tumors had very high levels of CK2 protein and activity, suggesting that CK2 activity was required for ongoing tumor growth. This is supported by experiments with a selective pharmacologic inhibitor of CK2, the flavonoid apigenin, which inhibits growth of the CK2α transgenic breast cancer cell lines *in vitro*. Experiments using other selective CK2 inhibitors to confirm these results are underway in the laboratory.

The variety of histologic patterns of the transgenic tumors is consistent with a variable and presumably stochastic activation of cooperating or downstream molecular pathways. The requirement for multiple additional steps to mammary transformation was also suggested by the long latency of tumor onset. We screened for such events based upon prior data about CK2-dependent pathways in growth control and cancer. In the CK2α transgenic breast tumors, upregulation of the Wnt/β-catenin pathway and of NF-κB occurred frequently. The role of the Wnt/β-catenin pathway is well-established in mouse mammary tumors [35–39], and in humans it is emerging as an increasingly important pathway to breast cancer [25, 37].

The NFκB pathway has recently been documented to also play an important role in early mammary tumorigenesis [7, 27]. Signal-dependent activation of NFκB occurs in the cytoplasm via phosphorylation of IκBα by the IκB kinases IKKα and IKKβ, which frees the NFκB subunits, allowing them to translocate to the nucleus and be transcriptionally active (reviewed in 19]. CK2 phosphorylates the C-terminal PEST domain of IκBα constitutively, and this has been found to also play a significant role in regulating IκB turnover, particularly with respect to regulation of basal levels of nuclear NF-κB [20–23, 41–44]. CK2 also phosphorylates the p65 subunit of NF-κB directly [45]. The transgenic CK2α mammary tumors have active nuclear NF-κB, suggesting cooperation between CK2 and NF-κB in mammary tumorigenesis.

c-*myc* is an important oncogene regulated by both Wnt/β-catenin signaling and by NF-κB [11, 29, 30]. Bi-transgenic expression of CK2α and c-*myc* in lymphocytes has demonstrated the potent transforming potential of co-expression of these genes [2]. c-Myc was well-expressed in the CK2α mammary tumors. On the other hand, although CK2α and *p53* loss are also collaborative in lymphoid transformation, we found evidence of *p53* loss in only 2 of the transgenic mammary tumors (data not shown).

Based on these findings and upon the prior report of CK2 involvement in the Wnt pathway, we have carried out studies on the role of CK2 in Wnt signaling. In Wnt-1 transfected or infected C57MG mouse mammary epithelial cells, the Wnt-dependent transcriptional co-factor β-catenin is upregulated as is CK2, at the level of mRNA, protein, and activity. CK2 was found associated with β-catenin and the upstream signaling molecule dishevelled in the mammary epithelial cells. β-catenin itself is a substrate for CK2 in this complex. This association appears to be important functionally, as CK2 activity is essential for maintenance of β-catenin and Dvl protein levels: treatment of the cells with the selective CK2 inhibitor apigenin dramatically accelerates the degradation of β-catenin and Dvl. Although apigenin has been reputed to also inhibit PI3K and MAPK, Wortmannin and PD98059, do not have the same effects. Studies are underway to identify the site of phosphorylation in β-catenin and determine how phosphorylation alters its stability and co-transcriptional activity in cells.

Acknowledgments

We thank Dr. Anthony M. C. Brown for kindly providing the pMV7-Wnt-1 and C57MG cells, and Dr. Bert Vogelstein for providing pCI-β-catenin construct. We thank Drs. Adrianne Rogers, Carol Rosenberg, and Abdul Traish in the Departments of Pathology and Medicine at Boston University School of Medicine for providing breast tumor and tissue specimens from rats and from human patients and Judy E. Walls in the

Transgenic Mouse Core Histopathology Laboratory at U.C. Davis School of Medicine for outstanding technical work on the histology and immunohistochemistry. This work was supported by the Massachusetts Department of Public Health Breast Cancer Training Program (E.L.-B., D.H.S.); by the Association pour la Recherche sur le Cancer and the Fondation Bettencourt-Schueller (R.R-M.); the Department of Defense DAMD 17-98-1 (G.E.S., D.C.S.); NIH RO1 CA 82742 (G.E.S., D.C.S.); NIH R01 CA63929 (D.J.S.) ;and by the University of California Breast Cancer Research Program grant #JB-0014 (R.D.C.).

References

1. Xu X, Toselli PA, Russell LD, Seldin DC: Globozoospermia in mice lacking the casein kinase II alpha' catalytic subunit. Nat Genet 23: 118–121, 1999
2. Seldin DC, Leder P: Casein kinase II alpha transgene-induced murine lymphoma: Relation to theileriosis in cattle. Science 267: 894–897, 1995
3. Kelliher MA, Seldin DC, Leder P: Tal-1 induces T cell acute lymphoblastic leukemia accelerated by casein kinase IIalpha. Embo J 15: 5160–5166, 1996
4. Landesman-Bollag E, Channavajhala PL, Cardiff RD, Seldin DC: p53 deficiency and misexpression of protein kinase CK2alpha collaborate in the development of thymic lymphomas in mice. Oncogene 16: 2965–2974, 1998
5. Allende JE, Allende CC: Protein kinases. 4. Protein kinase CK2: An enzyme with multiple substrates and a puzzling regulation. FASEB J 9: 313–323, 1995
6. Pinna LA, Meggio F: Protein kinase CK2 ('casein kinase-2') and its implication in cell division and proliferation. Prog Cell Cycle Res 3: 77–97, 1997
7. Kim DW, Sovak MA, Zanieski G, Nonet G, Romieu-Mourez R, Lau AW, Hafer LJ, Yaswen P, Stampfer M, Rogers AE, Russo J, Sonenshein GE: Activation of NF-kappaB/Rel occurs early during neoplastic transformation of mammary cells. Carcinogenesis 21: 871–879, 2000
8. Sinn E, Muller W, Pattengale P, Tepler I, Wallace R, Leder P: Coexpression of MMTV/v-Ha-ras and MMTV/c-myc genes in transgenic mice: Synergistic action of oncogenes *in vivo*. Cell 49: 465–475, 1987
9. http://www-mp.ucdavis.edu/tgmice/firststop.html
10. Kuenzel EA, Mulligan JA, Sommercorn J, Krebs EG: Substrate specificity determinants for casein kinase II as deduced from studies with synthetic peptides. J Biol Chem 262: 9136–9140, 1987
11. Duyao MP, Buckler AJ, Sonenshein GE: Interaction of an NF-kappa B-like factor with a site upstream of the c-myc promoter. Proc Natl Acad Sci U S A 87: 4727–4731, 1990
12. Arsura M, FitzGerald MJ, Fausto N, Sonenshein GE: Nuclear factor-kappaB/Rel blocks transforming growth factor beta1-induced apoptosis of murine hepatocyte cell lines. Cell Growth Differ 8: 1049–1059, 1997
13. Huggins C, Grand LC, Brillantes FP: Mammary cancer induced by a single feeding of polynuclear hydrocarbons, and its suppression. Nature 189: 204–207, 1961
14. Rogers, BH: Dimethylbenzanthracene-induced mammary tumorigenesis in ethanol-fed rats. Nutr Res 10: 143–146, 1990
15. Stewart TA, Pattengale PK, Leder P: Spontaneous mammary adenocarcinomas in transgenic mice that carry and express MTV/myc fusion genes. Cell 38: 627–637, 1984
16. Cardiff RD: The biology of mammary transgenes: Five rules. J Mammary Gland Biol Neoplas 1: 61–73, 1996
17. Mahler JF, Stokes W, Mann PC, Takaoka M, Maronpot RR: Spontaneous lesions in aging FVB/N mice. Toxicol Pathol 24: 710–716, 1996
18. Cardiff RD, Wellings SR: The comparative pathology of human and mouse mammary glands. J Mammary Gland Biol Neoplas 4: 105–122, 1999
19. Willert K, Brink M, Wodarz A, Varmus H, Nusse R: Casein kinase 2 associates with and phosphorylates dishevelled. Embo J 16: 3089–3096, 1997
20. Pando MP, Verma IM: Signal-dependent and -independent degradation of free and NF-kappa B-bound Ikappa Balpha. J Biol Chem 275: 21278–21286, 2000
21. Janosch P, Schellerer M, Seitz T, Reim P, Eulitz M, Brielmeier M, Kolch W, Sedivy JM, Mischak H: Characterization of IkappaB kinases. IkappaB-alpha is not phosphorylated by Raf-1 or protein kinase C isozymes, but is a casein kinase II substrate. J Biol Chem 271: 13868–13874, 1996
22. Schwarz EM, Van AD, Verma IM: Constitutive phosphorylation of IkappaBalpha by casein kinase II occurs preferentially at serine 293: Requirement for degradation of free IkappaBalpha. Mol Cell Biol 16: 3554–3559, 1996
23. McElhinny JA, Trushin SA, Bren GD, Chester N, Paya CV: Casein kinase II phosphorylates I kappa B alpha at S-283, S-289, S-293, and T-291 and is required for its degradation. Mol Cell Biol 16: 899–906, 1996
24. Foo SY, Nolan GP: NF-kappaB to the rescue: RELs, apoptosis and cellular transformation. Trends Genet 15: 229–235, 1999
25. Nakshatri H, Bhat-Nakshatri P, Martin DA, Goulet RJ Jr, Sledge GW Jr: Constitutive activation of NF-kappaB during progression of breast cancer to hormone-independent growth. Mol Cell Biol 17: 3629–3639, 1997
26. Romieu-Mourez R, Landesman-Bollag E, Seldin DC, Traish AM, Mercurio F, Sonenshein GE: Roles of IKK kinases and protein kinase CK2 in activation of NF-kB in breast cancer. Cancer Res (in press)
27. Sovak MA, Bellas RE, Kim DW, Zanieski GJ, Rogers AE, Traish AM, Sonenshein GE: Aberrant nuclear factor-kappaB/Rel expression and the pathogenesis of breast cancer. J Clin Invest 100: 2952–2960, 1997
28. Howard DK, Schlom J, Fisher PB: Chemical carcinogen-mouse mammary tumor virus interactions in cell transformation. In Vitro 19: 58–66, 1983
29. Kessler DJ, Duyao MP, Spicer DB, Sonenshein GE: NF-kappa B-like factors mediate interleukin 1 induction of c-myc gene transcription in fibroblasts. J Exp Med 176: 787–792, 1992
30. He TC, Sparks AB, Rago C, Hermeking H, Zawel L, da Costa LT, Morin PJ, Vogelstein B, Kinzler KW: Identification of c-MYC as a target of the APC pathway. Science 281: 1509–1512, 1998
31. Serres M, Grangeasse C, Haftek M, Durocher Y, Duclos B, Schmitt D: Hyperphosphorylation of beta-catenin on serine-threonine residues and loss of cell-cell contacts induced by calyculin A and okadaic acid in human epidermal cells. Exp Cell Res 231: 163–172, 1997
32. Agullo G, Gamet-Payrastre L, Manenti S, Viala C, Remesy C, Chap H, Payrastre B: Relationship between flavonoid structure and inhibition of phosphatidylinositol 3-kinase: A comparison with tyrosine kinase and protein kinase C inhibition. Biochem Pharmacol 53: 1649–1657, 1997
33. Munstermann U, Fritz G, Seitz G, Lu YP, Schneider HR, Issinger OG: Casein kinase II is elevated in solid human tumours and rapidly proliferating non-neoplastic tissue. Eur J Biochem 189: 251–257, 1990
34. Ohi Y, Yoshida H: Influence of estrogen and progesterone on the induction of mammary carcinomas by 7,12-dimethylbenz(a)anthracene in ovariectomized rats. Virchows Arch B Cell Pathol Incl Mol Pathol 62: 365–370, 1992
35. Tsukamoto AS, Grosschedl R, Guzman RC, Parslow T, Varmus HE: Expression of the int-1 gene in transgenic mice is associated with

mammary gland hyperplasia and adenocarcinomas in male and female mice. Cell 55: 619–625, 1988
36. Barker N, Huls G, Korinek V, Clevers H: Restricted high level expression of Tcf-4 protein in intestinal and mammary gland epithelium. Am J Pathol 154: 29–35, 1999
37. Bui TD, Rankin J, Smith K, Huguet EL, Ruben S, Strachan T, Harris AL, Lindsay S: A novel human Wnt gene, WNT10B, maps to 12q13 and is expressed in human breast carcinomas. Oncogene 14: 1249–1253, 1997
38. Dale TC, Weber-Hall SJ, Smith K, Huguet EL, Jayatilake H, Gusterson BA, Shuttleworth G, O'Hare M, Harris AL: Compartment switching of WNT-2 expression in human breast tumors. Cancer Res 56: 4320–4323, 1996
39. Huguet EL, McMahon JA, McMahon AP, Bicknell R, Harris AL: Differential expression of human Wnt genes 2, 3, 4, and 7B in human breast cell lines and normal and disease states of human breast tissue. Cancer Res 54: 2615–2621, 1994
40. Israel A: The IKK complex: An integrator of all signals that activate NF-kappaB? Trends Cell Biol 10: 129–133, 2000
41. Bren GD, Pennington KN, Paya CV: PKC-zeta-associated CK2 participates in the turnover of free IkappaBalpha. J Mol Biol 297: 1245–1258, 2000
42. Heilker R, Freuler F, Pulfer R, Di Padova F, Eder J: All three IkappaB isoforms and most Rel family members are stably associated with the IkappaB kinase 1/2 complex. Eur J Biochem 259: 253–261, 1999
43. Tran K, Merika M, Thanos D: Distinct functional properties of IkappaB alpha and IkappaB beta. Mol Cell Biol 17: 5386–5399, 1997
44. Lin R, Beauparlant P, Makris C, Meloche S, Hiscott J: Phosphorylation of IkappaBalpha in the C-terminal PEST domain by casein kinase II affects intrinsic protein stability. Mol Cell Biol 16: 1401–1409, 1996
45. Bird TA, Schooley K, Dower SK, Hagen H, Virca GD: Activation of nuclear transcription factor NF-kappaB by interleukin-1 is accompanied by casein kinase II-mediated phosphorylation of the p65 subunit. J Biol Chem 272: 32606–32612, 1997

Molecular and Cellular Biochemistry **227**: 167–174, 2001.

Response of cancer cells to molecular interruption of the CK2 signal

Huamin Wang, Alan Davis, Shihui Yu and Khalil Ahmed
Cellular and Molecular Biochemistry Research Laboratory (151), Department of Laboratory Medicine and Pathology and University of Minnesota Cancer Center, University of Minnesota and the Department of Veterans Affairs Medical Center, Minneapolis, MN, USA

Abstract

Protein kinase CK2 is one of the key cellular signals for cell survival, growth, and proliferation. It is has been observed to be elevated in various cancers that have been examined. Various observations suggest that moderate dysregulation of CK2 may profoundly influence the cell response. We have examined the effects of interfering with the CK2 signal in various cancer cell lines by employing antisense oligodeoxynucleotides (ODN) against the α and β subunits of CK2. Our results demonstrate that antisense CK2-α and antisense CK2-β ODNs markedly influence cell viability of these cancer cells in a dose and time-dependent manner. Antisense CK2-α was slightly more effective than antisense CK2-β in most of the cells tested. The efficacy of the antisense ODN seemed to vary with the cell type; however, in all cases potent induction of apoptosis was observed. Significantly, the effects of the antisense ODN on the CK2 activity in the nuclear matrix were relatively small compared to the much stronger induction of apoptosis in cells. This suggests that modest downregulation of CK2 can evoke a much greater apoptotic response in cancer cells. (Mol Cell Biochem **227**: 167–174, 2001)

Key words: protein kinase CK2, antisense, nuclear matrix, prostate cancer, head and neck cancer, gene therapy

Introduction

Protein kinase CK2 is a ubiquitous messenger-independent protein serine/threonine kinase that has been implicated in multiple functions in the cell including control of cell growth and proliferation and apoptosis (for reviews see e.g. [1–4]). CK2 is a highly conserved enzyme, and has been suggested to be essential for cell survival [6]. The kinase is a heterotetramer consisting of two catalytic subunits (α and/or α') complexed with two β subunits. A major mode of its regulation relating to the control of cell growth is through its signaling in the nucleus where nuclear matrix and chromatin appear to be its preferential targets (see e.g. [1]).

Protein kinase CK2 has been strongly implicated in the neoplastic process because of its elevation in the various tumors that have been examined [3, 4]. Studies involving transgenic expression of CK2 have provided evidence to suggest that modest overexpression of CK2 imparts an oncogenic potential to the cells such that the incidence of neoplasia is greatly enhanced by CK2-α in conjunction with the presence of another oncogenic signal [6]. In studies of the human squamous cell carcinoma of the head and neck and of prostate cancer, it was found that the kinase appeared to be associated with aggressive tumor behavior [7–9]. It is noteworthy that studies of the expression of the kinase in tumor cells in a given specimen have revealed that the elevation in the kinase is not solely a reflection of the proliferation status of tumor cells but may also relate to their pathobiological status [10]. The recent observation that overexpression of CK2 in the cells causes a reduction in the chemical-mediated apoptosis further underscores the importance of the increased levels of CK2 in tumor cells [3,11].

Given these various considerations, we investigated CK2 as a molecular target using an *in vitro* antisense oligodeoxynucleotide (ODN) approach in head and neck cancer cells [12]. These studies showed that treatment of head and neck cancer cells with antisense to the α subunit of CK2 resulted in a reduction of the chromatin-associated enzyme activity,

Address for offprints: K. Ahmed, Cellular and Molecular Biochemistry Research Laboratory (151), Veterans Affairs Medical Center, One Veterans Drive, Minneapolis, MN 55417, USA (E-mail: ahmedk@tc.umn.edu)

and inhibition of cell growth in a dose-dependent manner [12]. We have now expanded these studies to include different cell lines derived from prostatic adenocarcinoma and squamous cell carcinoma of the head and neck (SCCHN), and have determined the nature of effects of various antisense ODN to CK2 subunits on these cells. These data demonstrate that antisense ODN to CK2 can induce apoptosis in cancer cells. Both the antisense ODN to the α and β subunits are effective in inducing apoptosis although the antisense ODN to the α subunit appears to be a slightly more potent inducer of apoptosis than that to the β subunit, especially in some of the cell types examined. These results suggest that CK2 may serve as a target for induction of apoptosis in cancer cells and thus may offer a novel anticancer strategy.

Materials and methods

Chemicals

Synthetic dodecapeptide substrate (RRRADDSDDDDD) for assay of CK2 activity [11] was purchased from Sigma-Genosys (The Woodlands, TX, USA). Etoposide, and Mayer's hematoxylin solution, were purchased from Sigma Chemical Co. (St. Louis, MO, USA). Cell proliferation reagent WST-1 and the Cell Death Detection Kit (TUNEL staining) were supplied by Roche Molecular Biochemicals (Indianapolis, IN, USA). Anti-CK2α antibody was purchased from Santa Cruz Biotechnology (Santa Cruz, CA, USA). All reagents were of the highest purity available.

Oligodeoxynucleotides

The sequences of the CK2-α and CK2-β antisense ODN were based on the cDNA sequences of the α and β subunits of CK2 [13]. The sequence for the antisense CK2-α ODN was 5′-GTC CCG ACA TGT CAG ACA GG-3′. The corresponding sense and nonsense sequences used as a control were 5′-CCT GTC TGA CAT GTC GGG AC-3′, and 5′-CTC GAG GTC GAC GGT ATC GA-3′, respectively. The nonsense sequence was a random sequence synthesized to maintain the overall A + T/G + C composition of the antisense CK2-α ODN. The antisense and sense sequences for ODN to the β subunit were 5′-GAG CTG CTC ATC TTC ACG TC-3′ and 5′-GAC GTG AAG ATG AGC AGC TC-3′, respectively. Standard and phosphorothioate ODN were obtained from MWG Biotech (High Point, NC, USA).

Cell culture

The various cell lines employed in these studies were LNCaP, PC-3 and ALVA-41 derived from human prostate adenocarcinoma, and Ca9-22 and NA from human head and neck squamous cell carcinoma. LNCaP and PC-3 cells were purchased from American Type Culture Collection (Manassas, VA, USA), whereas ALVA-41 cell line was obtained from Dr. Richard C. Ostenson (University of Washington, Seattle, WA, USA), and Ca9-22 and NA cell lines from Dr. T. Kuroki (University of Tokyo, Tokyo, Japan). The LNCaP, PC-3, and NA cells were maintained in RPMI 1640 (Life Technologies, Grand Island, NY, USA) supplemented with 2 mM L-glutamine, 25 mM HEPES, and 10% defined FBS (Hyclone Laboratories, Logan, UT, USA) in an atmosphere containing 5% CO_2. Culture medium for LNCaP cells also included 10^{-10} M 5α-dihydrotestosterone. The ALVA-41 and Ca9-22 cell lines were maintained in RPMI 1640 supplemented with 2 mM L-glutamine, 25 mM HEPES, and 6% defined FBS in an atmosphere containing 5% CO_2.

Transfection with ODN

ODN was introduced into the cells by transfection using DOTAP liposomal transfection reagent from Roche Molecular Biochemicals (Indianapolis, IN, USA). The ODN stock solution was prepared according to the manufacturer's instructions. Culture medium was added to obtain a final ODN concentration of 2 μg/ml and 10 μl/ml DOTAP. The original cell culture medium was removed and replaced with this mixture for 6 h. After this time the ODN containing medium was removed and replaced with fresh original medium. The cells were allowed to incubate in this medium until the appropriate time for the cell viability or apoptosis evaluations.

Cell viability and proliferation assay

The cell proliferation reagent WST-1 (a tetrazolium salt that is cleaved by mitochondrial dehydrogenases in viable cells) was employed to determine cell viability and proliferation in cells treated with various antisense ODN. An aliquot (100 μl) of cell suspension (containing $0.5–2 \times 10^4$ cells) was placed in each well of a 96-well plates, and cells were allowed to reattach over a period of 24 h. Various ODN were added to the culture medium to achieve the desired concentration using the DOTAP transfection method. Following the treatment of cells with various ODN for the indicated periods of time, WST-1 (10 μl) was added to each well and incubation was carried out at 37°C for an additional 30 min. An automated plate reader was employed to measure OD_{450}.

Measurement of apoptosis

Apoptosis in cells following treatments with various ODN was determined by using the Cell Death Detection Kit (TUNEL staining) employed as per instructions of the supplier. After completing the TUNEL staining procedure the cells were counterstained by dipping the slides in Mayer's hematoxylin solution for 30 sec. They were then rinsed twice with Tris buffered saline (TBS, consisting of 10 mM Tris-HCl and 0.9% NaCl, pH 7.4), dipped in 0.1% NH_4OH for 2 sec, and finally washed several more times with TBS. The stained cells were mounted with Biomeda's Crystal/Mount obtained from Fisher Scientific (Chicago, IL, USA). Digital images of the cells were obtained using a Zeiss Axiovert 25 CFL microscope equipped with a color digital CCD camera. The images were prepared for presentation using Adobe's Photoshop 5.0 software.

Measurement of CK2 activity

CK2 activity was measured in the nuclear matrix (NM) fraction isolated from cells following various treatments. Isolation of the NM and CK2 associated with it was determined by employing the CK2-specific dodecapeptide substrate as detailed previously [11].

Western blot analysis

Samples were denatured by heating at 95°C for 5 min in a sample buffer consisting of 10 mM sodium phosphate buffer (pH 7.0), 4 M urea, 2.5% SDS, and 1% 2-mercaptoethanol, and were subjected to SDS/Urea/10% polyacrylamide gel electrophoresis as described previously [14, 15]. The separated proteins were transferred from the gel to a nitrocellulose sheet. After blocking the sheet with a medium consisting of 10 mM Tris-HCl (pH 7.4), 0.9% NaCl, and 3% dry milk, the blot was successively incubated with goat anti-CK2-α IgG and rabbit anti-goat IgG-alkaline phosphatase-conjugated antibody. Immobilized alkaline phosphatase was visualized using 5-bromo-4-chloro-indolyl phosphate and Nitro Blue Tetrazolium as described previously [14, 15].

Results

In previous work, employing the head and neck cancer cell line Ca9-22, we demonstrated that treatment of those cells with antisense CK2-α ODN resulted in inhibition of cell growth [12]. These studies were expanded to determine the effects of antisense ODN corresponding to the α and β subunits of CK2 on the growth response of several other tumor cell lines, and to characterize the nature of these effects. The results in Fig. 1 show that treatment of the androgen-dependent prostate cancer ALVA-41 cells with the CK2-α antisense ODN

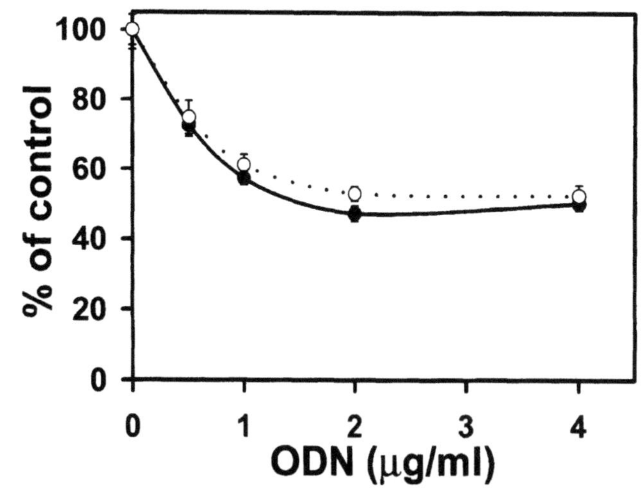

Fig. 1. Effect of antisense CK2-α and antisense CK2-β ODNs on ALVA-41 cell growth. Dose response of ALVA-41 cells employing antisense CK2-α ODN (●) and antisense CK2-β ODN (○) is shown. Cells were treated with the various ODN for a period of 6 h and then transferred to fresh media without the ODN and allowed to grow for 48 h at which time the cell viability was determined using the WST-1 assay. The results are expressed as percent of control (DOTAP treatment only).

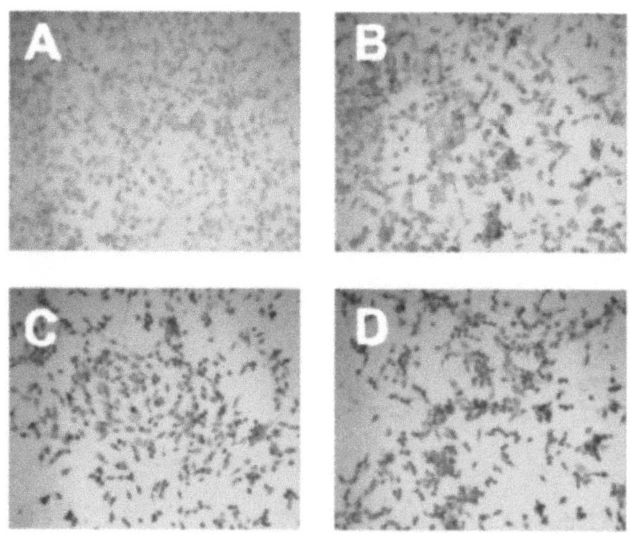

Fig. 2. TUNEL staining of ALVA-41 cells following treatment with antisense CK2-α ODN at different concentrations. Cells were treated with antisense CK2-α ODN at varying concentrations for a period of 6 h following which they were transferred to the fresh media (without the antisense ODN) and allowed to grow for 48 h. At this time, they were fixed and stained using the Cell Death Detection Kit (TUNEL assay), and then counterstained with Meyer's hematoxylin solution. (A) untreated control; (B) 1.5 μg/ml of antisense CK2-α ODN; (C) 3.0 μg/ml of antisense CK2-α ODN; (D) 6.0 μg/ml of antisense CK2-α ODN.

and CK2-β antisense ODN for 48 h resulted in a dose-dependent loss of cell viability such that the effect was near maximal for both type of antisense ODNs at a concentration between 1 and 2 μg/ml.

To determine the nature of the effects of the CK2 antisense ODNs on the ALVA-41 cells, they were subjected to TUNEL staining at 48 h following treatment with varied amounts of the CK2-α antisense ODN. The results in Fig. 2 (Panels A–D) demonstrate that there was a dose-dependent induction of apoptosis in cells treated with 1.5, 3.0 and 6.0 μg/ml of antisense CK2-α ODN as evidenced by the TUNEL staining. Thus, the change in the number of viable cells measured in Fig. 1 reflected the loss of cells through apoptosis as a result of the treatment with the antisense CK2 ODNs.

Further analysis of the ALVA-41 cells treated for varied times with CK2-α antisense ODN was carried out. For this, the cells were transfected with 2.0 μg/ml of the antisense CK2-α ODN for the initial period of 6 h, following which they were placed in fresh media without the antisense ODN. The results in Fig. 3 provide the evidence of progressive apoptosis in these cells when examined after 24, 48, and 72 h by TUNEL staining. Thus, the effect of the initial treatment of the cells with the antisense CK2-α ODN persists over time resulting in progressive induction of apoptosis.

In order to determine that the above described effects of the CK2 antisense were evoked by a reduction in the activity and level of the kinase, we examined the CK2 activity in the nuclear matrix preparations from ALVA-41 cells treated with the CK2-α antisense ODN. A representative result is presented in Table 1 which shows that the response of the

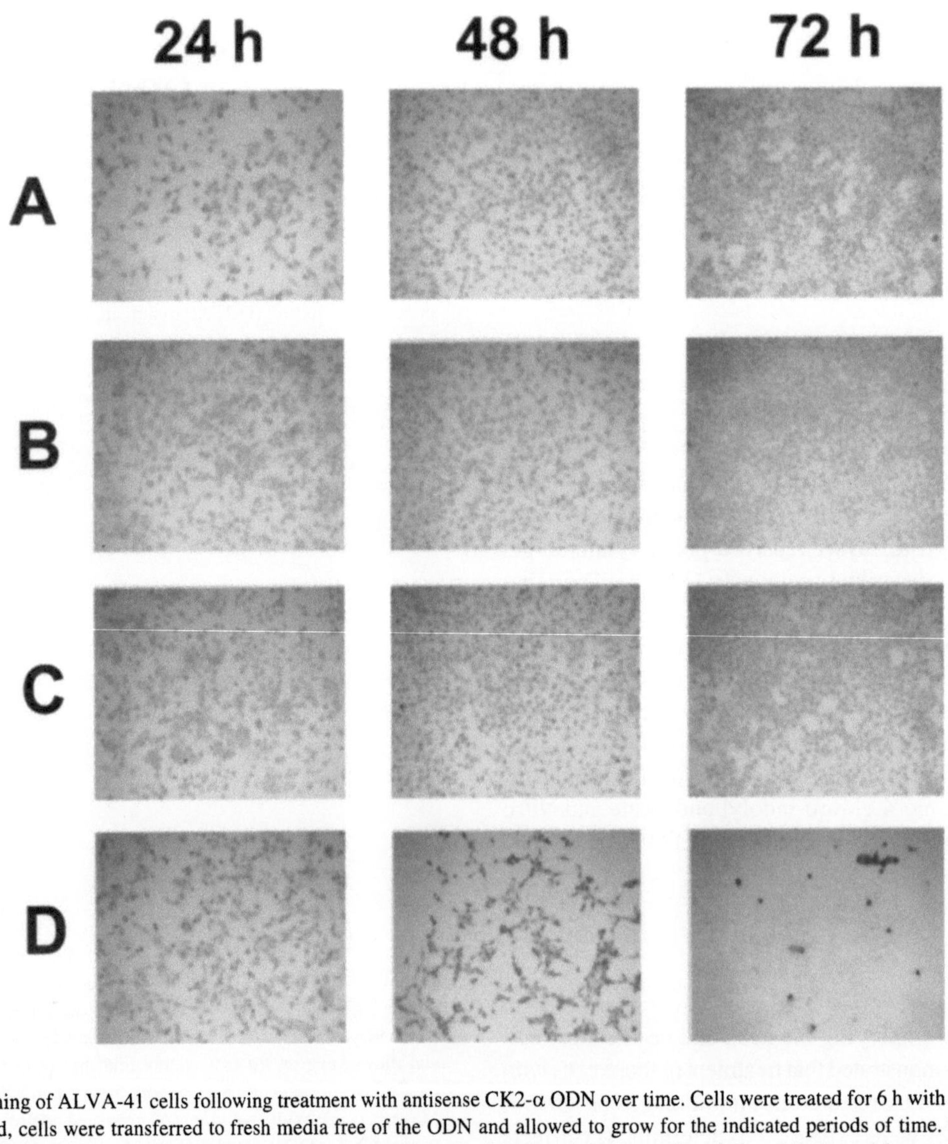

Fig. 3. TUNEL staining of ALVA-41 cells following treatment with antisense CK2-α ODN over time. Cells were treated for 6 h with the various ODN. At the end of this period, cells were transferred to fresh media free of the ODN and allowed to grow for the indicated periods of time. Cells were fixed and stained using the Cell Death Detection Kit (TUNEL) and then counterstained with Mayer's hematoxylin solution. (A) untreated control cells; (B) 2.0 μg/ml of nonsense ODN (C) 2.0 μg/ml of CK2-α sense ODN; (D) 2.0 μg/ml of CK2-α antisense ODN.

Table 1. CK2 activity in the nuclear matrix preparations from ALVA-41 cells treated with the CK2-α antisense ODN

Treatment of cells	CK2 activity
DOTAP only, control	29.8 ± 2.8
2 μg/ml nonsense ODN	28.6 ± 2.1
2 mg/ml CK2-α antisense ODN	18.2 ± 1.5

CK2 activity in the NM fractions from ALVA-41 cells represents nmol ^{32}P incorporated in the CK2 specific peptide/mg of protein/h. CK2 activity was determined at 48 h following the treatment with the indicated ODN, as described under Materials and methods.

CK2 activity to the antisense was relatively modest such that treatment with the ODN at a concentration of 5 μg/ml resulted in a loss of the NM-associated CK2 of the order of 33% after 48 h, whereas a much greater level of apoptosis was apparent even at concentrations of 1–2 μg/ml of the antisense CK2-α ODN. This is in accord with the previous observation that antisense CK2-α ODN produced a moderate effect on chromatin-associated CK2 activity in the Ca9-22 cells. In that case, the effect on the activity in the cytoplasmic fraction was affected to a lesser extent than that in the chromatin [12]. Immunoblot measurements of the level of CK2-α subunit protein following treatment with antisense CK2-α ODN was also carried out under similar conditions. The results show that concordant with the reduction in the enzyme activity shown in Table 1, there was a loss of the enzyme protein in the NM fractions after a period of 48 h, following the initial treatment of ALVA-41 cells with the 2 μg/ml of the antisense ODN (Fig. 4).

The nature of the effects of the antisense CK2-β ODN on the ALVA-41 cells was also examined. The results in Fig. 5 (Panels A–D) show that ALVA-41 cells treated with 2.0 μg/ml of the ODN gave clear evidence of induction of apoptosis as determined by TUNEL staining. The antisense CK2-β was, however, somewhat less effective than the antisense CK2-α in inducing apoptosis in these cells.

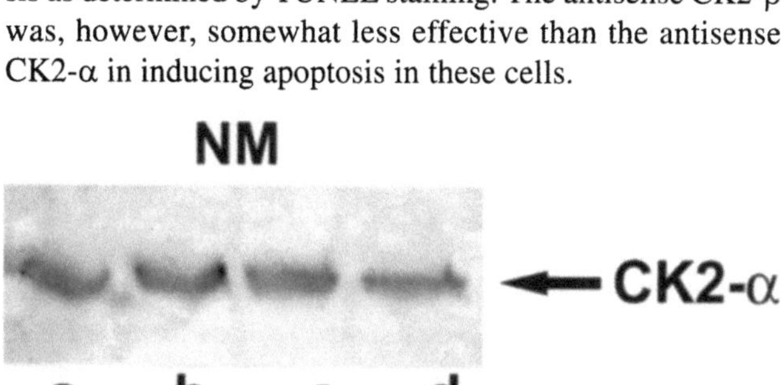

Fig. 4. Immunoassay of the Western blots of NM from LNCaP cells treated with antisense CK2-α ODN. NM was isolated from cells at 48 h following the treatment with the various ODN, as described in Fig. 2. Blots were incubated with the anti-CK2-α antibody and the immunostain was visualized as described in Materials and methods. Lane a = control (DOTAP only); lane b = 2.0 μg/ml of nonsense ODN; lane c = 2.0 μg/ml of CK2-α sense ODN; lane d = 2.0 μg/ml of CK2-α antisense ODN. The relative densitometric values for lanes a–c are 1.00, 0.79, 0.67, and 0.58, respectively.

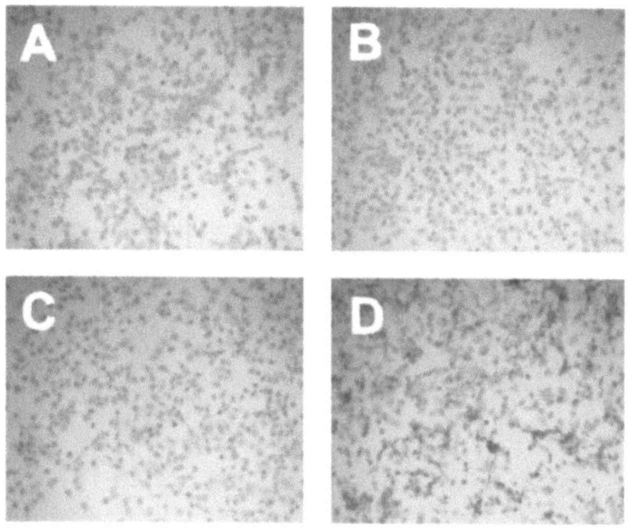

Fig. 5. TUNEL staining of ALVA-41 cells treated with antisense CK2-β ODN. TUNEL and Mayer's hematoxylin stainings were carried out at 48 h following the initial treatment with the ODN following the same experimental design as described above in legends to the other figures. (A) untreated control cells; (B) 2.0 μg/ml of the nonsense ODN; (C) 2.0 μg/ml of the CK2-β sense ODN; (D) 2.0 μg/ml of the CK2-β antisense ODN.

Analogous studies were undertaken on the effects of the antisense CK2-α ODN on several other cell lines. The results in Fig. 6 demonstrate the effects of antisense CK2-α ODN and antisense CK2-β ODN on LNCaP cells (another androgen-dependent prostate cancer cell line). Again, the viability of these cells was affected in a time-dependent manner in the presence of a single dose (5.0 μg/ml) of the antisense ODNs although these cells appeared to be somewhat more resistant to the effect of the antisense (Fig. 6, Panel A). Interestingly, these cells appeared to be equally sensitive to the antisense CK2-α or CK2-β ODN. That the effect on cell viability was a consequence of apoptosis in these cells was determined by TUNEL staining analysis which showed a progressive incidence of apoptosis in cells transfected with 2.0 or 5.0 μg/ml of the antisense CK2-α ODN for the initial period of 6 h, followed by maintenance in fresh medium in the absence of the ODN for 48 h (Fig. 6, Panel B, a–c). The effects of 2 μg/ml of antisense CK2-α or CK2-β ODN on the androgen-insensitive prostate cancer cells PC-3 were also examined at 48 h following the initial treatment (Fig. 6, Panel B, d–f). The TUNEL staining results clearly demonstrate the susceptibility of these cells to the antisense ODN treatment resulting in induction of apoptosis.

Similar experiments were undertaken with the squamous cell carcinoma of the head and neck cell lines Ca9-22 and NA. Representative results based on the effects of 5 μg/ml of antisense CK2-α and CK2-β ODNs at 48 h are presented in Fig. 7. The TUNEL staining of these cells clearly indicates the evidence of strong apoptosis in these cells.

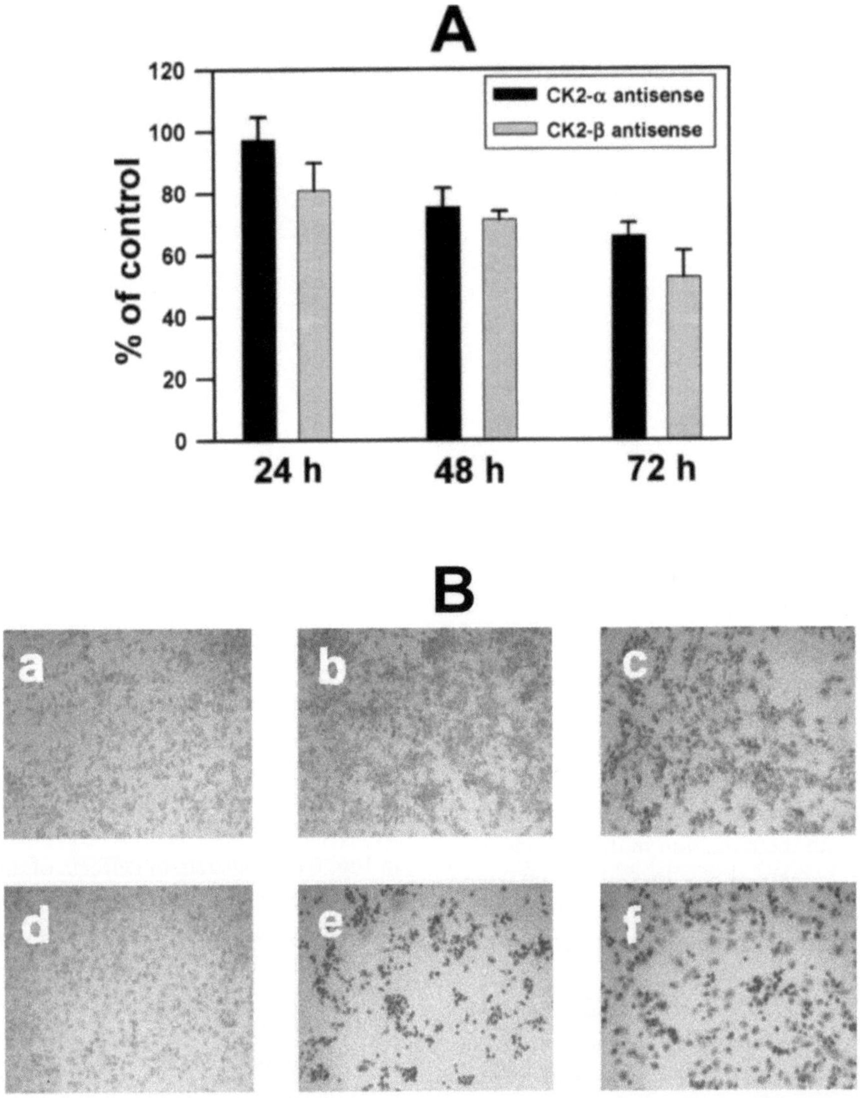

Fig. 6. Effects of antisense CK2 ODN on growth and apoptosis in LNCaP and PC-3 cells. The experimental design was the same as described in legends to the previous figures. Cells were treated with the antisense ODN for 6 h, and were transferred to fresh media without antisense, and allowed to grow for the periods of time indicated when they were subjected to WST-1 assay. (A) shows the effects of 5.0 μg/ml of CK2-α antisense (dark bars) and 5.0 μg/ml of CK2-β antisense (light bars). (B) (a–c) shows the TUNEL staining measured at 48 h in LNCaP cells treated at varying concentrations of antisense CK2-α ODN. a = untreated control cells; b = 2.0 μg/ml of antisense CK2-α ODN; c = 5.0 μg/ml of antisense CK2-α ODN. (B) (d–f) shows the TUNEL staining measured at 48 h in PC-3 cells. d = untreated control cells; e = 2 μg/ml of antisense CK2-α ODN; and f = 2 μg/ml of antisense CK2-β.

Discussion

Several reports have indicated the potential of employing antisense approaches as therapeutic modalities [16, 17]. In this regard, the biological functional characteristics of the protein kinase CK2 would make a particularly appealing target for investigation of the antisense approach to inducing death in cancer cells. The results described in the foregoing clearly demonstrate that the treatment of a variety of cancer cells with antisense CK2-α ODN or antisense CK2-β ODN can evoke a strong apoptotic response in these cells. The interesting feature of this effect is that the observed down-regulation of the CK2 activity (and protein) as a result of the treatment with the antisense ODN is relatively modest compared to the extent of the resulting apoptosis in these cells. The decrease in the NM-associated activity by antisense CK2-α ODN described here accords with the previously described moderate effect of this agent on the chromatin-associated CK2 activity in the head and neck cancer Ca9-22 cells [12]. Taken together, these observations suggest that the cancer cells (in which CK2 is known to be elevated) exhibit a very limited tolerance to a reduction in the level of the intrinsic CK2 in these cells. In previous work, it was shown that depletion of the α and β subunits of CK2 by antisense ODN

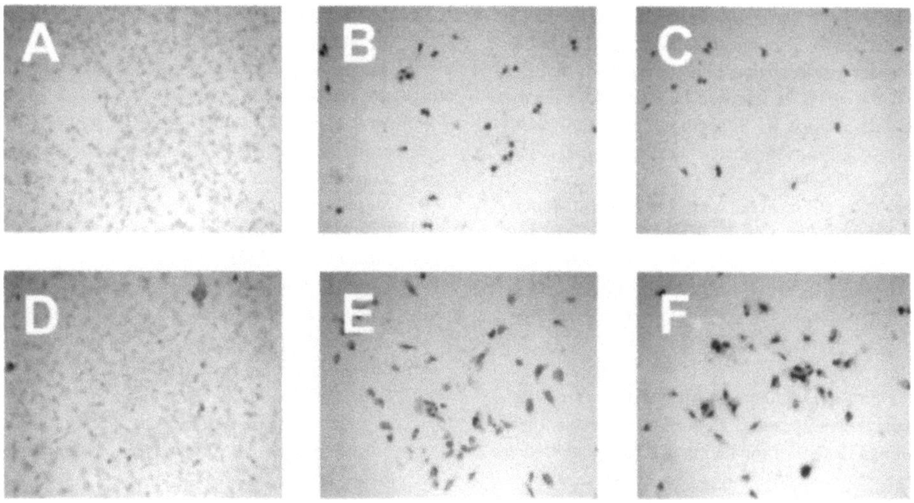

Fig. 7. TUNEL staining of various head and neck cancer cells following treatment with antisense CK2-α ODN and antisense CK2-β ODN. The TUNEL staining was carried out at 48 h following the initial treatment with various ODN, as described in the legends to the previous figures. (A) Ca9-22 cells treated with no ODN; (B) Ca9-22 cells treated with 5.0 μg/ml of antisense CK2-α ODN; (C) Ca9-22 cells treated with 5.0 μg/ml of antisense CK2-β ODN; (D) NA cells were treated with no ODN; (E) NA cells treated with 5.0 μg/ml of antisense CK2-α ODN; and (F) NA cells treated with 5.0 μg/ml of antisense CK2-β ODN.

treatment resulted in inhibition of neurite outgrowth in mouse neuroblastoma cells, and that the effect of the antisense β ODN was somewhat lower than that of the antisense α ODN [18]. However, the effects of the antisense CK2-α ODN were observed at relatively high concentrations (50–300 μg/ml) compared with those employed in the present experiments on various cancer cell lines. Likewise, doses of the order of 100 μg/ml were required to inhibit growth in fibroblasts [19]. The apparent differences in the efficacy of the antisense CK2 ODN in these and our studies may relate to differences in experimental design. However, it is tempting to postulate that the cancer cells (as studied in the present work) may be more amenable to induction of apoptosis through reduction of the CK2 by even a moderate level. This accords with the notion that modest overexpression of CK2 in transgenes imparts a significant oncogenic potential (see e.g. [3]). As suggested previously [12], the observed high sensitivity of the various cancer cell types to antisense CK2-α ODN could also have therapeutic implications as it would suggest that repeated treatment regimens or prolonged tumor exposure to antisense ODN may not be required to achieve a therapeutic response.

The present work has provided additional support to the notion that CK2 is critical for cell survival in addition to being involved in cell growth [1–3]. The possibility that antisense CK2 ODN may be employed to cause a modest reduction in CK2 in the nuclear compartment which would lead to an induction of apoptosis merits further investigation by employing *in vivo* experimental models. Developing novel approaches to achieve delivery of the antisense CK2 in a target tissue specific manner would be a key to the successful utilization of this approach in cancer therapy.

Acknowledgments

This work was supported in part by United States Public Health Service Research Grant CA-15062 awarded by the National Cancer Institute, Department of Health and Human Services, and in part by the Medical Research Fund of the United States Department of Veterans Affairs.

References

1. Ahmed K: Nuclear matrix and protein kinase CK2 signaling. Crit Rev Eukary Gene Exp 9, 329–336, 1999
2. Ahmed K, Davis AT, Wang H, Faust RA, Yu S, Tawfic S: Significance of protein kinase CK2 nuclear signaling in neoplasia. J Cell Biochem (suppl 35): 130–135, 2000
3. Tawfic S, Yu S, Wang H, Faust R, Davis A, Ahmed K: Protein kinase CK2 signal in neoplasia. Histol Histopathol 16: 573–582, 2001
4. Guerra B, Issinger O-G: Protein kinase CK2 and its role in cellular proliferation, development and pathology. Electrophoresis 20: 391–408, 1999
5. Padmanabha R, Chen-Wu JLP, Hanna DE, Glover CVC: Isolation, sequencing, and disruption of the yeast CKA2 gene: Casein kinase II is essential for viability in *Saccharomyces cerevisiae*. Mol Cell Biol 10: 4089–4099, 1990
6. Kelliher MA, Seldin DC, Leder P: Tal-1 induces T cell acute lymphoblastic leukemia accelerated by casein kinase IIα. EMBO J 15: 5160–5166, 1996
7. Yenice S, Davis AT, Goueli SA, Akdas A, Limas C, Ahmed K: Nuclear casein kinase 2 (CK2) activity in human normal, benign hyperplastic and cancerous prostate. Prostate 24: 11–16, 1994
8. Gapany M, Faust RA, Tawfic S, Davis A, Adams GL, Ahmed K: Association of elevated protein kinase CK2 activity with aggressive behavior of squamous cell carcinoma of the head and neck. Mol Med 1: 659–666, 1995

9. Faust RA, Gapany M, Tristani P, Davis A, Adams GL, Ahmed K: Elevated protein kinase CK2 activity in chromatin of head and neck tumors: Association with malignant transformation. Canc Lett 101: 31–35, 1996
10. Faust RA, Niehans GA, Gapany M, Hoistad D, Knapp D, Cherwitz D, Davis A, Adams GL, Ahmed K: Subcellular immunolocalization of protein kinase CK2 in normal and carcinoma cells. Int J Biochem Cell Biol 31: 941–949, 1999
11. Guo C, Yu S, Davis AT, Wang H, Green JE, Ahmed K: A potential role of nuclear matrix-associated protein kinase CK2 in protection against drug-induced apoptosis in cancer cells. J Biol Chem 276: 5992–5999, 2001
12. Faust RA, Tawfic S, Davis AT, Bubash LA, Ahmed K: Antisense oligonucleotides against protein kinase CK2-α inhibit growth of squamous cell carcinoma of the head and neck *in vitro*. Head Neck 22: 341–346, 2000
13. Ahmed K, Davis A, Hanten J, Lambert D, McIvor RS, Goueli SA: Cloning of cDNAs encoding the α and β subunits of rat casein kinase 2 (CK-2): Investigation of molecular regulation of CK-2 by androgens in rat ventral prostate. Cell Mol Biol Res 39: 451–462, 1993
14. Guo C, Yu S, Davis AT, Ahmed K: Nuclear matrix targeting of the protein kinase CK2 signal as a common downstream response to androgen or growth factor stimulation of prostate cancer cells. Cancer Res 59: 1146–1151, 1999
15. Yu S, Davis AT, Guo C, Green JE, Ahmed K: Differential targeting of protein kinase CK2 to the nuclear matrix upon transient overexpression of its subunits. J Cell Biochem 74: 127–134, 1998
16. Mercola D, Cohen JS: Antisense approaches to cancer gene therapy. Cancer Gene Ther 2: 47–59, 1995
17. Stein CA, Cohen JS: Oligodeoxynucleotides as inhibitors of gene expression: A review. Cancer Res 48: 2659–2668, 1988
18. Ulloa L, Diaz-Nido J, Avila J: Depletion of catalytic and regulatory subunits of protein kinase CK2 by antisense oligonucleotide treatment of neuroblastoma cells. Cell Mol Neurobiol 14: 407–414, 1994
19. Pepperkok R, Lorenz P, Jakobi R, Ansorge W, Pyerin W: Cell growth stimulation by EGF: Inhibition through antisense-oligodeoxynucleotides demonstrates important role of casein kinase II. Exp Cell Res 197: 245–253, 1991

Molecular and Cellular Biochemistry **227:** 175–183, 2001.

The genomic structure of two protein kinase CK2α genes of *Xenopus laevis* and features of the putative promoter region

Vivian Wilhelm, Guy Neckelman, Jorge E. Allende and Catherine C. Allende

Programa de Biología Celular y Molecular, Instituto de Ciencias Biomédicas, Facultad de Medicina, Universidad de Chile, Santiago, Chile

Abstract

Protein kinase CK2 is an enzyme that is ubiquitous in eukaryotes. This enzyme, composed of catalytic (α and α′) and regulatory (β) subunits, is responsible for the phosphorylation of a large number of proteins and is implicated in cell division. Genomic clones coding for the CK2α subunit of *Xenopus laevis* have been isolated. Initial restriction enzyme profiles and subsequent PCR analysis and DNA sequencing indicated that these genomic clones correspond to two different genes. The two genes are highly homologous in the regions of the coding sequence (only 3 amino acid differences) but differ considerable in their intron sequences and lengths. Gene 1 corresponds to the cDNA of *Xl*CK2α which had been previously isolated and described. The genomic clone for this gene was truncated. Gene 2 contains the entire coding region for CK2α subunit as well as a fragment of 6.4 kb of the 5′upstream region. The exon/intron boundaries of both genes obey the GT/AG rule with the exception of intron V where the less common GC/AG is seen. Comparison of the size of ten coding exons and sites where these are interrupted by introns shows strong conservation with respect to the human CK2α gene. RT-PCR analysis of mRNAs from *X. laevis* ovary, oocytes and early embryos using a specific primer for gene 2 demonstrated that this gene is expressed in these tissues and cells. Analysis of transcription start sites using 5′RACE and RNA from stage VI oocytes demonstrated that there are multiple start sites in the *Xl*CK2α mRNA. It was also seen that a noncoding exon 1 is present 4 kb upstream of the translation start site and that alternate splicing occurs in gene 2 to give exon 1 of different lengths. Sequencing of the entire upstream genomic region of gene 2 revealed that there are regions of homology to the sequence of exon 1 of the human CK2α gene. Other sequences with consensus to transcription factor binding sites that are seen in the promoter region of human CK2α are also found in the *X. laevis* CK2α gene 2. These sites include Ets1, E2F, CCAAT and GC rich regions. No canonical TATA motif is observed. (Mol Cell Biochem **227**: 175–183, 2001)

Key words: casein kinase 2, transcriptional regulation, CDK2 promoter, Ets1 site

Introduction

Protein kinase CK2 (also known as casein kinase 2) is ubiquitous in eukaryotes and is responsible for the phosphorylation of more than 200 different protein substrates [1–3]. There is a large body of evidence that indicates that CK2 is involved in cell division. The finding that CK2 content, both at the protein and mRNA levels, is considerable higher in dividing and tumor tissues is coincident with this functional involvement and raises the issue of how the transcription of the genes coding for the catalytic (α and α′) and the regulatory (β) subunits of this enzyme is regulated. On the other hand, it is clear that the regulation of the expression of these genes is important, as seen by the fact that overexpression of CK2α in lymphoid cells of transgenic mice [4] or cattle infected with *Theileria parva* [5] can lead to lymphomas in these animals.

Address for offprints: C.C. Allende, Programa de Biología Celular y Molecular, Instituto de Ciencias Biomédicas, Facultad de Medicina, Universidad de Chile, Casilla 70086, Santiago 7, Chile (E-mail: cconnell@machi.med.uchile.cl)

Pyerin's laboratory has contributed considerable to our knowledge of the genomic structure of the genes coding for CK2α and CK2β in humans and also regarding some of the features present in their promoter regions that may contribute to the transcriptional regulation of these genes [6–8]. However, there has been a dearth of information about the structure of the CK2 genes in species other than human and about the conservation or divergence of their promoter regions throughout evolution.

In this report, we present studies that describe the presence of two different genes that code for CK2α in *X. laevis* and that are expressed in ovary and early embryos. Analysis of genomic clones of these two genes has allowed us to determine their structural features. It is shown that while they are nearly identical in the amino acids of the protein coded by their exons, they differ significantly in the sequence and length of their introns. A study of the transcriptional start sites using mRNA isolated form early *X. laevis* embryos demonstrate that there are at least two start sites for gene 2 and that both genes contain a noncoding first exon and this exon 1 contains regions of great homology to the corresponding first exon in human CK2α. Another important finding described here is the presence of an Ets1 site in the close vicinity of the start site of both CK2α genes of *X. laevis*. Ets1 sites have been found by Krehan *et al.* [8] and shown to be important for human CK2α transcription. The 5′ region of the XlCK2α gene also contains elements present in the *X. laevis* CDK2 gene promoter region and in the human Cyclin B gene promoter region. Several consensus sequences for the E2F transcription factor are also seen. Similar to the case of the human gene, *Xl*CK2α does not contain a canonical TATA box but does contain upstream CCAAT boxes and GC-rich regions.

Materials and methods

Cloning of genomic fragments of XlCK2α

A genomic library of *Xenopus laevis,* constructed in bacteriophage lambda EMBO4 vector, was a kind donation of Dr. Igor Dawid (NIH). Inserts were 15–20 kb cloned in a BamHI site. The library was screened by plaque hybridization with a nick-translated ^{32}P-labeled 1110 kb fragment corresponding to the entire coding region of *Xl*CK2α and which also included 40 nucleotides of the 5′ noncoding region. A total of 10^6 pfu were screened, 6 positive clones were isolated and secondary and tertiary screening was performed with the same probe and with two additional labeled fragments obtained by PCR and which represented nucleotides 105–293 and 568–728 of the *Xl*CK2α cDNA. In order to estimate the size and the presence of putative promoter regions in these clones, the clones were analyzed using restriction enzymes *EcoR*I, *Pst*I, *Hind*III and *Kpn*I (Promega) and the fragments were Southern hybridized with probes representing the entire coding region of *Xl*CK2α and a second 5′ terminal probe (nucleotides –40 to 169).

Partial sequencing of clones 1 and 2 was performed initially using primers of the λ vector sequence together with reverse primer **pC2**, (5′-CACGTGTGACTCATAGTCCC-AA-3′), designed from the 5′ coding region of *Xl*CK2α cDNA, (nucleotides 69–90 from the transcription start site). For complete sequencing of the upstream region of clone 2 (gene 2), restriction fragments generated by digestion with *Sal*I were subcloned in vector PUC18 (Pharmacia) and sequenced in both directions using vector primers and then internal primers designed from determined sequences. A fragment of 6.4 kb upstream of the ATG translation start site and part of exon 2 was sequenced (NCBI Genbank Accession number AY032954). Sequencing was performed using Applied Biosystems Model 377 automatic sequencer. All PCR reactions with genomic clones were performed with Expand Long Template PCR System (Roche Molecular Biochemicals).

PCR amplification of genomic clones 1 and 2 and comparison with products obtained with cDNA of XlCK2α

Genomic DNA from liver was prepared as described by Gustincich *et al.* [9]. A series of synthetic oligonucleotides designed from the sequence of *Xl*CK2α cDNA [10] were used as primers for PCR for the comparison of genomic DNA (liver) and the cDNA of CK2α (see Fig. 2), as follows:

p1, 5′-GTCTACAAAACGCTCTGCCT-3′;
p2, 5′-TGGCTTCAAAGACTTCACTG-3′;
p3, 5′-CCAGGATGATTATCAGTTAGT-3′;
p4, 5′-GCCAGGGTGATAATGTTGG-3′;
p5, 5′-CCAACATTATCACCCTGGC-3′;
p6, 5′-GTGGCAATAATCCAGTGCC-3′;
p7, 5′-TGGTAATCCACAAGCAGCTC-3′;
p8, 5′-GTCCGTGTTGCTTCCCG-3′;
p9, 5′-ACCAGCTGGTCGTAGTTATC-3′;
p10, 5′-CCACGCTTTAATGATATTCT-3′;
p11, 5′-CTTGTCCAGGAAATCCAGTG-3′;
p12, 5′-GACTGACCTGACATCATAC-3′.

Oligonucleotide primers were synthesized in the core facility of the University of Chile using an automated sequencer (ABI377).

Boundaries of exons/introns were determined by comparing the cDNA and genomic sequences. The remaining coding sequence of gene 2 was determined by PCR as well as direct sequencing.

Analysis of genomic sequence fragments for characteristic promoter sites was performed with TFMATRIX transcription factor binding site profile database (E. Wingender, R. Knueppel, P. Dietze, H. Karas (GBF-Braunschweig). Searches for sequence similarities were performed using NCBI Blast 2 Sequences program.

Expression of XlCK2α

The primer used for detection of the specific expression of *Xl*CK2α gene 2 mRNA by RT-PCR using different tissues was **pαG2**: 5′-CTGCCCAGGATATCATTG-3′. Total RNA was prepared from *X. laevis* oocytes and embryos as given by Chomczynski and Sacchi [11] and was reverse transcribed using AMV reverse transcriptase (Promega) at 37° for 60 min using primer **pαG2** followed by PCR using primers **p8** and **pαG2**. In parallel and as control, reverse transcription was carried out with primer **p12**, followed by PCR with **p8** and **p12**. *X. laevis* oocytes and embryos were obtained as described [12].

Rapid amplification of 5′ mRNA ends (5′ RACE) was performed using a 5′,3′ RACE system (Boehringer Mannheim) and was used as suggested by the manufacturer. The reverse transcriptase step of the preparation used the CK2α gene-specific primer **pTrans** 5′-CAGAGGCAAAACTGGGCGGT-3′ and PCR performed with the supplied primer. RNA for these studies was prepared with Trisol (Gibco), using stage VI *Xenopus laevis* oocytes. The cDNA products were cloned in PUC19 and sequenced.

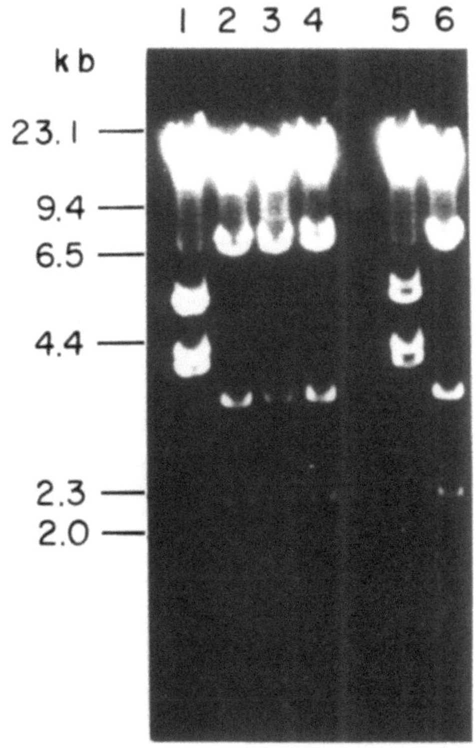

Fig. 1. Comparison of the EcoRI restriction patterns obtained using selected genomic clones which hybridize with *Xl*CK2α cDNA. DNA from 6 genomic clones positive for CK2α was purified and digested with EcoRI and analyzed by 0.7% agarose gel electrophoresis.

Results and discussion

Isolation of genomic clones coding for X. laevis *CK2α: Evidence for the presence of two different genes*

A genomic gene library constructed in λEMBL4 was screened as given in Materials and methods using as probe a 1100 base pair fragment which included the coding region and 40 nucleotides of the 5′ upstream region of the *X. l.* CK2α cDNA previously isolated. [10]. A total of six positive clones were obtained after tertiary screening in which specific probes representing the 5′ and 3′ regions were included in the analysis. Southern analysis with two distinct probes from the amino and carboxyl coding regions of CK2α showed that two of these clones (clones 1 and 5) did not interact with the probe representing the 3′ end of the gene (not shown). Digestion of all six clones with EcoRI demonstrated that clones 1 and 5 gave similar patterns but a different digestion pattern was seen with the other 4 clones (2, 3,4 and 6) which were all clearly related (Fig. 1).

The possibility that these two sets of genomic clones might represent two different CK2α genes was explored using PCR

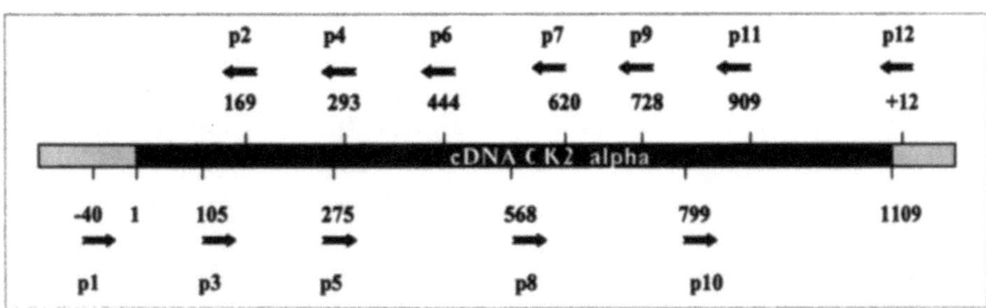

Fig. 2. Schematic representation of the nucleotide primers used for the analysis of the genomic structure of *Xl*CK2α and for sequencing of the coding regions. Primer sequences are given in Materials and methods.

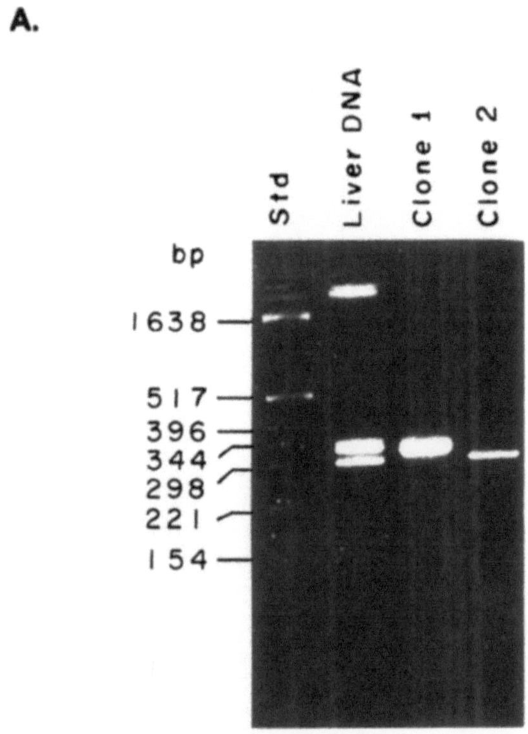

B.

```
cDNA     CCAGGATGATTATCAGTTAGTGAGGAAACTCGGTCGGGGCAAATACAGTG
gene 1   CCAGGATGATTATCAGTTAGTGAGGAAACTCGGTCGGGGCAAATACAGTG
gene 2   CCAGGATGATTATCAGTTAGTGAGGAAACTCGGCCGGGGTAAATACAGTG
         ********************************* ***** **********

cDNA     AAGTCTTTGAAGCCATTAACATCACCAACAACGAGAAGGTTGTGGTAAAA
gene 1   AAGTCTTTGAAGCCATTAACATCACCAACAACGAGAAGGTTGTGGTAAAA
gene 2   AAGTCTTTGAAGCCATTAACATCACCAACAACGAGAAGGTTGTGGTAAAA
         **************************************************

cDNA     ATTCTCAAG-----------------------------------------
gene 1   ATTCTCAAGgtaagcatgggctccagattctgaaccaggaatttctcccc
gene 2   ATCCTTAAGgtaagaatggtttgtaggttctgtgcgtgtgaa----cccc
         ** ** ***

cDNA     --------------------------------------------------
gene 1   accctagaagttccctatgagcctctccatatgtggtgggtgtccataat
gene 2   cccc--------------------------ataaattggtggtcattatt

cDNA     --------------------------------------------------
gene 1   aagggcactttgctgtgtgccttaccc-tatcttcttgtctttctttcca
gene 2   agtgcctctttgctgtgtgccttacccctatcttctcgtctttctttcca

cDNA     -CCTGTGAAGAAGAAGAAAATTAAGCGTGAGATAAAGATCCTGGAGAACC
gene 1   gCCTGTGAAGAAGAAGAAAATTAAGCGTGAGATAAAGATCCTGGAGAACC
gene 2   gCCTGTGAAGAAGAAGAAAATTAAGCGTGAGATAAAGATCCTGGAGAACC
         **************************************************

cDNA     TGCGGGGTGGGCCCAACATTATCACCCTGGC
gene 1   TGCGGGGTGGGCCCAACATTATCACCCTGGC
gene 2   TGCGGGGTGGGCCCAACATTATCACCCTGGC
         *******************************
```

Fig. 3. PCR products and nucleotide sequences obtained with primers p3 and p4 using *X. laevis* genomic DNA or genomic clones 1 and 2 for *Xl*CK2α. (A) Primers p3 and p4 were used with 300 ng liver DNA or 20 ng genomic clone 1 and 2 as templates for PCR amplifications and the products analyzed by agarose gel electrophoresis. The first lane shows DNA standards. (B) Nucleotide sequence of the two *X. laevis* CK2α genomic clone fragments are compared with the cDNA of *X. l.*CK2α. PCR amplification and sequence determination was performed with primers p3 and p4. Asterisks indicate identity between the three fragments. Capital and lowercase letters are exon and intron sequences, respectively.

with sets of primers designed from the coding regions of the CK2 cDNA. Figure 2 shows a scheme of the primers used. These PCR analyses were also carried out with genomic DNA isolated from *X. laevis* liver. Figure 3A shows that using primers 3 and 4, the PCR products obtained with rat liver DNA is a mixture that is resolved into two different products when either clone 1 or clone 2 DNA is used as template. Sequencing of the two PCR products obtained from clones 1 and 2 demonstrated that clone 1 is identical to the cDNA previously obtained and that the sequences of clones 1 and 2 are very similar within the coding regions but differed significantly in the base composition and length of the introns (Fig. 3B). PCR analyses with set of primers covering the entire length of the cDNA indicated that clone 2 contains the entire coding sequence for CK2α and that the coding exons and the 10 intervening introns correspond to 8 kb in length. The sequence observed for clone 1 corresponds to about one half of the CK2α coding region. These results strongly support the presence of two clones corresponding to two separate genes coding for *X. laevis* CK2α (henceforth referred to as genes 1 and 2, derived from clones 1 and 2, respectively). The coding region of the two genes as present in these two clones is shown in Fig. 4. Comparison of the sequence of these two forms to that of the cDNA previously isolated [10] demonstrated that this cDNA was clearly equivalent to the truncated gene 1. The deduced amino acid sequences coded for by gene 1 and gene 2 are shown in Fig. 5, where a total of 3 conserved amino acid changes are found within the studied segments.

It is interesting to point out that, contrary to what has been observed with most eukaryotes studied, no CK2α′ species has been found in *Xenopus laevis*. It seems possible, however, that the second CK2α gene which lacks the distinctive features of the CK2α′ and is clearly a CK2α form, may be fulfilling the function of the CK2α′ gene which has been shown to be specifically involved in spermiogenesis through experiments in knock-out mice [13]. However, the tetraploidy of *X. laevis* frequently results in the existence of 2 genes coding for a single protein [14].

Exon-intron structure of the coding region of XlCK2α *genes*

Partial sequencing of PCR products using *X. laevis* liver DNA as template and primers representing the entire *Xl*CK2α cDNA enabled us to define the overall exon-intron structure of the coding region of the *Xl*CK2α genes. Table 1 summarizes the data for gene 2 and it may be seen that the intron/exon boundaries conform to the GT/AG rule, with the single exception of intron V where GC/AG is found. As seen in Figs 4 and 5 above, there is nearly complete identity between the coding regions of genes 1 and 2. These genes, which are very different in intron size and sequence, possess absolute

```
cDNA      ATGTCAGGACCTGTGCCAAGTCGGGCCCGTGTCTACACGGATGTGAACACACACAGACC
clone 2   ATGTCAGGACCTGTGCCAAGTCGGGCCCGTGTCTACACGGATGTGAACACACACAGACC
clone 1   ATGTCAGGACCTGTGCCAAGTCGGGCCCGTGTCTACACGGATGTGAACACACACAGACC

cDNA      CCGAGACTATTGGGACTATGAGTCACACGTGGTAGAATGGGG    CAACCAGGATGATT
clone 2   CCGAGAATATTGGGACTACGAGTCACACGTTGTAGAGTGGGG II GAACCAGGATGATT
clone 1   CCGAGAATATTGGGACTATGAGTCACACGTGGTAGAATGGGG    CAACCAGGATGATT
                •           •           •     •

cDNA      ATCAGTTAGTGAGGAAACTCGGTCGGGGCAAATACAGTGAAGTCTTTGAAGCCATTAAC
clone 2   ATCAGTTAGTGAGGAAACTCGGCCGGGGTAAATACAGTGAAGTCTTTGAAGCCATTAAC
clone 1   ATCAGTTAGTGAGGAAACTCGGTCGGGGCAAATACAGTGAAGTCTTTGAAGCCATTAAC
                                •     •

cDNA      ATCACCAACAACGAGAAGGTTGTGGTAAAAATTCTCAAG     CCTGTGAAGAAGAAGAA
clone 2   ATCACCAACAACGAGAAGGTTGTGGTAAAAATCCTTAAG III CCTGTGAAGAAGAAGAA
clone 1   ATCACCAACAACGAGAAGGTTGTGGTAAAAATTCTCAAG     CCTGTGAAGAAGAAGAA
                                          •  •

cDNA      AATTAAGCGTGAGATAAAGATCCTGGAGAACCTGCGGGGTGGGCCCAACATTATCACCC
clone 2   AATTAAGCGTGAGATAAAGATCCTGGAGAACCTGCGGGGTGGGCCCAACATTATCACCC
clone 1   AATTAAGCGTGAGATAAAGATCCTGGAGAACCTGCGGGGTGGGCCCAACATTATCACCC

cDNA      TGGCAGACATAGTGAAAGACCCGGTG    TCCAGAACGCCAGCACTTGTCTTCGAACAT
clone 2   TGGCAGACATTGTGAAAGACCCTGTG IV TCCAGAACGCCAGCACTCGTCTTCGAACAT
clone 1   TGGCAGACATAGTGAAAGACCCGGTG    TCCAGAACGCCAGCACTTGTCTTCGAACAT
                                                         •

cDNA      GTCAACAATACAGATTTTAAG   CAGTTATATCAGACTCTGACAGACTACGACATTAG
clone 2   GTCAACAATACAGATTTTAAG V CAGTTATATCAGACTCTGACTGACTACGACATTAG
clone 1   GTCAACAATACAGATTTTAAG   CAGTTATATCAGACTCTGACAGACTACGACATTAG

cDNA      ATTCTACATGTATGAAATTCTTAAG    GCACTGGATTATTGCCACAGTATGGGTATAA
clone 2   ATTCTACATGTATGAAATCCTTAAG VI GCACTGGATTATTGCCACAGTATGGGTATAA
clone 1   ATTCTACATGTATGAAATTCTTAAG    GCACTGGATTATTGCCACAGTATGGGTATAA

cDNA      TGCACAGAGATGTGAAACCTCACAATGTAATGATCGACCATGAGCACAGAAAG     CTA
clone 2   TGCACAGGGATGTGAAACCTCACAATGTAATGATTGACCATGAGCACAGAAAG VII CTA
clone 1   TGCACAGAGATGTGAAACCTCACAATGTAATGATTGACCATGAGCACAGAAAG     ---
                                            •

cDNA      CGTCTCATAGACTGGGGCTTGGCCGAATTCTACCACCCAGGACAGGAGTACAATGTCCG
clone 2   CGTCTCATAGACTGGGGCCTGGCTGAATTCTACCACCCAGGACAGGAGTACAATGTCCG
clone 1   -----------------------------------------------------------

cDNA      TGTTGCTTCCCGATACTTCAAGGGGCCGGAGCTGCTTGTGGATTACCAA      ATGTACG
clone 2   TGTTGCTTCCCGATACTTCAAGGGCCCGGAGCTGCTTGTTGATTACCAA VIII ATGTATG
clone 1   ---------------------------------------------------------------
                                  •              •                    •

cDNA      ACTACAGCCTGGACATGTGGAGCCTGGGGTGCATGCTGGCCAGTATGATTTTCAGAAAA
clone 2   ACTACAGTCTGGATATGTGGAGCTTGGGCTGCATGCTGGCCAGTATGATCTTCAGAAAA
clone 1   -----------------------------------------------------------
                 •     •          •    •                   •

cDNA      GAGCCATTTTTCCATGGCCATGATAACTACGACCAG IX CTGGTGAGAATAGCGAAGGT
clone 2   GAGCCATTTTTCCATGGGCATGATAACTACGACCAG    CTGGTGAGAATAGCCAAGGT
clone 1   ------------------------------------------------------------
                           •                                    •

cDNA      GCTGGGCACAGAAGACCTCTACGATTATATTGACAAGTACAACATAGAACTGGATCCAC
clone 2   TCTGGGCACAGAGGACCTATACGATTATATTGACAAGTACAACATAGAACTGGATCCAC
clone 1   ----------------------------------------------------------
          •           •     •

cDNA      GCTTTAATGATATTCTTGGGAG X GCACTCGCGCA¤ACGGTGGGAAAGATTTGTACAC
clone 2   GCTTCAATGATATCCTGGGCAG   GCACTCGCGCAAACGGTGGGAAAGATTTGTACAC
clone 1   ----<==================-----------------------------------
              •        •  •  •

cDNA      AGCGAGAACCAACATCTAGTCAGTCCTGAGGCACTGGATTTCCTGGACAAGCTGCTGCG
clone 2   AGTGAGAACCAACATCTAGTCAGTCCTGAGGCACTGGATTTCCTGGACAAGCTGCTGCG
clone 1   -----------------------------------------------------------
            •

cDNA      ATATGACCATCAGACCCGACTGACTGCCCGTGAAGCTATGGACCACCCATACTTCT XI
clone 2   ATATGACCATCAGACCCGACTGACTGCCCGTGAGGCTATGGACCACCCATACTTCT
clone 1   --------------------------------------------------------
                                             •

cDNA      ATCCCATCGTAAAGGACCAGTCCCGGATGGCGGCTCTAATATGCCCAGTAGCAGCACAC
clone 2   ACCCCATCGTAAAGGACCAGTCCCGGATGGCGGCTCTAATATGCCCAGTGGCAGCACAC
clone 1   ----------------------------------------------------------
           •                                               •

cDNA      CCGTCAGTAGCGCCAGTATGA
clone 2   CAGTCAGTAGCGCCAGTATGA
clone 1   ---------------------
           •
```

Fig. 4. Comparison of the coding sequences of the genomic clones 1 and 2 with the *Xl*CKα cDNA. Roman numeral inserts indicate the positions of intron segments; dots below the sequence indicate variations in the clone 2 sequence with respect to the *Xl*CK2α cDNA; dashed lines indicate the region deleted in clone 1. The arrow indicates the position of primer pαG2 designed to detect exclusively gene 2 and used to analyze the expression of this gene.

Table I.

Exon-intron structure and splice sites of CK2α of X. laevis.

Exon No.	Exon size (bp)	SEQUENCES AT JUNCTIONS 5' donor 3' acceptor	Intron No.	Intron size (bp)
1.	85	CAG/gtgtgg--//--cag/CGA	I	4203
	115	CTT/gtacgt--//--cag/CGA		4173
2.	189	GGG/gtaagt--//--cag/CAA	II	(4000)
3.	112	AAG/gtaaga--//--cag/CCT	III	112
4.	102	GTG/gtaagt--//--cag/TCC	IV	81
5.	51	AAG/gcgagt--//--cag/CAG	V	(1700)
6.	60	AAG/gtaatt--//--cag/GCA	VI	100
7.	84	AAG/gtaagt--//--cag/CTA	VII	80
8.	111	CAA/gtatgt--//--cag/ATG	VIII	155
9.	102	CAG/gtgaga--//--cag/CTG	IX	(1700)
10.	101	CAG/gtaagt--//--cag/GCA	X	96
11.	149	TCT/gtaagt--//--cag/ACC	XI	249

Capital letters denote exon sequences and lowercase letters intron sequences. Intron sizes were determined by sequence analyses except (in parenthesis) when estimated by the size of the PCR product.

```
x.l.αcDNA     MSGPVP-SRARVYTDVNTHRPRDYWDYESHVVEWGNQDDYQLVRKLGRGKY 50
x.l.Gene2     ......................E............................
H.s.CK2α      ......................E............................
H.s.CK2α'     .P..AAG......AE..SL.S.E.....A..PS..................

x.l.αcDNA     SEVFEAINITNNEKVVVKILKPVKKKKIKREIKILENLRGGPNIITLADI 100
x.l.Gene2     ..................................................
H.s.CK2α      ..................................................
H.s.CK2α'     .........NS..R.................V.........T...K.I.T

x.l.αcDNA     VKDPVSRTPALVFEHVNNTDFKQLYQTLTDYDIRFYMYEILKALDYCHSM 150
x.l.Gene 2    ..................................................
H.s.CK2α      ..................................................
H.s.CK2α'     ......K.......YI..........I...F........L..........

x.l.αcDNA     GIMHRDVKPHNVMIDHEHRKLRLIDWGLAEFYHPGQEYNVRVASRYFKGP 200
x.l.Gene 2    ..................................................
H.s.CK2α      ..................................................
H.s.CK2α'     ................QQK...............A..........SY...

x.l.αcDNA     ELLVDYQMYDYSLDMWSLGCMLASMIFRKEPFFHGHDNYDQLVRIAKVLG 250
x.l.Gene 2    ..................................................
H.s.CK2α      ..................................................
H.s.CK2α'     ............................R......Q..............

x.l.αcDNA     TEDLYDYIDKYNIELDPRFNDILGRHSRKRWERFVHSENQHLVSPEALDF 300
x.l.Gene 2    ........E.........................................
H.s.CK2α      ..................................................
H.s.CK2α'     ..E..G.LK..H.D...H......Q.......N.I....R.........L

x.l.αcDNA     LDKLLRYDHQTRLTAREAMDHPYFYPIVKDQSRMGGSNMPSGSSTPVSSA 350
x.l.Gene 2    ..................................................
H.s.CK2α      ...................................S.S..G..-......
H.s.CK2α'     ..........Q....K...E......V..E..------Q.CADNAVL..G

x.l.αcDNA     SMMSGISTVPTPSALGSLAGSPVISATNTLGTPVAAAAGATQ-392
x.l.Gene 2    ...........Q.................
H.s.CK2α      N......S.....-..P.......A.A.P..M..P.....Q.
H.s.CK2α'     LTAAR
```

Fig. 5. Comparison of the predicted amino acid sequences of *Xl*CK2α cDNA with *Xl*CK2α gene 2 and with CK2α and α′ of *H. sapiens*. Dots indicate identity with *X. laevis* and dashes indicate that the amino acid is absent from the corresponding sequence.

homology in the coding exon sizes and in all of the donor and acceptor splice signals, as far as has been determined (gene 1 clone being truncated at exon 8 and gene 2 at exon 11). A comparison of the exon size of the coding structure of *Xl*CK2α with that found in the human CK2α gene [7] shows that the positions of RNA processing are also conserved between these two species, giving rise to exons of the same size and with similar boundaries.

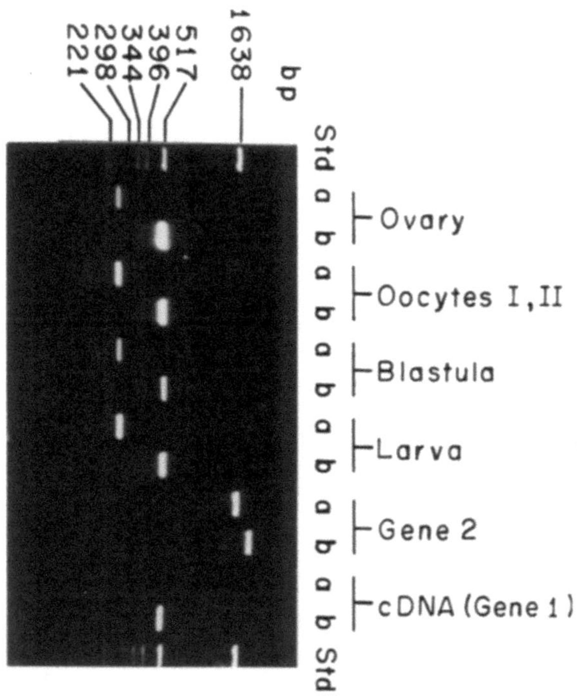

Fig. 6. Expression of *Xl*CK2α gene 2 in *X. laevis*. Total *X. laevis* RNA from oocytes, ovary and embryos of stages 8 (blastula) and 34 (larva) was used for reverse transcription using primer pαG2 (specific for gene 2, see Fig. 4) or with primer 12 (common for cDNA of both gene 1 and 2) followed by PCR using: lanes a – primers p8 and pαG2; lanes b – primers p8 and p12. Controls were performed using cDNA of gene 1 and DNA from gene 2.

Expression of the two X l*CK2α genes*

The human genome contains a silent pseudogene for CK2α [15]. It was pertinent therefore to determine whether the two *Xl*CK2α genes observed were expressed. Since gene 1 is identical to the cDNA isolated from oocytes, it was clear that this gene is actively expressed in this type of cell. To detect the presence of the specific mRNA corresponding to gene 2, a primer was designed for a region which contained a cluster of structural differences between gene 1 and gene 2. The horizontal arrow in Fig. 4 indicates the region used for the primer sequence. Using this primer together with a reverse primer (primer 8 of Fig. 2), gene 2 gave a PCR product of the expected length (1700 bp) while gene 1 did not act as a template for amplification. RT-PCR of total mRNA isolated from ovary, oocytes an early embryos (blastula and larval stages) demonstrated that RNA transcribed from gene 2 was present in all of these cells and tissues (Fig. 6). Sequence determinations of the RT-PCR products confirmed that the RNA corresponded to gene 2 of *Xenopus laevis* CK2α (not shown).

The 5′ upstream region of the X. laevis *gene 2*

A 6.4 kb fragment corresponding to the region upstream of the 5′ coding end of gene 2 was sequenced as described in Materials and Methods. Using mRNA from X. laevis stage VI oocytes and RT-PCR (5′RACE), experiments were carried out to determine the transcription start sites of the CK2α genes and to relate this information to the genomic structure. Figure 7A. shows the results of these studies and indicates the transcription start sites obtained by the RT-PCR analysis. It is seen that there are two principle start sites (designated as nucleotides 1 and 28) and that a nontranslated exon is present within this fragment. In addition, examination of the RT-PCR products also indicated that two distinct 5′-donor sites exist within exon 1, a fact that indicates that alternative splicing of the transcripts would produce exon 1 of different lengths (see Table 1). Examination of the sequences present in a region 2000 bases upstream of the transcription start sites of gene 2 indicates that this gene is similar in structure to the so-called housekeeping genes in that it lacks the typical TATA box and CCAAT signals proximal to a unique start site. Searches for canonical sequences in this segment reveal several elements that match those of transcription factor binding sites, including, besides those indicated in the figure 7A, sites for factors YY1 (–85), SRY (–103, –97: Sex Determination Region Y gene) and upstream CCAAT boxes. The interpretation of the importance of these elements awaits further study.

Identity of the sequence of one of the 5′ RACE cloned products with that of a sequence present in the nontranslated segment near the initiator ATG codon present in the cDNA of gene 1 indicate that this RT-PCR product corresponded to gene 1 mRNA. This RT-PCR product also contains regions with strong homology to those found in exon 1 of gene 2 (Fig. 7B).

In Fig. 7B a comparison is made of the noncoding exon 1 of human CK2α [7] with its counterparts from *X l*CK2α genes 1 and 2. Significant stretches of homology exist in this CK2α exon sequence of these two species (double-underlined in the human sequence). It should be noted that the homology includes the Ets-1 site which has been identified by Pyerin's laboratory [8] as a major determinant of the human CK2α promoter strength. *X. laevis* gene 1 transcription starts precisely in the Ets1 site while *Xl*CK2α gene 2 contains that signal at –20 nucleotides of the first detected start site. It is very interesting that the human CK2β gene also has an Ets1 site very close to the transcription start site [6]. The proximity of Sp1 sites to the Ets1 site has been judged also to have relevance in the promoter function [8]. In the *Xl*CK2α genes there are several GC-rich segments localized near the transcription start site and these regions are both within and near the regions of homology of the Sp1 sites seen in the human gene. These findings suggest a possible coordination

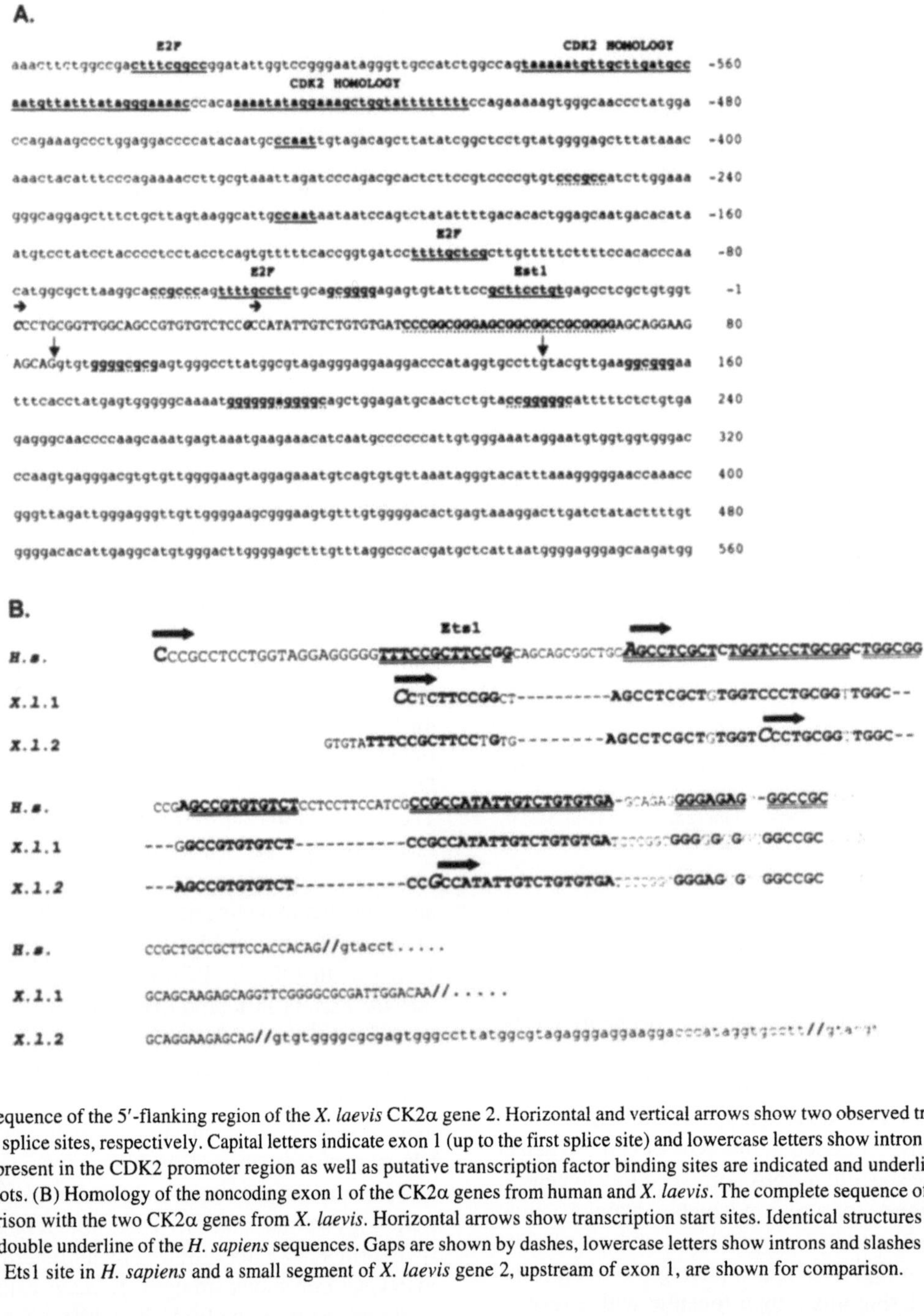

Fig. 7. (A) DNA sequence of the 5′-flanking region of the *X. laevis* CK2α gene 2. Horizontal and vertical arrows show two observed transcription start sites and two alternative splice sites, respectively. Capital letters indicate exon 1 (up to the first splice site) and lowercase letters show intron I. A region of homology to a sequence present in the CDK2 promoter region as well as putative transcription factor binding sites are indicated and underlined. GC-rich regions are underlined by dots. (B) Homology of the noncoding exon 1 of the CK2α genes from human and *X. laevis*. The complete sequence of exon 1 of *H. sapiens* is given for comparison with the two CK2α genes from *X. laevis*. Horizontal arrows show transcription start sites. Identical structures are indicated by bold capital letters with double underline of the *H. sapiens* sequences. Gaps are shown by dashes, lowercase letters show introns and slashes represent splice sites. The presence of an Ets1 site in *H. sapiens* and a small segment of *X. laevis* gene 2, upstream of exon 1, are shown for comparison.

in the regulation of the transcription of the two subunits of CK2 and conservation of the regulatory signals that control the expression of CK2 throughout evolution. It may also be relevant that within the region that is homologous to the human CK2α promoter, there is a 90 to 95% homology in a 22 base segment (nucleotides +4 to +25 and –1 to –23 of the *Xl*CK2α gene 1 and gene 2, respectively) to a sequence in the promoter region of human cyclin B1 gene. It is also noteworthy that the putative promoter region of XlCK2α gene 2 contains a region of high homology to a region that is present in the promoter region of *X. laevis* CDK2 (region –505 to –541 Fig. 7A.). This observation suggests that these two kinase which are closely related structurally in the protein kinase family [16] and which are both involved in cell division, might also share similarities in the regulation of the transcription of their genes.

Acknowledgements

This work was supported by FONDECYT-Chile grant # 1000624 and by ICGEB (Trieste, Italy) grant # CRP/CHI99-004 (1).

References

1. Allende JE, Allende CC: Protein kinase CK2: An enzyme with multiple substrates and a puzzling regulations. FASEB J 9: 313–323, 1995
2. Pinna LA, Meggio F: Protein kinase CK2 and its implication in cell division and proliferation. Prog Cell Cycle Res 3: 77–97, 1997
3. Guerra B, Issinger OG: Protein kinase CK2 and its role in cellular proliferation, development and pathology. Electrophoresis 20: 391–408, 1999
4. Seldin DC, Leder P: Casein kinase II alpha is a novel lymphoid oncogene in a mouse model of tropical theileriosis of cattle. Science 267: 894–897, 1995
5. Ole-MoiYoi OK, Brown WC, Iams KP, Nayar A, Tsukamoto T, Mackin MD: Evidence for the induction of casein kinase II in bovine lymphocytes transformed by the intracellular protozoan parasite *Theileria parva*. EMBO J 12: 1621–1631, 1993
6. Voss H, Wirkner U, Jakobi R, Hewitt NA, Schwager C, Wimmermann J, Ansorge W, Pyerin W: Structure of the gene encoding human casein kinase II subunit β. J Biol Chem 266: 13706–13711, 1991
7. Wirkner U, Voss H, Ansorge W, Pyerin W: Genomic organization and promoter identification of the human protein kinase CK2 catalytic subunit α (CSNK2A1). Genomics 48: 71–78, 1998
8. Krehan A, Ansuini H, Bocher O, Grein W, Wirkner U, Pyerin W: Transcription factors Ets1, MF-κB, and Sp1 are major determinants of the promoter activity of the human protein kinase CK2α gene. J Biol Chem 275: 18327–18336, 2000
9. Gustincich S, Manfioloetti G, Del Sal G, Schneider C, Carnici P: A fast method for high quality genomic DNA extraction from whole blood. Biotechniques 11: 298–302, 1991
10. Jedlicki A, Hinrichs MV, Allende CC, Allende JE: The cDNAS coding for the α and β subunits of *Xenopus laevis* casein kinase II. FEBS Lett 297: 280–284, 1992
11. Chomczynski P, Sacchi N: Single-step method of RNA isolation by acid guanidimium thiocyanate-phenol-chloroform extraction. Anal Biochem 162: 156–159, 1987
12. Wilhelm V, Rojas P, Gatica M, Allende CC, Allende JE: Expression of the subunits of protein kinase CK2 during oogenesis in *Xenopus laevis*. Eur J Biochem 232: 671–676, 1995
13. Xu X, Toselli PA, Russell LD, Seldin DC: Globozoospermia in mice lacking the casein kinase II alpha' catalytic subunit. Nat Genet 23: 118–121, 1999
14. Plant KE, Hair A, Morgan GT: Genes encoding isoforms of transcription elongation factor TFIIS in *Xenopus* and the use of multiple unusual RNA processing signals. Nucleic Acids Res 24: 3514–3521, 1996
15. Wirkner U, Voss H, Lichter P, Weitz S, Ansorge W, Pyerin W: Human casein kinase II subunit α: Sequence of a processed (pseudo)gene and its localization on chromosome 11. Biochim Biophys Acta 1131: 220–222, 1992
16. Hanks SK, Hunter T: The eukaryotic protein kinase superfamily: Kinase (catalytic) domain structure and classification. FASEB J 9: 576–596, 1995

Molecular and Cellular Biochemistry **227:** 185–187, 2001.

Index to Volume 227

MIX
Papier aus verantwortungsvollen Quellen
Paper from responsible sources
FSC® C105338

If you have any concerns about our products,
you can contact us on
ProductSafety@springernature.com

In case Publisher is established outside the EU,
the EU authorized representative is:
Springer Nature Customer Service Center GmbH
Europaplatz 3, 69115 Heidelberg, Germany

Printed by Libri Plureos GmbH
in Hamburg, Germany